Boston Studies in the Philosophy and History of Science

Volume 316

More information about this series at http://www.springer.com/series/5710

Joyce van Leeuwen

The Aristotelian *Mechanics*

Text and Diagrams

Springer

Joyce van Leeuwen
Max Planck Institute
for the History of Science
Berlin, Germany

Dissertation zur Erlangung des Doktorgrades an der Philosophischen Fakultät I der Humboldt-Universität zu Berlin. Tag der mündlichen Prüfung: 9. Juli 2012.

ISSN 0068-0346 ISSN 2214-7942 (electronic)
Boston Studies in the Philosophy and History of Science
ISBN 978-3-319-25923-9 ISBN 978-3-319-25925-3 (eBook)
DOI 10.1007/978-3-319-25925-3

Library of Congress Control Number: 2015958726

Springer Cham Heidelberg New York Dordrecht London

Printed on acid-free paper

Springer International Publishing AG Switzerland is part of Springer Science+Business Media (www.springer.com)

Acknowledgments

This book is a revised version of my dissertation, which was accepted at the Humboldt-Universität zu Berlin in 2012. It was written in the Excellence Cluster Topoi in Berlin to which I am grateful for research support between the years 2009 and 2011. First and foremost, I would like to thank my advisers, Jonathan Beere, Reviel Netz, and Henry Mendell, for their continued encouragement and help at various stages of my project. I was particularly fortunate to have the opportunity to spend a semester at Stanford University to work with Reviel Netz, to whom I am grateful for much illuminating conversation. It should also be evident from this book how much his own work on diagrammatic reasoning in Greek mathematics has influenced me. I acknowledge the Dean of the Faculty of Philosophy I at the Humboldt-Universität zu Berlin, Michael Seadle.

For granting a Mellon Postdoctoral Fellowship in 2012–2013, I am grateful to the Department of History and Philosophy of Science at the University of Pittsburgh. I benefited from Jim Lennox's vast knowledge of Aristotle, for which I am especially thankful. I am grateful to Department 1 of the Max Planck Institute for the History of Science in Berlin for its generous support in 2013–2015. I owe a great debt to Jürgen Renn and Matteo Valleriani for their faith in this project and for expanding my horizon beyond the ancient world. I greatly benefited from the Department's research on long-term developments of mechanical knowledge and, especially, the relationships between practical and theoretical knowledge. I thank Peter McLaughlin for many discussions concerning Aristotelian mechanics.

Earlier versions of parts of chapters have appeared in other publications. I thank Cambridge University Press for permission to reproduce my paper "The Text of the Aristotelian *Mechanics*," published in *Classical Quarterly* 63.1 (2013), which is included in an extended version in Chap. 3. I also thank Brill for permission to reuse material from "Thinking and Learning from Diagrams in the Aristotelian *Mechanics*," published in *Nuncius* 29 (2014), which appears as parts of Chaps. 5 and 6. These chapters were improved by the valuable comments of Markus Asper, Marko Malink, Jacob Rosen, Ken Saito, and the anonymous readers of *CQ* and *Nuncius*.

Parts of my work were presented in seminars at the Humboldt-Universität zu Berlin, Lund University, Stanford University, Ludwig-Maximilians-Universität München, University of Pittsburgh, and the Max Planck Institute for the History of Science in Berlin. I am grateful for various comments and encouragements offered by all these audiences. The collation of the manuscripts of the *Mechanics* was conducted at the Aristoteles-Archiv of the Freie Universität Berlin. I benefited from its extensive microfilm collection as well as paleographical and codicological materials. I thank Dieter Harlfinger for instruction on the Aristotelian manuscript tradition. I wish to thank Urte Brauckmann for helping to obtain digital reproductions of the images and the related permissions. I am indebted to Lindy Divarci and Lucy Fleet for assisting me in the publishing process. Joshua Crone and Simon Mills proofread the manuscript for me and saved me from many mistakes.

Berlin Joyce van Leeuwen
May 2015

Contents

1 Introduction 1
1.1 Collections of *Problēmata* 4
1.2 Authenticity 7
1.3 History of the Text in Antiquity 18
Bibliography 23

2 General Characteristics of the Transmission 25
2.1 On the Method 25
2.2 List of Manuscripts 27
Bibliography 33

3 The Textual Transmission 37
3.1 Manuscript Families 39
3.1.1 Family *a* 39
3.1.2 Family *b* 43
3.1.3 O^a and Its Descendants 48
3.1.4 Family *c* 50
3.1.5 Contaminated and Incomplete Manuscripts 54
3.2 The Paraphrase by Pachymeres 57
3.3 The Latin Tradition 61
3.4 The Arabic Tradition 62
3.5 Critical Editions 64
3.5.1 The *Editio Princeps* 64
3.5.2 Later Editions 65
3.6 Results 67
3.6.1 *Stemma Codicum* 67
3.6.2 On a New Text 68
Bibliography 70

4 Editing Diagrams 73
4.1 Transmission of Diagrams 75
4.1.1 Manuscript Families 75
4.1.2 Correspondence of Diagrams 79
4.2 Practices of Diagrams 89
4.2.1 Terminology 89
4.2.2 The Lettered Diagram 91
4.2.3 Characteristics of Diagrams 96
4.2.4 Byzantine Origins 97
4.3 Critical Edition 101
4.3.1 Introductory Remarks 101
4.3.2 Case Study: Problems 3 and 23 102
4.3.3 Critical Edition 114
4.4 Notes on the Critical Edition 119
4.5 Independent Diagrams 134
Bibliography 136

5 Diagrammatic Reasoning 139
5.1 Diagrams in Modern Critical Editions 140
5.2 Validating the *Stemma Codicum* 144
5.3 Cognition of Diagrams 148
5.3.1 Rectilinear Motion 148
5.3.2 Circular Motion 151
5.3.3 Circular Motion Continued 154
Bibliography 157

6 Early Modern Transformations of Mechanics 159
6.1 The Mechanical Discipline 160
6.2 Latin Translations and Their Sources 164
6.3 Niccolò Leonico Tomeo's *Mechanical Questions* 166
6.4 Practical Knowledge 170
6.5 Selected Problems 178
6.5.1 Problem 20 178
6.5.2 Problem 25 182
Bibliography 187

7 Conclusion 189
Bibliography 191

Appendix: Reproductions and Reconstructions of the Diagrams 193
Introduction 193
Problem 1.1 199
Problem 1.2 201
Problem 1.3 204
Problem 1.4 206

Problem 2 210
Problem 3 215
Problem 5 219
Problem 17 221
Problem 21 224
Problem 22 228
Problem 23 230
Problem 24.1 235
Problem 24.2 240
Problem 25 243
Problems 26–27 248
Problem 30 251

Chapter 1
Introduction

The text of the Aristotelian *Mechanics* has elicited different evaluations and opinions across the centuries. Interest in the treatise reached its climax in the early modern period, when people from many and diverse backgrounds worked on the text.[1] Among them were humanists, mathematicians and engineers, who wrote translations and paraphrases of the *Mechanics* and published their commentaries on the treatise. Well-known figures like Niccolò Tartaglia, Alessandro Piccolomini, Bernardino Baldi and Galileo Galilei were all influenced and inspired by the work. These scholars covered a wide spectrum of activities ranging from translations of the Greek text in Latin or vernacular languages to extended commentaries in which the mechanical theories from antiquity were related to contemporary technologies. In his in 1547 published Latin paraphrase of the *Mechanics*, Piccolomini summarizes the relevance of the treatise for mechanical developments in early modern times:

> I will merely add that although this work is short, it can be thought of as boundless since it allows one to understand the force of intellect and incredible learning of Aristotle, in that it investigates the true reasons for almost all the wonderful machines which not only have already been discovered but which will be invented in the future.[2]

The early modern period saw many innovations in the field of practical knowledge, especially in architecture and construction. In order to account for these developments, the Aristotelian *Mechanics* played a significant role in that it explained the underlying causes and principles of the working of machines and other mechanical devices. The text of the *Mechanics* thus offered a theoretical framework within which the practical activities could be explained. This strong connection between theoretical and practical knowledge is characteristic of the mechanical discipline in the early modern period.[3]

[1] See Rose and Drake (1971), 65–104 for a detailed discussion of the reception of the *Mechanics* in Renaissance culture.

[2] Transl. Rose and Drake (1971), 82–83.

[3] See Valleriani (2009, 2010), particularly 117–190.

J. van Leeuwen, *The Aristotelian Mechanics*, Boston Studies in the Philosophy and History of Science 316, DOI 10.1007/978-3-319-25925-3_1

In the centuries that followed, interest in the Aristotelian *Mechanics* waned. Due to advances in the natural sciences, the treatise no longer attracted scholars working in this field, nor were classical philologists focusing on Greek literature interested in the subject matter of the text. This decline in interest was further accelerated by general doubts that began to surface in the nineteenth century regarding the authenticity of the text. Today this situation has drastically changed, as shown by the increased number of publications on the *Mechanics* in recent years. Scholars now generally agree on the great significance of the treatise for the history of science, as it is the first extant theoretical treatment of machines. It is, however, impossible to appreciate the value of the treatise without first having acquired a solid knowledge of the principles of the text, especially considering that many publications rely on certain false assumptions about the transmission of the text and its diagrams. A critical examination of the complete manuscript material of the Aristotelian *Mechanics* is therefore necessary. The present book aims to perform this task in order to establish a sound basis for further research.

The manuscript tradition of the Aristotelian *Mechanics* is an exclusively Byzantine affair. The greater part of the manuscripts that have been handed down to us can be dated to the fifteenth century; only a very few were copied as early as the beginning of the fourteenth. My manuscript analysis has shown that all extant manuscripts of the treatise can be traced back to a common source, a lost codex from around the tenth century. The Byzantine tradition of the *Mechanics* is of great importance, since it provides the basis for all later reception of the text. The manuscripts from the Byzantine period are our earliest sources for the Aristotelian *Mechanics*; to date, no Latin translation of the text from the Middle Ages has been found, and the Arabic tradition of the treatise does not add to our knowledge of the transmission, since it consists solely of an abridged version of the introduction and first problem of the text.

At the end of the fifteenth century, the *Mechanics* became widely available through the *editio princeps* of the text, printed by Aldus Manutius in Venice. This intensified the examination of the text in early modern times, even though many humanists still displayed a particular philological interest in Greek manuscripts. Piccolomini for example, in the introduction to his paraphrase, mentions that he took the effort to consult manuscripts in several Italian libraries.[4] On the one hand, we see that early modern scholars aimed to reconstruct the authentic text, while, on the other hand, the contents of the work were adapted to and appropriated by the context of sixteenth-century mechanics. How this actually worked can be demonstrated by the above-quoted citation, in which Piccolomini asserts that the *Mechanics* investigates the causes of both current and future machines. The mechanical knowledge present in the Aristotelian text can be extended to different practical activities. By merging the ancient theories with contemporary technologies, transformations necessarily occur in both fields of knowledge.[5] Such processes of transfer and transformation of mechanical knowledge will not only become apparent from the word-

[4] See Piccolomini (1565), f.6v.

[5] See Valleriani (2009), 185.

ing of the commentaries, but can also be demonstrated by means of the visualizations contained in these texts.

The diagrams in the early modern printed editions of the *Mechanics* are of a strikingly different nature from the diagrams in the Byzantine manuscript tradition. The manuscript diagrams offer geometrical representations of the mechanical devices. They are contained in a greater part of the manuscripts and are prescribed by the author of the text. These diagrams are crucial to our understanding of the mechanical principles, however, as I will argue, in their present form they are later Byzantine additions. On the basis of a great number of similarities they can be traced back to the Byzantine archetype of the *Mechanics*. The diagrams reflect certain ancient diagrammatic practices, but for several reasons they are not likely to be identical with the author's diagrams. Whereas the textual tradition is thus primarily concerned with the conservation of the treatise, the Byzantine diagrams present a commentary on the text. For this reason the diagrams are a powerful tool to study processes of transfer and transformation of knowledge. This becomes especially relevant when we compare the Byzantine manuscript diagrams with those in the printed editions and commentaries from the early modern period. On the one hand, part of the early modern diagrams still preserve the geometrical properties known from the Greek manuscript traditions; on the other hand, we see a different type of diagrams that is connected with the Renaissance scholars' interests in contemporary technologies. These diagrams are no longer strictly geometrical, but rather illustrate the wide range of practical activities.

The aim of the present book is twofold: it focuses both on aspects of tradition and innovation in the transmission of the Aristotelian *Mechanics*. Whereas the Byzantine manuscript tradition is characterized by meticulous copying of the text on the part of the scribes, the diagrams in the manuscripts and printed texts present various modifications and transformations of the mechanical contents that are characteristic of the respective stages in the transmission process. Chapters 2 and 3 set out the Byzantine manuscript tradition of the *Mechanics*, resulting in a newly established *stemma codicum* which illustrates the affiliations of the manuscripts. Most importantly, I will show the urgent need for a new critical edition of the text. Chapter 4 presents a first critical edition of the diagrams contained in the Greek manuscripts of the treatise. Whereas these chapters are of interest primarily to classical philologists as they offer the basis for a new critical edition of the *Mechanics*, the remainder of the book addresses a broader audience with interests in the history of science. The examination of the diagrams bridges these two areas of interest, and connects questions on the transmission of the text and diagrams to broader issues on the conception and development of mechanical knowledge over time. I will argue in Chap. 5 that the diagrams are significant for a reconstruction of the authentic text; yet at the same time they can be considered as a commentary on the text and inform us about the state of mechanical knowledge during a specific period. Finally, Chap. 6 will make a further contribution to our understanding of the history of mechanics. I will claim that the changed scope of mechanics in the sixteenth century is reflected in the diagrams and images that are contained in the translations, paraphrases and commentaries on the *Mechanics* from the early modern period. In this context, I will focus in particular on the Latin translation of the text by the Italian humanist Niccolò Leonico Tomeo, published in the year 1525. Apart from Leonico Tomeo's translation of the treatise, we also possess a Greek manuscript by his hand, which enables us to closely observe the altered diagrammatic practices in the process from handwritten manuscript to printed edition. Before turning to my analysis of the manuscript tradition of the *Mechanics*, in the present chapter I will present some important issues relating to the genre of *problēmata*, the authenticity of the text and its history in antiquity.

1.1 Collections of *Problēmata*

The text of the *Mechanics* resembles the style of the pseudo-Aristotelian treatise titled the *Problems* and other ancient collections of *problēmata*. The genre arises from a Socratic background in which problems and questions are posed and answered. Causal explanations of a wide range of phenomena are at the core of these *problēmata* texts. In the *Topics*, Aristotle himself describes the nature of a problem in a dialectical context. He asserts that not all kinds of problems count as dialectical (105a3-9):

> Not every problem, nor every thesis, should be examined, but only one which might puzzle one of those who need argument, not punishment or perception. For people who are puzzled to know whether one ought to honour the gods and love one's parents or not need punishment, while those who are puzzled to know whether snow is white or not need perception. The subjects should not border too closely upon the sphere of demonstration, nor yet be too far removed from it; for the former cases admit of no doubt, while the latter involve difficulties too great for the art of the trainer.[6]

Problems thus consist in asking something on which there is no general agreement. According to Aristotle, these involve questions that are difficult, but nevertheless solvable. In the following passage we are told that dialectical problems are either ethical, logical or natural (105b20-21). Since many of the questions in the *Problems* and the *Mechanics* concern natural phenomena, we would have expected them to belong to the final class of dialectical problems mentioned here by Aristotle. However, the type of questions and answers in these *problēmata* texts are different from a dialectical context. A fragment from Aristotle's lost work *On Problems*, that has been preserved in Alexander of Aphrodisias' commentary on the *Topics* instructs us on the different context of these problems (*in Top.* 63.11-19):

> Are these not physical problems, as Aristotle has said in his work *On Problems*? Physical phenomena whose causes are unknown constitute physical problems. Still, there are dialectical problems even about physical matters, as there are about ethical and logical matters; those of one kind are dialectical, those of another physical. All dialectical problems will be reducible to the inquiry whether the connexion of an attribute with a thing is a fact, and the inquiry whether a things exists, which are two of the four questions enumerated at

[6] Transl. Pickard-Cambridge in Barnes (1984), 174.

> the beginning of the second book of the *Posterior Analytics*; for the questions 'What is the reason of a connexion?' and 'What is the nature of a thing?' are *not* dialectical problems.[7]

This passage makes clear that there are two types of problems on physical matters, those that are dialectical and those that are physical. It is the form of the question that places a problem in the framework of dialectical or scientific explanation.[8] Dialectical problems concerning natural objects involve "that it is" and "whether it is" questions, whereas there are also natural problems that raise questions of the type "why it is" and "what it is". The latter belong to scientific contexts and take the form in which they are found in the *Problems* and the *Mechanics*. These questions ask for causal demonstrations and demand a sort of response that is different from the "yes" or "no" answers to the dialectical problems.[9] The "why it is" questions display a syllogistic structure and search for demonstration via "middles".[10] A question of the type "why is every A a C" is answered in the form "every A is C because B".

Causal understanding of phenomena by means of "why it is" (διὰ τί) questions are characteristic of the treatise the *Problems*, which is a large collection of problems on diverse topics compiled by the Peripatetic School. It developed from an authentic Aristotelian core to which many questions and answers were added over time until it probably reached its present form at some point in the fifth or sixth century.[11] Because of its lack of internal coherence, it is generally assumed that the *Problems* were not designed for publication, but rather derived in a pedagogical process from lecture notes containing problems for discussion.[12] Just as formulaic as the words διὰ τί that introduce each new question, is the response format of the problems. The answer begins with the words ἢ ὅτι, ἢ διότι, or an occasional πότερον… ἤ.[13] Questions on similar topics are grouped together in the *Problems*, for example Book 15 contains problems that are connected with mathematical theory. Question 10 in this context inquires about the length of shadows (912b4-10):

> Why are the shadows thrown by the moon longer than those thrown by the sun, though both are thrown by the same perpendicular object? Is it because the sun is higher than the moon, and so the ray from the higher point must fall within that from the lower point? Let ΑΔ be the gnomon, Β the moon, and Γ the sun. The ray from the moon is ΒΖ, so that the shadow will be ΔΖ; but the ray from the sun is ΓΕ, and its shadow therefore will necessarily be less, viz. ΔΕ.[14]

---

[7] Ross (1952), 104.

[8] See Lennox (2001), 77–78.

[9] See Lennox (2001), 79–80.

[10] See Lennox (2001), 80.

[11] See Blair (1999), 171.

[12] See Blair (1999), 174.

[13] See Flashar (1962), 341.

[14] Transl. Forster in Barnes (1984), 1419.

This short problem presents a geometrical demonstration in order to answer a question arising from observation.[15] This kind of mathematical argument is only occasionally used in the *Problems*, it is, however, characteristic of the questions that are presented in the *Mechanics*, as we shall now see. Many of the problems refer to corresponding diagrams by means of letter labels in the text in the same way as in the question above on the length of shadows.

The text of the *Mechanics* deploys a similar structure to the *Problems*, in which we see that the διὰ τὶ question is followed by a rather tentative answer "is it because" (ἢ διότι). In this context, the author sometimes suggests several possible answers, but usually limits himself to one particular explanation. The *Mechanics* comprises 35 separate problems preceded by an introduction to the mechanical discipline in which the author sets out the explanatory structure of the text. In many of the problems, the author is concerned with explaining the working of simple machines, as for example the balance (problem 2), the lever (problem 3), the rudder (problem 5), the wedge (problem 17), the dental forceps (problem 21), or the nutcracker (problem 22). The text also includes problems on diverse topics, such as problems 32–34 on projectile motion or problem 25 on the construction of beds. The problems are of varying length, ranging from a few lines to several pages. Problem 3, for instance, starts with a typical question of the aforementioned type διὰ τὶ (850a30-32): "Why is it that small forces move great weights with the help of a lever, as was said at the beginning, although the weight of the lever is added?"[16] In the answer, starting with ἢ ὅτι, the principle of the lever is explained by referring to the balance (850a33-35): "Is it because the lever is the reason, being a beam with its cord fixed below and divided into two unequal parts?"[17] The author's description in the next few lines is of especial significance, since here we find a pre-Archimedean formulation of the law of the lever (850a39-850b2): "Therefore, as the weight moved is to the weight moving it, so, inversely, is the length of the one arm to the length of the other."[18] Finally, the author completes his explanation using a diagram (850b6-9): "Let AB be the lever, Γ the weight, Δ the mover, and E the fulcrum; and Δ will be moved to H, and Γ, the weight moved, to K."[19] We find here the same mathematical vocabulary as in the above quoted example from the *Problems*. The explanation of the principle proceeds by the use of letter labels in the text to refer to the corresponding parts in a diagram. The diagrams to problem 3 as well as those to other questions of the *Mechanics* are in many cases preserved in the Byzantine manuscript tradition of the treatise.

We have seen in the third problem that the author refers the principle of the lever to that of the balance. This move is part of the overall deductive strategy of the

[15] See Taub (2008), 24.

[16] διὰ τί κινοῦσι μεγάλα βάρη μικραὶ δυνάμεις τῷ μοχλῷ, ὥσπερ ἐλέχθη καὶ κατ' ἀρχήν, προσλαβόντι βάρος ἔτι τὸ τοῦ μοχλοῦ;

[17] ἢ ὅτι αἰτιόν ἐστιν ὁ μοχλός, ζυγὸν [ὢν] κάτωθεν ἔχον τὸ σπαρτίον καὶ εἰς ἄνισα διῃρημένον;

[18] ὃ οὖν τὸ κινούμενον βάρος πρὸς τὸ κινοῦν, τὸ μῆκος πρὸς τὸ μῆκος ἀντιπέπονθεν.

[19] ἔστω μοχλὸς ἐφ' οὗ AB, βάρος δὲ ἐφ' ᾧ τὸ Γ, τὸ δὲ κινοῦν ἐφ' ᾧ τὸ Δ, ὑπομόχλιον ἐφ' ᾧ τὸ E, τὸ δὲ ἐφ' ᾧ τὸ Δ κινῆσαν ἐφ' ᾧ τὸ H, κινούμενον δὲ τὸ ἐφ' οὗ Γ βάρος ἐφ' οὗ K.

work, which ultimately aims to derive all mechanical phenomena from the principle of the circle (848a11-14): "The things that take place in the balance can be referred to the circle, and those in the lever to the balance, while almost all other mechanical motions refer to the lever."[20] The author does not only define the circle as the underlying principle of all mechanical phenomena, but also dwells upon its peculiar properties in the introduction to the treatise. The most remarkable feature of the circle is the fact that a point on a radius that is further from the centre moves faster than a point that is closer (848a15-17). This principle forms an important aspect of the argument of the *Mechanics*, and it returns amongst others in the author's discussion of problem 1. Although not all problems are explained by means of the lever, balance or circle,[21] the explanatory model applied in the *Mechanics* sets this text apart from the treatise the *Problems.* Whereas the questions in the *Problems* are very loosely connected to each other, the overall deductive structure of the *Mechanics*, as set out by the author in the introduction, combines the separate problems into a cohesive whole. From this strategy as well as from the mathematical practice of applying letter labels to diagrams, we notice that the *Mechanics* is not only connected with the genre of *problēmata* through the same question-and-answer format, but also shares some important characteristics with ancient mathematical texts.[22]

1.2 Authenticity

Although it was long thought that the *Mechanics* was written by Aristotle, today the treatise is generally considered to be a later product of the Peripatetic School. Strato of Lampsacus, the second successor to Aristotle as the head of the Lyceum, is the author most often suggested to have written our text. He was an obvious candidate, since the title Μηχανικὸν appears on the ancient lists of works by both Aristotle and Strato.[23] However, as István Bodnár has argued, doctrinal considerations make it unlikely that Strato was responsible for the *Mechanics* in its present form.[24]

The negative opinion on Aristotle's authorship does not have a very long history. Until the nineteenth century, it was not seriously doubted that Aristotle had written the *Mechanics*. In the Renaissance there were some skeptics, for example Gerolamo

[20] τὰ μὲν οὖν περὶ τὸν ζυγὸν γινόμενα εἰς τὸν κύκλον ἀνάγεται, τὰ δὲ περὶ τὸν μοχλὸν εἰς τὸν ζυγόν, τὰ δ' ἄλλα πάντα σχεδὸν τὰ περὶ τὰς κινήσεις τὰς μηχανικὰς εἰς τὸν μοχλόν.

[21] E.g. problems 32–34 on projectile motion do not fit the explanatory scheme of the text, but rather involve general questions pertaining to Aristotle's natural philosophy. See Schiefsky (2009), 43–67 for a detailed analysis of the reduction program of the *Mechanics*.

[22] See Asper (2007), 75.

[23] See Bodnár (2011), 443.

[24] See Bodnár (2011), 443–455. In this paper Bodnár compares the views in the *Mechanics* on projectile motion and circular motion with what we know about the doctrines of Strato. Although it cannot be excluded that the text was written by Strato, Bodnár argues that there is not a shred of evidence to substantiate that anything comparable to what we find in the *Mechanics* was undertaken by Strato.

Cardano, but he stood alone in his opinion. Moreover, this sixteenth-century skepticism concerned all Aristotelian writings not transmitted through Arabic and available in Latin translations from the Middle Ages onwards.[25] Cardano specifically mentions the obscurity of the text as reason for doubting that the text was written by Aristotle.[26] The obscurity of the *Mechanics* was noted by many early modern scholars in relation to several of the problems. However, this did not result in any negative opinions on the authorship; rather, they attempted to resolve it by looking for good manuscripts and interpreting the text in terms of Aristotle's known works. This can be exemplified by Alessandro Piccolomini's Latin paraphrase of the *Mechanics* (1547), in which Piccolomini strongly argues for the authenticity of the treatise. A comparison between the *Mechanics* and other works in which Aristotle makes use of mathematical demonstrations clearly shows, Piccolomini asserts, that the phrasing is peculiar to Aristotle.[27] Moreover, Piccolomini revealed many parallels to Aristotle's writings. In the prologue, he expounds on the place of mechanics in Aristotle's classification of the sciences. Mechanics is mentioned along with optics, astronomy and harmonics as one of the sciences that is subordinate to mathematical principles.[28] This description of mechanics as one of the sciences that relies on mathematics for its demonstrations agrees with Aristotle's definition at several points in the corpus.[29] At times Piccolomini is aware of possible contradictions within Aristotle's works, for instance when mechanics is defined in our text as both an art and a science. In order to demonstrate the authenticity of the *Mechanics*, Piccolomini points to several passages from other texts in which Aristotle did not make a clear distinction between an art and a science.[30]

The first substantial criticism of the *Mechanics* was uttered by Jean Étienne Montucla in the *Histoire des mathématiques* (1799).[31] Montucla's judgement on the text is harsh: "A la vérité, ils y sont encore tellement défigurés par l'erreur, qu'on ne peut les regarder que comme une grossière ébauche de ces sciences. […]. Ils trouveront sans doute que la plupart des explications qu'il donne sont entièrement fausses, et que la principale et la première est tout-à-fait ridicule."[32] He dismisses the content of the text as "completely wrong" and "totally ridiculous". However, Montucla does not cast any doubt on the authenticity of the *Mechanics*. In the wake of Montucla's publication, we find the first scholars in the nineteenth century who

[25] See Krafft (1970), 17.

[26] I thank Peter McLaughlin for allowing me to read his manuscript *The Question of the Authenticity of the Mechanical Problems*, in which he quotes amongst others this passage from Cardano.

[27] See Piccolomini (1565), f.6r.

[28] See Piccolomini (1565), f.5r.

[29] See e.g. *APo.* 76a22-25 or 78b35-39.

[30] See Piccolomini (1565), f.8r.

[31] See Bottecchia Dehò (2000), 27–51 for a detailed presentation of the *status quaestionis* on the authorship of the *Mechanics* in the nineteenth and twentieth centuries. In my treatment of the question, I will especially analyse the arguments against the authenticity that are most frequently cited in the literature.

[32] Montucla (1799), 187.

question the authenticity of the text. Valentine Rose (1854) simply sets the work aside as being "trivial" and "confused".[33] His main argument against the authenticity is that Aristotle did not write any mathematical texts. Brandis (1862) and Zeller (1879) follow Rose's example and argue against the authenticity of the *Mechanics*.[34] They represent the view that the text does not fit into the philosophical system of Aristotle. Brandis even goes so far as to argue that in the exceptional case that some Aristotelian thoughts can be demonstrated in the *Mechanics* or in other minor works, it would not enhance our knowledge of Aristotle's philosophy.[35]

There is, however, no general agreement among nineteenth-century scholars about the authorship of the *Mechanics*. In the introduction to his German translation, Friedrich Theodor Poselger (1829) advocates the authenticity of the text. He mentions that the style and approach of the text are typically Aristotelian, and that, therefore, the *Mechanics* unquestionably belongs to the authentic treatises.[36] Another positive voice on the *Mechanics* is Moritz Cantor (1862). In opposition to Montucla's rejection, Cantor is deeply impressed by the subject matter of the text and the solutions offered to the mechanical questions. He holds that as a mathematician he does not possess the linguistic knowledge required to render a judgement on the authenticity of the text. Nevertheless, he makes the following observation:

> So viel dürfte indessen zu behaupten sein, dass die Mechanik im aristotelischen Geiste verfasst ist, dass ein innerer Widerspruch gegen andere Schriften des großen Gelehrten nicht nachgewiesen ist. Behaupten darf man auch, dass die Möglichkeit einer aristotelischen Mechanik ebensowenig geleugnet werden kann als die geistige Bedeutsamkeit der unter diesem Titel auf uns gekommenen Schrift.[37]

For that reason the *Mechanics* could very well have been written by Aristotle, as Cantor concludes: "Die sogenannte Mechanik des Aristoteles würde, sagen wir, seines Namens nicht unwürdig sein."[38]

The differing positions held by these scholars depend in particular on whether or not they value the *Mechanics*. In general, we see that nineteenth-century philologists are not impressed by the text and come to a negative judgement on its authorship, whereas mathematicians and historians of science recognize the significance of the text and thus deem Aristotle a suitable author. All critics merely address the style of the treatise, without raising any strong objections against Aristotle's authorship. The absence of substantiated arguments could result in completely unfounded and unconvincing opinions, such as the view expressed by Edward Seymour Forster at the beginning of his Oxford translation of the *Mechanics* (1913): "Whilst the scientific standpoint of the *Mechanica* is certainly Peripatetic, the writer's interest in the practical application of the problems involved is quite un-Aristotelian."[39] This

---

[33] See Rose (1854), 192.

[34] See Brandis and Zeller as cited in Nobis (1966), 265–276.

[35] See Nobis (1966), 266.

[36] See Poselger (1829), 65.

[37] Cantor (1862), 240.

[38] Cantor (1862), 241.

[39] Forster (1913), 12.

statement was easily countered by Marshall Clagett (1959): "The attribution to Aristotle has been questioned mainly on the ground that the treatise's attention to practical problems is 'quite un-Aristotelian,' which is doubtful reasoning at best, considering the enormous range of intellectual activity of Aristotle during his last years at the Lyceum."[40] Moreover, the author is not in the first place interested in the practical applications themselves, but rather in referring these back to underlying mathematical principles.

From the second half of the twentieth century onwards, we find an increasing number of scholars who are in favour of Aristotle's authorship. Through a close examination of the text, many parallels with other works from the Aristotelian corpus could be discovered. Paul Gohlke (1957), for example, in the introduction to his German translation, states that the *Mechanics* and the *Meteorology* might be near contemporary, since Aristotle mentions the parallelogram of motions in his *Meteorology* (342a24).[41] In his 1970 book *Dynamische und statische Betrachtungsweise in der antiken Mechanik*, Fritz Krafft presents a detailed comparison between the *Mechanics* and Aristotle's dynamics.[42] In this context, he pays special attention to *Physics* VII.2, a passage that will be considered in more detail below. Despite these scholars' efforts, the view that Aristotle is in fact the author of our treatise could not prevail. Therefore, I will now examine the arguments against the authenticity claim that are based on the actual subject matter of the text. There are three serious objections that are repeatedly found in the literature. Firstly, the use of mathematical argument in the *Mechanics* is highly unusual for Aristotle. This applies especially to the long geometrical proof in the first problem of the text. I would argue, however, that it is only logical for Aristotle to use a mathematical argument here, since this is consistent with how he defines mechanics in the introduction to the text and in the *Posterior Analytics*. Mechanics is a science that is subordinate to geometry, and in order to explain mechanical phenomena, a mathematical demonstration is sometimes required. Moreover, this kind of mathematical argument is also present in other works by Aristotle. In the *Meteorology* we find long geometrical discussions about haloes and rainbows.[43] These meteorological phenomena belong to the field of optics, and as such they are also appropriately studied under geometry.

The second consideration concerns the terminology of the *Mechanics*; part of the terminology used is post-Euclidean, and for this reason the text must be dated to a period later than Aristotle's own time. According to Johan Ludvig Heiberg, Euclid was responsible for introducing the term τετράπλευρον (quadrilateral) into the mathematical vocabulary.[44] Definition 19 in Book I of the *Elements* can be attributed to Euclid: "*Rectilineal figures* are those which are contained by straight lines, *trilateral* figures being those contained by three, *quadrilateral* those contained by

[40] Clagett (1959), 4.

[41] See Gohlke (1957), 6.

[42] See Krafft (1970), 48–78.

[43] See e.g. *Mete.* 373a5-19 for a geometrical description of the circular shape of the halo.

[44] See Heiberg (1904), 15.

four, and *multilateral* those contained by more than four straight lines."[45] The term τετράπλευρον is not found in Plato or Aristotle, except for its use in the *Mechanics* at 848b20 and the *Problems* at 911b3.[46] It is, however, the only term that speaks against Aristotle's authorship, as Heiberg admits.[47] The Euclidian terminology is generally more common in the *Mechanics* than in other works attributed to Aristotle, but the origins of these terms can be traced back to an earlier time than Euclid, for example ῥόμβος (rhombus). Furthermore, there are many parallels between the terms in the *Mechanics* and those in Aristotle's corpus that are not found in Euclid.[48] Characteristic of the *Mechanics* is the fact that there is less consistency in the terminology than there is in Euclid. The variety and interchangeability of the terms employed is especially visible in the practice of applying letter labels to mechanical objects. The various techniques used here are similar to those applied by Aristotle elsewhere, among which the *Meteorology* is a particularly rich case. For Heiberg, this all points to two options regarding the date of origin of the *Mechanics*:

> Hieraus könnte man schließen wollen, dass die Mechanik entstanden sei, ehe Euklid die mathematische Terminologie konsequenter und bequemer machte, und eine solche Annahme bleibt möglich; denn dass τετράπλευρον erst von Euklid geschaffen sei, ist nur eine Vermutung, wenn auch eine sehr wahrscheinliche. Aber die Tatsachen der Terminologie lassen sich ebensogut durch die Annahme erklären, dass das Werkchen nach Euklid in Kreisen verfasst worden, die von der älteren, bzw. der aristotelischen, Terminologie beherrscht nur teilweise sich der euklidischen anbequemt hatten.[49]

Krafft agrees with Heiberg on a possible return to an older vocabulary at a time when a new terminology was already available.[50] He asserts, however, that it is implausible to maintain this with respect to the practice of applying letter labels to mechanical objects. It would have been highly unlikely for an author to use an older expression like τὸ εφ' οὗ AB, when he had to hand the direct identification of a letter with an object: τὸ AB. Although we see that authors writing on mechanics after Euclid preponderantly use direct identification of a letter with an object, it is conceivable that the author of the *Mechanics* adopted an archaic style. Similar periphrastic expressions by means of *epi* are for example also found in the works of the engineers Philo of Byzantium (latter half of the third century BC) and Hero of Alexandria (first century).[51]

In their arguments, both Heiberg and Krafft assume that Euclid was responsible for a systematization of the mathematical vocabulary. However, the development of the mathematical language should not be understood as being fixed at a specific point in time. Aristotle's use of mathematics is for the most part similar in style to Euclid, and the marginal differences between the two can be explained by the fact

[45] Transl. Heath (2002), 2.

[46] See Heiberg (1904), 15.

[47] See Heiberg (1904), 31.

[48] See Heiberg (1904), 31.

[49] Heiberg (1904), 32.

[50] See Krafft (1970), 92.

[51] For this periphrastic use of *epi* see e.g. Philo's *Bel.* (over 20 times) or Hero's *Aut.* (about 5 times).

that Aristotle used mathematics for his own special purposes, as Reviel Netz points out.[52] We should not suppose that the mathematical lexicon was set by Euclid, for it was already used by Aristotle, and many new terms were still being introduced into the lexicon after Euclid's time. For example, Archimedes, as he wrote *On Floating Bodies*, needed to introduce some terms into the mathematical lexicon for his specific purposes.[53] He borrowed these terms from everyday Greek, which is a characteristic feature of the mathematical lexicon, namely, that it is built up of ordinary words with different technical meanings.

Because the mathematical lexicon as a whole was not fixed by Euclid, we cannot, in a strict sense, distinguish mathematical terminology that is pre- or post-Euclidian. Moreover, the language in the *Mechanics* is not entirely comparable with the mathematical vocabulary. Although Aristotle deploys mathematical terminology in the *Mechanics*, the text does not belong to the mathematical discourse. Instead of singling out certain mathematical terms as pre- or post-Euclidian, we should focus instead on the fact that the various practices used in applying letter labels to objects in the *Mechanics* are consistent with the techniques in the authentic texts by Aristotle, and it is this that speaks for his authorship.[54]

The third argument most scholars raise against the authenticity of the *Mechanics* concerns the idea that the division into natural and unnatural motion as it is presented in the first problem of the text, is incompatible with Aristotelian doctrine that downward motion is natural to heavy bodies.[55] I believe that this objection is based on the mistaken notion that Aristotle's description of natural motion in the *Physics*, as the motion of an object to its proper place in the universe, has to be equivalent to his analysis in the *Mechanics*. Furthermore, this claim is not based on the text and diagrams as they are contained in the Greek manuscripts, but rather on an emendation of the text by Johannes van Cappelle (1812) and a modern diagram of circular motion. I will first describe the interpretation of this difficult passage as it is given by most commentators on the text, and then offer an alternative analysis of natural motion in the context of an understanding of mechanics based on the text as it is contained in the manuscripts.

In the first problem of the text, the author divides rectilinear and circular motion into different components. Lines 849a14-16 of the text further specify these components in the case of circular motion as being either natural or unnatural. Van Cappelle in his edition reads this passage as follows: "This happens with every line which describes a circle, and it moves along the circumference, according to nature towards the tangent, but contrary to nature towards the centre."[56] Upon this reading of the text, the component of circular motion that is κατὰ φύσιν (*kata phusin*) is tangential,

---

[52] See Netz (1999), 275.

[53] See Netz (1999), 118.

[54] See Sect. 4.2.2 for a description of the different techniques used by the author of the *Mechanics* in applying letter labels to mechanical objects.

[55] For a recent statement of this opinion, see Berryman (2009), 109–110.

[56] πάσῃ μὲν οὖν κύκλον γραφούσῃ τοῦτο συμβαίνει, καὶ φέρεται κατὰ τὴν περιφέρειαν, τὴν μὲν κατὰ φύσιν εἰς τὸ πλάγιον, τὴν δὲ παρὰ φύσιν εἰς τὸ κέντρον.

whereas the component that is παρὰ φύσιν (*para phusin*) is directed towards the centre. Most scholars base their interpretation on this reading of the text and note that this natural tangential motion is unrelated to the Aristotelian notion that downward motion is natural to heavy bodies. In their commentaries they refer to the corresponding diagram by van Cappelle (Fig. 1.1), where the tangential motion is depicted as a lateral motion.[57] Given that the first problem of the *Mechanics* describes the motion of the arms of a balance, it would make more sense to depict the tangential motion as a downward motion. As is apparent from Fig. 1.2, this is exactly how natural motion was understood in the Byzantine period. In the manuscript diagrams the tangential motion is directed downwards, and is therefore consistent with a heavy body's downward motion to its proper place.

However, as both Christiane Vilain and Mark Schiefsky rightly note, this analysis of circular motion is understood by the author in a much more general sense.[58] It not only describes the specific case of circular motion in the arms of a balance, but should be applied to circles in any orientation. If we look at problem 8 of the text, which questions why circular bodies are easiest to move, we find a different orientation of circular motion. Here, in the case of a potter's wheel, the tangential motion is sideways, and not in a downward direction. But in the case of the balance too the situation is more complicated. Even if one motion is directed downwards on one side of the circle, there is an upward motion on the other side. And they cannot both be natural movements of heavy bodies, as Vilain asserts.[59]

Schiefsky offers a solution to the problem posed by the fact that natural motion as it is defined in the *Mechanics* has different directions. He understands the motion

Fig. 1.1 Edition by van Cappelle

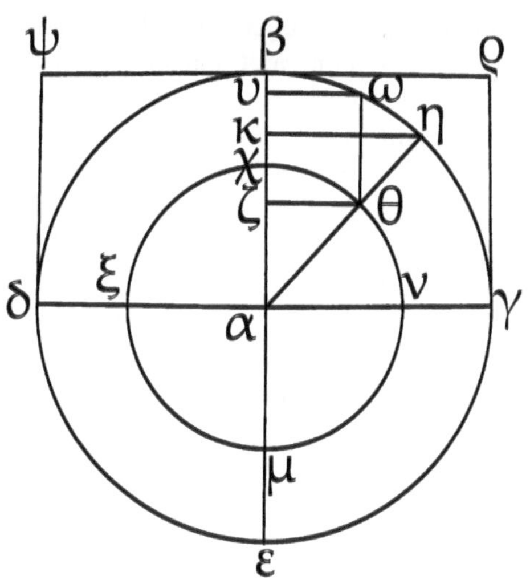

[57] See e.g. Vilain (2008), 153 or Schiefsky (2009), 57 n.15.

[58] See Vilain (2008), 154 and Schiefsky (2009), 57 n.15.

[59] See Vilain (2008), 154.

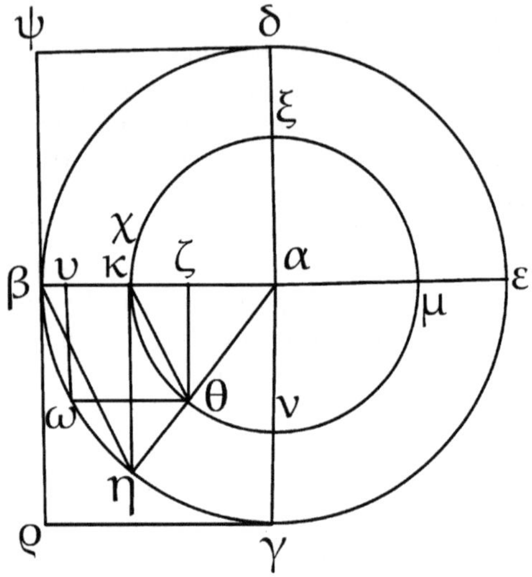

Fig. 1.2 Reconstruction of the archetype

that is *kata phusin* not in the strict sense as the motion of an object towards its proper place. Instead, the contrast between *kata phusin* and *para phusin* should be interpreted as a contrast between free and constrained motion.[60] An object moving in a circle, for example, a stone in a sling, would fly off in a straight line if it moved freely; it is, however, constrained by a motion to the centre of the circle, and therefore the resulting motion is curvilinear. We can see how this interpretation works, not only in the case of the arms of a balance, but also for circles with different orientations, like the potter's wheel in problem 8 of the text. Schiefsky's analysis is, however, based on van Cappelle's emendation of the text in lines 849a14-16 and does not take into account the reading of the manuscripts, to which I will turn below.

The notion that circular motion consists of a natural component in the direction of the tangent, and an unnatural one towards the centre, is not presented in the text of any of the extant manuscripts. The codices contain the following reading of lines 849a14-16: "This happens with every line which describes a circle, and it moves according to nature along the circumference, but contrary to nature towards the tangent and the centre."[61] Here a natural circular motion is composed of two unnatural components: one that is tangential, and another that is directed towards the centre. A natural motion made up of unnatural constituents complicates an interpretation of this passage, but the primary reason van Cappelle gives for emending the text is that the author of the *Mechanics* is mistaken in his explanation of the free motion of a radius in a circle. There cannot be a completely free motion of a radius along the

[60] See Schiefsky (2009), 57 n.15.

[61] πάσῃ μὲν οὖν κύκλον γραφούσῃ τοῦτο συμβαίνει, καὶ φέρεται τὴν μὲν κατὰ φύσιν κατὰ τὴν περιφέρειαν, τὴν δὲ παρὰ φύσιν εἰς τὸ πλάγιον καὶ τὸ κέντρον.

circle's circumference, since it is constrained and held back due to the fact that one end of it is fixed.[62]

Gianni Micheli is the only commentator who retained the manuscript reading of these lines in his interpretation of natural motion. The starting point of his analysis is based on the distinction between ἰσχύς (*ischus*) and ῥοπή (*ropē*) as it is expounded by Krafft.[63] Both *ischus* – along with the use of δύναμις (*dunamis*) in Aristotle – and *ropē* are distinct types of forces. *Ischus* is a force that acts on a body from without, and can be understood in the modern sense as the product of weight and speed, while *ropē* is a tendency that is inherent in natural bodies, through which, for example, heavy bodies move in a downward direction. Krafft translates *ropē* by the term "Bewegungskraft", in which the motion stands for a body's proper motion.[64] Upon his interpretation of lines 849a14-16, Micheli connects the motion of the radius along the circumference with *ropē*, the natural tendency of bodies.[65] The tangential motion is related to *ischus*, and is unnatural, since it is effected from without by the weight on the balance. The curvilinear and tangential motions are influenced by another motion, namely that towards the centre of the circle. By this unnatural motion the radius is attracted to the centre of the circle.

According to Micheli's analysis, the author of the *Mechanics* is not primarily interested in the contrast between *kata phusin* and *para phusin* motions. It is the distinction between *ropē* and *ischus* that is central to the author's argument. The tangential component of motion is therefore not per se against nature, but the motion that is caused by *ischus* is unnatural.[66] As an example, Micheli mentions problem 8 of the text, where the tangential component of motion in a potter's wheel is caused by a force working from the outside (*ischus*). All motions that are generated by *ischus* are unnatural, irrespective of their orientation; sometimes this motion is directed downwards (e.g. in a balance), sometimes sideways (e.g. in a potter's wheel). In this way, Micheli solves the difficulty of differently orientated tangential motions without resorting to the interpretation of *kata phusin* and *para phusin* as free and constrained motion, which is based on the modern notion of centrifugal and centripetal forces. Nevertheless, one further difficulty still remains with this interpretation. Micheli connects the motion along the circle's circumference with *ropē*, the tendency that is inherent in natural bodies. The diagram from the Byzantine manuscripts depicts the motion along the circumference as a downward motion, which is the natural motion of heavy bodies. We have already seen that the author's analysis of circular motion should, however, be understood in a more general sense. Micheli's interpretation does not offer an explanation for the general case, in which the curvilinear motion has a different orientation, – sideways, for example, in case of the potter's wheel.

[62] See van Cappelle (1812), 157.

[63] See Krafft (1970), 74.

[64] See Krafft (1970), 74.

[65] See Micheli (1995), 65.

[66] See Micheli (1995), 64–65.

In what follows, I would like to present a new analysis of this problematic passage based on the text as it is contained in the manuscripts. All of the commentators on the text depart from the strange starting point that natural motion in the *Mechanics* should be understood in terms of an object's motion towards its proper place as it is presented in Aristotle's natural philosophy. The author of the *Mechanics* is not concerned, however, with the natural motion of physical objects. He does not question, for instance, the motion of a wheel, that, in this respect, would naturally fall towards the earth. The author's inquiry is an entirely different one, since he is interested in finding out which motion is natural to the specific properties of a mechanical object.[67] If we look at circular motion in a wheel, we notice that it is natural to the properties of a round object to move along its circumference. The primary concern in the text is to examine natural motion in the specific case of mechanical objects, and not to study the natural position of, for example, a wheel in the physical world. Hence, the concept of natural motion as it is expounded in the *Mechanics* is rather different from Aristotle's natural philosophy, but not necessarily incompatible with the ideas expressed in the *Physics*.

If we look again at lines 849a14-16 of the text, the terms *kata phusin* and *para phusin* acquire a different meaning. The motion that is natural to the properties of round objects, is a motion along the circumference. It is the resulting motion that is composed of two basic motions: one towards the tangent and the other towards the centre. Whereas the resulting curvilinear motion is natural, both components of which this motion is made up, are unnatural.[68] These two unnatural components differ from each other in that the tangential motion is forced from without, for example, by a weight placed on one side of a balance, while motion towards the centre is inherent in the object itself and draws back to the centre. The motions towards the tangent and the centre of a circle are *para phusin* in the sense that they do not belong to the properties of a circle, whilst the motion that is *kata phusin* along the circle's circumference is proper to the nature of a circle.

A passage in *Physics* VII.2 might enlighten our understanding of circular motion in machines.[69] Here, Aristotle describes forcible circular motion as being produced by a combination of pushing and pulling (lines 244a2-4): "And twirling is a compound of pulling and pushing; for that which is twirling a thing must be pulling one part of the thing and pushing another part, since it impels one part away from itself and another part towards itself."[70] The person who imparts circular motion to an object, for example, in a potter's wheel, simultaneously pushes and pulls at different sides of the wheel. These motions of pushing and pulling occur at the tangents of the circle, and are forced upon the wheel from without by the person handling it. The

[67] I am grateful to Ido Yavetz for pointing this out to me.

[68] *Ph.* VIII.4 offers a parallel in Aristotle of a natural motion consisting of unnatural components. Lines 254b17-19 describe the case of animal motion, in which the animal as a whole moves itself naturally, whereas its parts may be in motion in an unnatural way. I am grateful to Henry Mendell for this reference.

[69] See Krafft (1970), 64 for a reference to this passage.

[70] Transl. Hardie and Gaye in Barnes (1984), 410.

identification of tangential motion for the case of spinning an object in *Physics* VII.2 agrees with the unnatural component of circular motion towards the tangent in the *Mechanics*. This understanding of circular motion as the result of a combination of forced rectilinear motions, finds parallels in problem 8 of the *Mechanics*. Here the author describes the tangential component of circular motion for round objects like a wheel, pulley or potter's wheel. The moving force pushes the object in the direction of the tangent (852a7-13), which can again be identified as an unnatural motion that is forced from without.

The analysis of circular motion in the *Mechanics* has shown the importance of a critical examination of the manuscript material for this treatise. The text of the Greek manuscripts together with the corresponding diagrams are very different from the textual basis on which most commentators rely in their interpretation of natural motion. Starting from the reading in the authentic manuscripts, we find a different notion of natural motion when compared with Aristotle's *Physics*. In the specific context of mechanics, a natural circular motion can be defined as a motion that is natural to the properties of a round object. This explanation stands apart from the idea of an object's motion to its proper place and thus does not pose any unavoidable obstacles within the framework of Aristotle's natural philosophy.

None of the aforementioned arguments against the authenticity of the treatise necessarily precludes Aristotle's authorship. The evidence is far from conclusive, and the opinions of most scholars are more or less based on a gut feeling that the character and style of the text are un-Aristotelian. Although the arguments involving the use of mathematical vocabulary in the *Mechanics* or its description of natural motion cannot plausibly sustain the case of those who argue against its authenticity, there is no positive evidence either that this text must have been written by Aristotle. There are certain indications of the dating of the treatise, we know for example that the text was probably written before Archimedes' time, since certain concepts, such as the idea of the centre of gravity, are absent from the text.[71] We have also seen parallels between the *Mechanics* and the works by Aristotle. Most notably, the connection between mechanics and mathematics in the introduction of our text is defined along the same lines as in the Aristotelian corpus. Both emphasize the importance of mathematics for mechanics and describe the mechanical discipline as being subordinate to mathematical principles. Moreover, we cannot simply ignore the fact that most early modern scholars did not doubt the authenticity of the treatise. Among them, Piccolomini fervently argued for the authorship of Aristotle by demonstrating that there were no internal inconsistencies between the *Mechanics* and Aristotle's works. This would all make it probable that we have to look for our author in the circle of Peripatetics. Recent research on the *Mechanics* also points in this direction, as, for example, Michael Coxhead's paper on Aristotle's use of poets for the purpose of persuasion.[72] Without taking sides on the authorship question, Coxhead shows that the citation of the tragic poet Antiphon in the introduction to the *Mechanics* is similar to Aristotle's quotations in other works. While

[71] See van Cappelle (1812), 277.

[72] See Coxhead (2012), 300–306.

showing parallels between the Aristotelian corpus and the *Mechanics* does not offer any convincing proof in favour of Aristotle being the author of the text nor so much as hints at a certain Peripatetic philosopher, my analysis of the counter-arguments has shown that his authorship cannot be precluded. I believe the best way to consider the issue of authorship is to approach it with an open mind, which might eventually lead to suggesting a different candidate as the author of the *Mechanics* than those previously mentioned.[73]

1.3 History of the Text in Antiquity

In this section I will trace the history of the *Mechanics* in antiquity by examining the references and allusions to the text that precede the Byzantine manuscript tradition. But first, I will briefly relate a story that was current in antiquity on the fate of Aristotle's library. Strabo and Plutarch represent our main sources for this story.[74] After his death in 322 BC, Aristotle left his own library to Theophrastus, his successor as head of the Lyceum. Then it passed to Neleus, who took the books to the city of Scepsis in Asia Minor and left them to his successors, as Strabo reports. The books were stored carelessly underground in some sort of tunnel, where they disappeared for almost two centuries. At some point early in the first century BC, the family sold the books of Aristotle and Theophrastus to Apellicon of Teos, who lived in Athens. Apellicon is characterized by Strabo as a bibliophile rather than a philosopher, which means that he prepared new books that were full of errors, as he tried to repair the damage done to the books by being stored underground for such a long time. After capturing Athens in 86 BC, Sulla took the library to Rome, where the scholar Tyrannio prepared the books, and Andronicus of Rhodes subsequently made an edition of Aristotle.

We do not know how much of this story is true, but it is all the evidence we have. As Barnes rightly points out, the ancient story states that Aristotle's *books* disappeared for two centuries; it does not say anything about his *works*.[75] It is likely that there were still some copies left at the Lyceum. Although the Peripatetic School declined after Theophrastus' death and Aristotle's works might not have been easily accessible for two centuries, it was not altogether impossible to study Aristotle.

[73] Winter (2007), iii–ix argues that the fourth-century BC Pythagorean philosopher and mathematician Archytas of Tarentum is the author of the *Mechanics*. Winter does not give any further arguments for this attribution other than that Archytas is the most likely author among those mentioned by Vitruvius in a list of sources on mechanics. Nevertheless, the suggestion in itself might be worthwhile to consider. Archytas' conception of natural motion as it can be reconstructed from the fragments does seem to bear similarities to the theory we find in the *Mechanics*. See Huffman (2005), 526–529 on Archytas' account of natural motion, especially his distinction between natural motion and instrumental motion.

[74] For the texts by Strabo and Plutarch, as well as commentaries on these texts, see Barnes (1997), 1–69.

[75] See Barnes (1995), 10.

There is reason to assume that later Peripatetics knew at least some of Aristotle's writings.[76] What we know for certain, then, is that Andronicus prepared an edition of the texts of Aristotle in the first century BC. Andronicus' activities merely consisted in the grouping of the different texts, and even his work as arranger of the treatises was not epoch-making, as Barnes has shown.[77] Andronicus' text of Aristotle exerted little influence on later authors, other than that we can still distinguish some of his achievements in the ancient catalogue of Aristotle's works by Ptolemy al-Gharīb. There are three such ancient lists, all of which include a work on mechanics written by Aristotle.

The earliest references in this connection are found in two ancient lists, one in Diogenes Laërtius, and the other the *Vita Hesychi*, an anonymous list also known in the literature under the name of *Vita Menagiana*.[78] These lists can be traced back to a common source somewhere in the third century BC. The author of the list is still disputed; it was either Hermippus of Smyrna, a learned Alexandrian librarian who catalogued the holdings of the library of Alexandria,[79] or it was compiled by Aristo of Ceos, a Peripatetic philosopher who probably succeeded Lyco as head of the Lyceum.[80] In the catalogue of Diogenes Laërtius we encounter a mechanical treatise at number 123 under the title Μηχανικὸν ᾱ. As Paul Moraux notes, this work is likely to be the same text that we have now in the Aristotelian corpus.[81] In the *Vita Hesychi* we find the exact same entry under number 114. In both lists the text on mechanics is contained along with other works that belong to the ancient genre of *problēmata*. Different collections of *problēmata* and *aporēmata* are mentioned in this connection, such as the *Homeric Questions*, *Physical Problems*, and *Problems from Democritus*. Thus, a mechanical treatise which might be identified with our *Mechanics*, is already listed at a very early time among the collections of Aristotelian problems.

A third ancient catalogue of Aristotle's writings has been transmitted to us in the Arabic tradition and is attributed to Ptolemy al-Gharīb.[82] It is posterior to the other two lists and must be dated after the edition of Aristotle by Andronicus of Rhodes in the first century BC.[83] This is evident from Ptolemy's proem, in which he mentions Andronicus several times.[84] Ptolemy aims to give a compendium of Aristotle's works in contrast to Andronicus' list, which counted about 1000 writings. A work titled *Mechanical Problems* is mentioned in the first part of the list at number 22: Μηχανικὰ προβλήματα. This title represents without a doubt the same work as in

[76] See Barnes (1997), 14–16.

[77] See Barnes (1997), 66.

[78] These lists are found in Düring (1957), 29–56 and 82–89.

[79] Düring (1957) defends this thesis at 67–69.

[80] Moraux (1951), 211–247 argues for this attribution.

[81] See Moraux (1951), 120.

[82] See Hein (1985), 415–439 for the Arabic text and Greek translation of Ptolemy's catalogue, including a proem to Gallus.

[83] See Moraux (1951), 289.

[84] See Hein (1985), 416–419.

the other ancient catalogues, as Moraux notes.[85] It is mentioned together with other collections of problems and dialogues, many of which are now lost.

Another ancient reference can be found in the *Vita Aristotelis Marciana*. This biography of Aristotle is preserved on two pages of *Marc.gr.* 257, a codex from the thirteenth century.[86] Here, a treatise on mechanics is again associated with the ancient genre of *problēmata*. It is referred to as *Mechanical Problems*, μηχανικὰ προβλήματα, along with other problems on physics (φυσικὰ προβλήματα) and optics (ὀπτικὰ προβλήματα). The connection between mechanics and the branch of optics is found at several places in the *Corpus Aristotelicum*. Both sciences are characterized by Aristotle as being subordinate to mathematical principles.[87] The *Vita Marciana* is a descendant of an original epitome of Ptolemy's *Life of Aristotle*, which existed in the latter half of the fifth century.[88] Since it was used in oral instruction, it cannot be attributed to one single author, but is rather the collective product of several generations of Neoplatonic students of Aristotle.[89] Once the school tradition had been broken, it was used as an introduction to any manuscript containing the writings of Aristotle. But it is still most commonly found in manuscripts containing the text of, or commentaries on the *Categories*, as Ingemar Düring asserts.[90]

Because the *Vita Aristotelis Marciana* was part of the Neoplatonic school tradition, there is some interplay with commentaries on the works of Aristotle from the sixth century. The commentaries on the *Categories* by Simplicius, Olympiodorus and Elias all list mechanical writings under mathematics. Simplicius makes the connection between treatises on geometry and mechanics: τὰ δὲ μαθηματικά, οἷα γεωμετρικά τε αὐτῷ γέγραπται καὶ μηχανικὰ βιβλία.[91] Olympiodorus in his commentary mentions optics and mechanics together in relation to mathematics: μαθηματικὰ δὲ οἷα τὰ Μηχανικὰ καὶ 'Οπτικὰ προβλήματα.[92] A similar classification is found in Elias' commentary on the *Categories*: τὰ δὲ μαθηματικὰ αὐτοῦ ὡς τὰ 'Οπτικὰ καὶ Μηχανικὰ αὐτῷ βιβλία γεγραμμένα.[93] Christel Hein refers in this context to the Arabic classification of the works of Aristotle, in which not only the treatises on optics and mechanics, but also the text *On Indivisible Lines* are included in the mathematical writings.[94] This compilation of non-authentic or non-extant treatises is apparently a characteristic of the Alexandrian tradition, as Hein asserts.[95]

In addition to the ancient catalogues, it is important to scrutinize the circulation of Aristotle's works in antiquity. There is some evidence for this in the papyri find-

[85] See Moraux (1951), 120.
[86] See Gigon (1962), 8.
[87] See n.29.
[88] See Düring (1957), 117.
[89] See Düring (1957), 118–119.
[90] See Düring (1957), 119.
[91] Simplicius, *in Cat.*, 4.25–26.
[92] Olympiodorus, *in Cat.*, 7.33.
[93] Elias, *in Cat.*, 116.11–12.
[94] See Hein (1985), 304.
[95] See Hein (1985), 305.

ings: 13 papyri fragments of Aristotle have been preserved.[96] These papyri can for the most part be dated to the end of the first century and contain fragments of the following works: *Posterior Analytics*, *On the Heavens*, *Categories* (2), *Nicomachean Ethics*, *History of Animals*, *Politics*, *Protrepticus*, *Constitution of Athens* (3) and *Topics* (2). Insofar as it is possible to generalize on the basis of only 13 fragments, these texts do not seem to be completely representative of Aristotle's works. Since five of these fragments belong to Aristotle's logical works, and four to his political texts, we might say that there may have been considerable interest in Aristotle's logic and politics in antiquity, perhaps in the context of school exercises. Aristotle's works on natural philosophy, which occupy more than half of all Bekker pages, are somewhat underrepresented in the papyri findings. This might be connected to the conclusion Netz draws from the papyrological evidence that there was no general interest in scientific texts as such in antiquity.[97] They were merely collected for other purposes, one of them being their literary value.

Aristotle was not widely read in antiquity; this becomes evident when we compare the number of extant papyri of Aristotle to those of Plato: 13 to 104. The main reason for Aristotle's unpopularity is probably related to the style and difficulty of his esoteric works. The division of Aristotle's works into esoteric and exoteric is an ancient one.[98] The exoteric works were written for a larger public and were meant for publication, but unfortunately very few fragments of them have survived. The esoteric works can be regarded as unpolished drafts that were made for the use of philosophers and students within the school; these texts are the ones that have been preserved and that we now read.

The meagre papyrological evidence of Aristotle's works does not give cause for hope that we will find many allusions to the *Mechanics* in other ancient treatises on mechanics. In several different contexts it is stated that Aristotle is not mentioned by Vitruvius in Book VII of his *De Architectura* among the authors of books on machines. This might be an unfortunate coincidence, but it can by no means serve as conclusive evidence that such a text by Aristotle does not exist.[99] We have to look further for references to the *Mechanics*. Hero of Alexandria, who was active in the first century, wrote several books on mechanics that are only preserved in Arabic manuscripts. In these books Hero was clearly influenced by the *Mechanics*. The most conspicuous example of this is the inclusion of a list of mechanical questions of the same type as in the Aristotelian *Mechanics* at the end of Book II.[100] Some of

[96] I searched for papyri through the online database CEDOPAL.

[97] I am grateful to Reviel Netz for allowing me to read his manuscript *Space, Scale, Canon: Parameters of Ancient Literary Culture*.

[98] See Barnes (1995), 12.

[99] Vitruvius in *De Architectura* VII mentions numerous mechanicians from the Hellenistic period and does not aim to give an exhaustive list of all ancient authors on mechanics. Therefore, we should not take his list as a starting point and simply assume that the author of the *Mechanics* must have been mentioned by Vitruvius among the persons writing on machines, as Winter (2007), iii–ix does.

[100] See Nix and Schmidt (1900), 174–189.

these questions are exactly similar to those that are found in our treatise, only the answers given by Hero are different, and sometimes more correct. Question *i* for example: "Why, in extracting teeth, does a dentist use a forceps instead of the hand?" is comparable to our problem 21, or question *p*: "Why is it that a small rudder displaces a large ship?" is similar to problem 5 of the *Mechanics*. At no point in the text does Hero mention Aristotle by name, although he usually refers to his sources, as he for instance names Archimedes several times in his works. A possible explanation for this might be that Hero refers to an ancient genre of mechanical questions rather than to the *Mechanics* specifically, but the analogies to our text are striking.

There is actually one ancient mechanician who refers to Aristotle in the context of mechanics, namely Athenaeus. Controversy surrounds the dating of the works of Athenaeus, but he was most likely active in the first century BC.[101] In his book *De Machinis*, Athenaeus mentions Strato, Hestiaeus, Archytas and Aristotle as authors who wrote lengthy and unnecessary treatises:

> But those writing something or giving us advice and appearing to do this for our own good, not unreasonably, write at length in unnecessary treatises and spend their time displaying how much they know; for they leave their books filled with digressions; and this despite the fact that it was well said by the ancient philosophers that one should know how to measure opportunity, as being a defining-point of wisdom, This, you see, is how anyone setting about a practical treatise should have benefited: having carefully understood himself on the basis of the famous Delphic precept, rather than the works of Strato and Hestiaeus and Archytas and Aristotle and the others who have written works similar to theirs. For younger devotees of knowledge they would be useful <as a training> in elementary principles; but for those already wanting to *do* something they would be altogether irrelevant and detached from practical thinking.[102]

The works of these authors are criticized by Athenaeus for being detached from practical thinking. Even though we cannot exclude the possibility that Athenaeus here refers to some other text by Aristotle, the connection between Aristotle and Strato in particular, given that Strato is also included in the ancient lists as the author of a work on mechanics, makes it likely that Athenaeus had in mind a mechanical treatise by Aristotle. The fact that we can find fewer references in early mechanical texts than in a somewhat later author like Hero might be related to the diffusion of Aristotle's works before the edition by Andronicus.

The evidence in antiquity for the Aristotelian *Mechanics* is not overwhelming. Although we do not possess any ancient commentaries on the text, a mechanical treatise – whether or not we identify it with the *Mechanics* – was known in antiquity to have been written by Aristotle. It is mentioned in several ancient commentaries on the *Categories* and included in all important ancient lists. However, in all of these cases a text on mechanics is listed in connection with other collections of *problēmata* or mathematical works that are either lost, or commonly regarded as spurious, like the treatise the *Problems*. A number of ancient mechanicians were influenced by the

[101] See Whitehead and Blyth (2004), 15–20.

[102] Transl. Whitehead and Blyth (2004), 45.

form and subject matter of the treatise, as, for example, Hero, although he does not refer to Aristotle by name. Here, I could find only the mechanician Athenaeus, who actually mocks a treatise by Aristotle for being too theoretical and of no practical use. In the next chapter, we will shift our focus from these ancient references to the actual Byzantine manuscript material of the *Mechanics*. When compared with its ancient classification among texts that are today generally not considered to have been written by Aristotle, in the Greek manuscript tradition the text of the *Mechanics* is often contained alongside authentic Aristotelian treatises.

Bibliography

Asper, Markus. 2007. *Griechische Wissenschaftstexte: Formen, Funktionen, Differenzierungsgeschichten*. Stuttgart: Franz Steiner Verlag.

Barnes, Jonathan. 1984. *The complete works of Aristotle*, 2 vols. Princeton: Princeton University Press.

Barnes, Jonathan. 1995. Life and work. In *The Cambridge companion to Aristotle*, ed. Jonathan Barnes, 1–26. Cambridge: Cambridge University Press.

Barnes, Jonathan. 1997. Roman Aristotle. In *Philosophia Togata II: Plato and Aristotle at Rome*, eds. J. Barnes and M. Griffin, 1–69. Oxford: Clarendon Press.

Berryman, Sylvia. 2009. *The mechanical hypothesis in ancient Greek natural philosophy*. Cambridge: Cambridge University Press.

Blair, Ann. 1999. The *Problemata* as a natural philosophical genre. In *Natural particulars: Nature and the disciplines in Renaissance Europe*, eds. A. Grafton and N. Siraisi, 171–204. Cambridge, MA: MIT Press.

Bodnár, István. 2011. The pseudo-Aristotelian *Mechanics*: The attribution to Strato. In *Strato of Lampsacus: Text, translation and discussion*, eds. M.L. Desclos and W. Fortenbaugh, 443–455. New Brunswick/London: Transaction Publishers.

Bottecchia Dehò, Maria Eisabetta. 2000. *Aristotele, Problemi meccanici: introduzione, testo greco, traduzione italiana, note*. Catanzaro: Rubbettino.

Busse, Adolf (ed.). 1900. Eliae in Porphyrii Isagogen et Aristotelis Categorias Commentaria. In *Commentaria in Aristotelem Graeca* 18,1. Berlin: Reimer.

Busse, Adolf (ed.). 1902. Olympiodori Prolegomena et in Categorias Commentarium. In *Commentaria in Aristotelem Graeca* 12,1. Berlin: Reimer.

Cantor, Moritz. 1862. *Vorlesungen über Geschichte der Mathematik*, vol. 1. Leipzig: Teubner.

Clagett, Marshall. 1959. *The science of mechanics in the Middle Ages*. Madison: University of Wisconsin Press.

Coxhead, Michael A. 2012. A close examination of the pseudo-Aristotelian *Mechanical Problems*: The homology between mechanics and poetry as *technē*. *Studies in History and Philosophy of Science* 43: 300–306.

Düring, Ingemar. 1957. *Aristotle in the ancient biographical tradition*. Göteborg: Almqvist & Wiksell.

Flashar, Hellmut. 1962. *Aristoteles: Problemata Physica*. Berlin: Akademie Verlag.

Forster, Edward Seymour. 1913. *The works of Aristotle*. Oxford: Clarendon Press.

Gigon, Olof. 1962. *Vita Aristotelis Marciana*. Berlin: De Gruyter.

Gohlke, Paul. 1957. *Aristoteles: Kleine Schriften zur Physik und Metaphysik*. Paderborn: Schöningh.

Heath, Thomas L. 2002. *Euclid's Elements*, ed. D. Densmore. Santa Fe: Green Lion Press.

Heiberg, Johan Ludvig. 1904. Mathematisches zu Aristoteles. In *Abhandlungen zur Geschichte der mathematischen Wissenschaften*, 3–49. Leipzig: Teubner.

Hein, Christel. 1985. *Definition und Einteilung der Philosophie: von der spätantiken Einleitungsliteratur zur arabischen Enzyklopädie*. Frankfurt am Main: Peter Lang.

Huffman, Carl A. 2005. *Archytas of Tarentum: Pythagorean, philosopher and mathematician king*. Cambridge: Cambridge University Press.

Kalbfleisch, Karl (ed.). 1907. Simplicii in Aristotelis Categorias Commentarium. In *Commentaria in Aristotelem Graeca* 8. Berlin: Reimer.

Krafft, Fritz. 1970. *Dynamische und statische Betrachtungsweise in der antiken Mechanik*. Wiesbaden: Franz Steiner Verlag.

Lennox, James G. 2001. *Aristotle's philosophy of biology: Studies in the origins of life science*. Cambridge: Cambridge University Press.

Leonico Tomeo, Niccolò. 1525. *Nicolai Leonici Thomaei Opuscula nuper in lucem aedita quorum nomina proxima habentur pagella*. Venice: Bernardino Vitali.

Micheli, Gianni. 1995. *Le origini del concetto di macchina*. Florence: Olschki.

Montucla, Jean-Étienne. 1799. *Histoire des mathématiques*. Paris: Agasse.

Moraux, Paul. 1951. *Les listes anciennes des ouvrages d'Aristote*. Louvain: Ed. universitaires de Louvain.

Netz, Reviel. 1999. *The shaping of deduction in Greek mathematics: A study in cognitive history*. Cambridge: Cambridge University Press.

Nix, Leo Michael Ludwig and Wilhelm Schmidt (eds.). 1900. *Herons von Alexandria Mechanik und Katoptrik*. Leipzig: Teubner.

Nobis, Heribert M. 1966. Die wissenschaftshistorische Bedeutung der peripatetischen 'Quaestiones Mechanicae' als Anlass für die Frage nach ihrem Verfasser. *Maia* 18: 265–276.

Piccolomini, Alessandro. 1565. *In Mechanicas Quaestiones Aristotelis, Paraphrasis paulo quidem plenior*, 2nd ed. Venice: Traianus Curtius.

Poselger, Friedrich Theodor. 1829. Über Aristoteles Mechanische Probleme. *Abhandlungen der Königlichen Akademie der Wissenschaften zu Berlin*: 57–92.

Rose, Valentine. 1854. *De Aristotelis Librorum Ordine et Auctoritate Commentatio*. Berlin: Reimer.

Rose, Paul Lawrence and Stillman Drake. 1971. The pseudo-Aristotelian *Questions of Mechanics* in Renaissance culture. *Studies in the Renaissance* 18: 65–104.

Ross, David. 1952. *The works of Aristotle*. Oxford: Clarendon Press.

Schiefsky, Mark. 2009. Structures of argument and concepts of force in the Aristotelian *Mechanical Problems*. *Early Science and Medicine* 14: 43–67.

Taub, Liba. 2008. *Aetna and the moon: Explaining nature in ancient Greece and Rome*. Corvallis: Oregon State University Press.

Valleriani, Matteo. 2009. The transformation of Aristotle's *Mechanical Questions*: A bridge between the Italian Renaissance architects and Galileo's *First New Science*. *Annals of Science* 66(2): 183–208.

Valleriani, Matteo. 2010. *Galileo engineer*. Dordrecht: Springer.

Van Cappelle, Johannes Petrus (ed.). 1812. *Aristotelis Quaestiones Mechanicae*. Amsterdam: Den Hengst.

Vilain, Christiane. 2008. Circular and rectilinear motion in the *Mechanica* and in the 16th century. In *Mechanics and natural philosophy before the scientific revolution*, eds. W.R. Laird and S. Roux, 149–172. Dordrecht: Springer.

Whitehead, David and P. Henry Blyth. 2004. *Athenaeus mechanicus, on machines*. Stuttgart: Franz Steiner Verlag.

Winter, Thomas Nelson. 2007. *The mechanical problems in the corpus of Aristotle*. University of Nebraska-Lincoln: Faculty Publications, Classics and Religious Studies Department.

Chapter 2
General Characteristics of the Transmission

2.1 On the Method

For many treatises in the *Corpus Aristotelicum*, the earliest extant Greek manuscripts can be dated to the ninth or tenth century. The textual evidence is further complemented by amongst others late antique commentaries on the works of Aristotle or early translations of several of his treatises into Syriac or Arabic. However, for some texts in the corpus, the circumstances are less fortunate. With regard to the *Mechanics*, *Eudemian Ethics* and *On Indivisible Lines*, the textual tradition does not begin until the end of the thirteenth or the beginning of the fourteenth century.[1] In part, it might be a coincidence that we do not possess older manuscripts of these treatises, but it may also inform us about the relative unpopularity of these texts compared to other works by Aristotle in the Byzantine period.

The material transmitted to us from Byzantine times is extremely rich; Wartelle lists more than 1000 Greek manuscripts until the sixteenth century that contain the works of Aristotle.[2] In the case of the Aristotelian *Mechanics*, 31 Greek manuscripts have been handed down to us. As mentioned above, the textual transmission starts fairly late; the oldest manuscripts date from the beginning of the fourteenth century. A transmission that starts this late entails many difficulties, since it is hard to assess the earlier phases of the textual tradition with any certainty. The situation is further complicated by the presence of an independent branch in the tradition, that of the paraphrase by the Byzantine Georgius Pachymeres, which went on to influence some of the manuscripts and all critical editions of the treatise. Throughout the book, I will demonstrate the different influences exerted by Pachymeres upon the tradition of the *Mechanics*, and establish directives for a new critical edition of the text.

[1] See Harlfinger (1980), 456.

[2] See Wartelle (1963); although Wartelle's inventory is not exhaustive and contains some errors and repetitions, it clearly shows the popularity of Aristotle in Byzantine times.

J. van Leeuwen, *The Aristotelian Mechanics*, Boston Studies in the Philosophy and History of Science 316, DOI 10.1007/978-3-319-25925-3_2

The first transmission process of Aristotle's works was concentrated in Constantinople, where the transcription of manuscripts from majuscule into minuscule took place.[3] Due to the activities of the Byzantine scholars in the ninth and tenth centuries, we still have today our Greek minuscule manuscripts containing the works of Aristotle. The earliest tradition of the *Mechanics* can be connected to Constantinople as well; *Marc.gr.* 214 and *Vat.gr.* 253, the oldest extant manuscripts of the treatise, were both copied here.

At the beginning of the fifteenth century, contacts between Italy and Byzantium increased, and the transmission of Greek texts gradually moved westward. The Council of Florence-Ferrara in 1438–1439 was an important event in the intellectual exchange between the East and the West.[4] One of the prominent participants was the later Cardinal Bessarion, who played an important role in the transmission of the works of Aristotle in the fifteenth century. The fifteenth century was the most prolific century in the transmission of Greek manuscripts, and most manuscripts of Aristotle were written in this century (see Sect. 2.2 for a list of manuscripts of the *Mechanics*). Bessarion was the commissioner of several Aristotelian manuscripts, e.g. *Marc.gr.* 200 and *Marc.gr.* 216, both of which include the text of the *Mechanics*. After the sack of Constantinople in 1453, the Greek manuscript tradition definitively moved to Italy, where Venice became the centre of transmission in the second half of the fifteenth and the first half of the sixteenth century.[5] In Venice the complete edition of Aristotle was printed by Aldus Manutius between 1495 and 1498. By the sixteenth century, the appearance of the first printed editions had altered the Greek manuscript tradition. It was no longer necessary to copy texts in order to curate them, and we notice a rise in specialized manuscripts copied due to a concern for textual criticism, namely to produce a reliable text, or due to an interest in a particular subject matter. The beginning of the sixteenth century was further characterized by an increased preoccupation among humanists with works by Aristotle. These humanistic activities culminated in 1545 in the foundation of the "Aristotelian Academy" at the Council of Trent.[6] The Academy aimed to reconstruct on the basis of the Aldine the most authentic possible versions of various treatises in the *Corpus Aristotelicum*. Important members of this Academy include Diego Hurtado de Mendoza and Juan Páez de Castro, who worked together on the text of the *Mechanics*. Some traces of their philological activities are still visible in codex *New Haven Phill.* 7488, which contains notes by the hand of Páez.[7] Moreover, whilst attending the Council of Trent, Mendoza produced a Spanish translation of the *Mechanics*.[8] With the Council of Trent and the Academy, the manuscript tradition of Aristotle's

---

[3] See Harlfinger (1980), 452.

[4] See Harlfinger (1980), 462.

[5] See Harlfinger (1980), 465.

[6] See Harlfinger (1980), 469.

[7] See Harlfinger (1971), 83.

[8] See Rose and Drake (1971), 86.

works comes to an end, since after this time handwritten manuscripts were rarely composed.[9]

The aforementioned developments in the transmission of Greek texts are reflected in the manuscripts of the *Mechanics*, especially when we analyse which other works are contained in the manuscripts that have been handed down to us. As for the *Mechanics*, we observe a variety of reasons for copying this text. Most frequently we encounter manuscripts where the *Mechanics* is included along with other Aristotelian texts on diverse topics. These manuscripts, for example *Marc.gr.* 214, *Vat.gr.* 253, and *Marc.gr.* 200, were copied as part of a plan to compile all the texts by Aristotle. The *Mechanics* is generally contained at the end of these manuscripts, together with the treatises *On Indivisible Lines* and *On Breath*. Apart from this general philosophical interest in the *Mechanics*, we can distinguish a group of manuscripts, the subject matter of which was of greater interest to the commissioner or scribe. Here, the *Mechanics* is contained along with texts specializing in a certain topic, as, for instance, *Par.gr.* 2507, which contains various astronomical texts. Finally, there are some specialized codices that contain only the *Mechanics*, for example *New Haven Phill.* 7488 and *Par.gr.* 2115.

Before presenting my analysis of the manuscript families and single manuscripts, I will first give a summary of the 31 manuscripts of the Aristotelian *Mechanics*. All of the manuscripts were fully collated by me with the use of microfilms; they are ordered alphabetically according to their sigla.[10] The summary includes the most important codicological details, relevant information for the *Mechanics* and a selection of the available literature. For complete codicological data on most of the manuscripts and further information on their historical background, I will refer to the exhaustive descriptions in *Aristoteles Graecus* (*AG*)[11] and Harlfinger's important work on the textual transmission of the treatise *On Indivisible Lines*.[12]

2.2 List of Manuscripts

B[a] Vat. Pal.gr. 162 – 15th century, before 1445 – parchment – in 4° – ff. 216
Xen., *Mir.*, *Mu.*, *Mech.* (ff. 31–42), *Spir.*, – Theophrastus, *Ign.*, *Metaph.*, *Lap.*, *Sud.*, *Vert.*, *Lass.*, *Pisc.*, *Vent.*, *Od.*, *HP* Book X, – *Plant.* Book II.
Copyist: Johannes Scutariotes
Provenance: Florence
Lit.: Stevenson (1885), 88–89; Burnikel (1974), XXIX; Roselli (1992), 36.

[9] See Harlfinger (1980), 472.

[10] I used the sigla under which the codices have generally been known since Bekker, and as they were employed by Bottecchia in her critical edition of the *Mechanics*.

[11] Those descriptions of *Aristoteles Graecus* that remained unpublished are indicated in the summary by an asterisk.

[12] See Harlfinger (1971).

B^e Bern. 402 – 15th century, before 1497 – paper – 235 × 155 mm – ff. 144
Theophrastus, *Ign.*, *Metaph.*, *Lap.*, *Sud.*, *Vert.*, *Lass.*, *Pisc.*, *Vent.*, *Od.*, – *Xen.*, *Mir.*, *Spir.*, *Mech.* (ff. 99–114, with diagrams), *Mu.*, – Alexander Aphrodisiensis, *Fat.*
Copyist: Niccolò Leonico Tomeo
Provenance: Veneto
Lit.: *AG*, 53–55 (Moraux); Burnikel (1974), XXXII–XXXIII; Roselli (1992), 37; Vendruscolo (1996), 549–550; Andrist (2007), 188–196.

B^u Burney 67 – 17th century – paper – in 8° – ff. 400
Compendia of *Ph.*, *Mete.*, *HA*, *Mir.*, *Mech.* (ff. 153–157v), – Porphyrius, *Intr.*, – *Phgn.*, *Pr.*, *Cat.*, *Int.*, *APr.*, *APo.*, *Top.*, *SE*, *EN*, – excerpts from Pindarus.
Lit.: Forshall (1840), 22.

D^1 Berol.Phill. 1507 – 15th century, circa 1455 – paper – 286 × 218 mm – ff. 353
PA, *IA*, *Sens.*, *Mem.*, *Somn.Vig.*, *Insomn.*, *Div.Somn.*, *MA*, *GA*, *Long*, *Juv.*, *Resp.*, *Somn.Vig.*, *Insomn.*, *Div.Somn.*, *MA*, *LI*, *Spir.*, *HA*, *Col.*, *Mech.* (ff. 347–353v, with diagrams).
Copyist: Anonymous AR; Konstantin Laskaris; Anonymous KB; final pages of *Mech.* (ff. 350–353v) by Matthaios Kamariotes; other collaborating copyists.
Provenance: probably Constantinople
Lit.: *AG*, 40–42 (Harlfinger); Harlfinger (1971), 247–261; Escobar (1990), 45; Berger (2005), 59.

F^1 Laur.Acq. 65 – 15th century, 1450–1475 – paper – 298 × 207 mm – ff. 16
Mech. (ff. 1–15, with diagrams).
Copyist: Michael Lygizos
Provenance: Florence
Lit.: *AG*, 335 (Harlfinger).

F^2 Laur.Plut. 28.45 – 15th century, 1445 – paper – 193 × 142 mm – ff. 118
Mech. (ff. 1–53v, with diagrams), – Theophrastus, *Metaph.*, – Michael Psellos, *De operatione daemonum*.
Copyist: Demetrius Sguropulos
Provenance: Milan
Lit.: Bandini (1768), 67; *AG*, 194–195 (Harlfinger); Burnikel (1974), XXVII.

H^a Marc.gr. 214 – 13th–14th century, circa 1290–1303 – parchment – 330×230 mm – ff. 238
EN, *Metaph.*, *Ph.*, *GC*, *de An.*, excerpt from *Rh.*, – Cleomedes, *Caelestia*, – *Sens.*, *Mem.*, *Somn.Vig.*, *Insomn.*, *Div.Somn.*, *MA*, *Long*, *Juv.*, *Resp.*, *Col.*, *LI*, *Mech.* (ff. 203–210), *Spir.*, *Cael.*
Provenance: probably Constantinople
Lit.: Siwek (1961), 30–31; Harlfinger (1971), 159–173; Mioni (1981), 328–329; Escobar (1990), 73–74; Rashed (2001), 31 and 244–254.

H^d Ambr. 174 sup. (67) – 15th century, circa 1470 – paper – 320×230 mm – ff. 338
Mu., *Mir.*, – Theophrastus, *Sign.*, – *Vent.*, *Ph.*, *LI*, *PA*, *IA*, *de An.*, *Sens.*, *Mem.*, *Somn.Vig.*, *Insomn.*, *Div.Somn.*, *MA*, *Mech.* (ff. 209–218v), *Col.*, *Phgn.*, *Pr.*
Copyist: Johannes Rhosos; Manuel, pupil of Konstantin Laskaris (*Mech.*)
Provenance: possibly Rome
Lit.: Martini-Bassi (1906), 80; Siwek (1961), 68; Harlfinger (1971), 269–285; Nussbaum (1976), 115; Escobar (1990), 51–52; Vogt (1999), 215.

L Vat.gr. 253 – 14th century, beginning – paper – 241×164 mm – ff. 269
Cael., *GC*, *Mete.*, *de An.* Book III, *Sens.*, *Mem.*, *Somn.Vig.*, *Insomn.*, *Div.Somn.*, *MA*, *Long*, *Juv.*, *Resp.*, *Col.*, *LI*, *Mech.* (ff. 246v–262v, with diagrams), *Spir.*
Provenance: Constantinople
Lit.: Mercati (1923), 330–331; Siwek (1961), 29–30; Harlfinger (1971), 159–166; Escobar (1990), 63; Rashed (2001), 28 and 54–59.

L^v Voss.gr. Q.25 – 15th century, 1450–1500 – paper – 230×160 mm – ff. 165
Theophrastus, *Ign.*, *Metaph.*, *Lap.*, *Sud.*, *Vert.*, *Lass.*, *Pisc.*, *Vent.*, *Od.*, – *Xen.*, *Mir.*, *Spir.*, *Mech.* (ff. 128v–148v), – Alexander Aphrodisiensis, *Fat.*
Copyist: Johannes Rhosos
Lit.: *AG*, 399–400 (Victor); Burnikel (1974), XXXII; Roselli (1992), 37.

M^1 Matr. 4563 – 15th century, 1470 – paper – 285×215 mm – ff. 340
Ph., *de An.*, *Sens.*, *Mem.*, *Somn.Vig.*, *Insomn.*, *Div.Somn.*, *MA*, *Long*, *Juv.*, *Resp.*, *Col.*, *Spir.*, *Cael.*, *GC*, *Mete.*, *Metaph.*, *LI*, *Mech.* (ff. 286v–294v, with diagrams), *HA*.
Copyist: Manuel, pupil of Konstantin Laskaris; Konstantin Laskaris (*Mech.*)
Lit.: Harlfinger (1971), 285–290; Escobar (1990), 51; Rashed (2001), 22; Berger (2005), 60.

M^v Mosq. 453 – 15th century, circa 1450 – paper – 220 × 147 mm – ff. 316
de An., *Sens.*, *Mem.*, *Somn.Vig.*, *Insomn.*, *Div.Somn.*, *MA*, *Long*, *Juv.*, *Resp.*, *Mech.* (ff. 69–78v, with diagrams), *Spir.*, *PA*, *GA*, *IA*, *HA*, *Col.*, *LI.*
Copyist: Matthaios Kamariotes
Provenance: probably Constantinople
Lit.: Harlfinger (1971), 247–261; Escobar (1990), 54; Berger (2005), 61.

N^a Marc.gr. 215 – 15th century, circa 1470 – parchment – 258 × 180 mm – ff. 300
Rh., *Rh.Al.*, *Po.*, *Plant.*, *Mech.* (ff. 131–143), *Mir.*, *Mu.*, *LI*, *Col.*, *Spir.*, *Phgn.*, - Theophrastus, *Sign.*, – *Vent.*, *Xen.*, *Pr.*
Copyist: Demetrius Triboles; Anonymus KB (*Mech.*); other copyists.
Provenance: probably Rome
Lit.: Harlfinger (1971), 309–318; Mioni (1981), 329–330; Vogt (1999), 215.

N^b Nap. IIIAA14bis – 15th century, 1450–1500 – parchment – 237 × 175 mm – ff. 70
Hippocrates, *Ep.*, – Diogenes, *Ep.*, – Brutus, *Ep.*, – *Mech.* (ff. 48–70v, with diagrams).
Provenance: Naples
Lit.: Diels (1905), 38; Napolitano et al. (1977), 26–27; **AG* (Harlfinger); Formentin (1997), 207–216.

N^c Nap. IIIC8 (266), 16th century, beginning – paper – in 8° – ff. 173
Theol.Ar., – *Mech.* (ff. 88–128v).
Lit.: Cirillo (1826), 352.

N^h New Haven Phill. 7488 – 15th century, middle – parchment – 210 × 130 mm – ff. 35
Mech. (ff. 1–35, with diagrams).
Copyist: Manuel (marginalia added by Juan Páez de Castro, circa 1545)
Lit.: Faye and Bond (1962), 45; Harlfinger (1971), 83; Shailor (1984), 355–356.

O^a Marc.gr. 216 – mid 15th century, 1445 – paper – 220 × 135 mm – ff. 348
Phgn., *Vent.*, *Mir.*, *Mu.*, *Xen.*, *Mech.* (ff. 90–110v), *Spir.*, *Plant.*, *Pr.*
Copyist: Johannes Scutariotes
Provenance: Florence
Lit.: Harlfinger (1971), 283–285; Mioni (1981), 330–331; Roselli (1992), 37; Vogt (1999), 213.

P Vat.gr. 1339 – 14th century, 2nd half – parchment – 280×210 mm – ff. 460
PA, *GA*, *IA*, *de An.*, *Sens.*, *Mem.*, *Somn.Vig.*, *Insomn.*, *Div.Somn.*, *MA*, *Long*, *Juv.*, *Resp.*, *Col.*, *LI*, Mech. (ff. 281v–294, with diagrams), *Spir.*, *Mu.*, *HA*.
Copyist: Joseph
Lit.: Vogel and Gardthausen (1909), 221; Siwek (1961), 129–136; Harlfinger (1971), 247–261; Nussbaum (1976), 128–131; Escobar (1990), 68; Berger (2005), 64; Bloch (2008), 51–56.

P^4 Par.Suppl.gr. 333 – 15th–16th century, before 1515 – paper – in 4° – ff. 244
PA, *GA*, Mech. (ff. 163–176), *IA*, *Sens.*, *Mem.*, *Somn.Vig.*, *MA*, *Long.*
Copyist: Johannes Serbopulos; Demetrius Chalkondyles (*Mech.*)
Lit.: Omont, III (1888), 250; *RGK*, II 138, II 240; Escobar (1990), 62–63.

Par Par.gr. 2115 – 15th–16th century – paper – 168×120 mm – ff. 39
Mech. (ff. 1–39v, with diagrams).
Lit.: Omont, II (1888), 197; **AG* (Harlfinger).

P^s Par.Suppl.gr. 541 – 15th century – paper – in 8° – ff. 353
Ancyranus Nilus, *Capita Paraenetica*, – Hero, *Geodaesia*, – Isaac Argyrus, *De Triangulis*, – Hero, *Op.*, – Michael Psellus and Basilius Megalomita, *Aenigmata*, – Phocylides, *Carmina*, – Albinus, *Introductio in Platonem*, – Mech. (ff. 155v–169, with diagrams), – Boethius, *De Consolatione Philosophiae*, – Macrobius, *Somnium Scipionis*, – Plutarchus, *Vit.Hom.*
Copyist: Georgius Gregoropulos; Andronikos Kallistos (*Mech.*)
Lit.: Omont, III (1888), 274–275; *RGK*, II 25, II 78.

P^t Par.gr. 2507 – 14th century, 1370–1380 – paper – 216×144 mm – ff. 206
Heliodorus, *In Paulum Alexandrinum Commentarium*, – Hephaestio, *Apotelesmatica*, – anonymous astronomical texts, – Albumasar, *De Revolutionibus Nativitatum*, – Mech. (ff. 135–139v, with diagrams), – *Commentarius in Ptolemaei Tetr.*
Copyist: possibly Johannes Ambramios (Harlfinger)
Lit.: Omont, II (1888), 274; Pingree (1968), XI; **AG* (Harlfinger); Mondrain (2007), 167.

q Urb.gr. 76 – 15th century, 1400–1450 – paper – 208×146 mm – ff. 109
Cleomedes, *Caelestia*, – *Dicta Septem Sapientum*, – Hesiodus, *Op.*, – Mech. (ff. 80–99, with diagrams), – Constantinus, *De Caelo.*
Copyist: several copyists; Johannes Chionopulos (ff. 1–58v, dated 1421)
Lit.: Stornajolo (1895), 99–103; Todd (1990), IX; Tziatzi-Papagianni (1994), 346–348; *RGK*, III 314.

Q Marc.gr. 200 – 15th century, 1457 – parchment – 425×285 mm – ff. 596
Ph., *Cael.*, *GC*, *Mete.*, *de An.*, *Sens.*, *Mem.*, *Somn.Vig.*, *Insomn.*, *Div.Somn.*, *Long*, *Juv.*, *Resp.*, *HA*, *PA*, *IA*, *MA*, *GA*, *Plant.*, *Spir.*, *Col.*, *LI*, *Phgn.*, – Theophrastus, *Sign.*, – *Vent.*, *Mir.*, *Mu.*, *Xen.*, *Mech.* (ff. 311–316v), *Pr.*, *Metaph.*, *MM*, *EE*, *EN*, *Oec.*, *Pol.*, *Rh.*, *Rh.Al.*, *Po.*
Copyist: Johannes Rhosos
Provenance: Rome
Lit.: Harlfinger (1971), 183–191; Mioni (1981), 311–313; Escobar (1990), 70–71; Vogt (1999), 214; Rashed (2001), 30; Berger (2005), 65.

u Reg.gr. 124 – circa 1500 – paper – 217×131 mm – ff. 369
Pr., – Alexander Aphrodisiensis, *Pr.*, – *Mech.* (ff. 206–221), *Metaph.*, – Theophrastus, *Metaph.*
Lit.: Stevenson (1888), 88–89; **AG* (Eleuteri); Burnikel (1974), XXVIII.

U^m Marc.gr. IV.57 – 15th century, before 1446 – parchment – 203×148 mm – ff. 59
Mech. (ff. 1–24, with diagrams), – Michael Psellus, *De Operatione Daemonum*, – Petrus Peregrinus de Maricourt, *Tractatus De Magnete*, – Flaccus Africus discipulus Belbenis, *Tractatus de septem herbis.*
Copyist: Johannes Symeonakis
Provenance: Crete
Lit.: Mioni (1972), 246–247; Zorzi (1993), 40.

V^1 Vat.gr. 905 – 15th century, ca. 1430 – paper – 218×146 mm – ff. 160
Johannes Tzetzes, *Εἰς τὴν Ὁμήρου Ἰλιάδα ἐξήγησις*, – *LI*, *Mech.* (ff. 35v–51, with diagrams), *Mu.*, *HA.*
Copyist: Demetrius Branas (Harlfinger)
Lit.: Harlfinger (1971), 266–268; Berger (2005), 64.

V^2 Vat.gr. 2231 – 14th century, 1317–1338 – paper – 206×140 mm – ff. 281
Pythagoras, *Carmen Aureum*, – Hierocles, *in CA*, – Epictetus, *Ench.*, – Simplicius, *in Epict.*, – Marcus Aurelius, *De Rebus Suis*, – Maximus Planudes, *Disticha Catonis*, – Plutarchus, *Op.*, – Michael Psellus, *de An.*, – Theophrastus, *Sign.*, – excerpt from *Mech.* (ff. 275–281).
Copyist: several anonymous copyists
Lit.: Hadot (1978), 89–92; Lilla (1985), 328–333.

V[3] Barb.gr. 22 – 16th century, beginning – paper – 169×114 mm – ff. 304
Mech. (ff. 1–27v), – Georgius Gemistus Pletho, *Tractatus De Legibus*, – Zoroaster, *Oracula Magica*, with commentary by Pletho, – Sophocles, *Aj.*, verse 1–13, – Aeschines, *In Timarchum*, *De Falsa Legatione*, *In Ctesiphontem*, – Polybius, *Historiarum Reliquiae*.
Copyist: Demetrius Damilas
Provenance: probably Florence
Lit.: Capocci (1958), 24–27; Canart (1977–1979), 281–347.

V[ph] Vind.Phil.gr. 231 – 15th century, 1458 – paper – 215×145 mm – ff. 196
Phgn., *Vent.*, *Mir.*, *Mu.*, *Spir.*, *Mech.* (ff. 121–159v), *Xen.*, *LI.*
Copyist: Emmanuel ὁ καλούμενος Φυσκόμηλος (?)
Provenance: Naples
Lit.: Hunger (1961), 340–341; Harlfinger (1971), 269–285; Roselli (1992), 37; Vogt (1999), 214.

W[u] Urb.gr. 44 – 15th century, circa 1465 – paper – 293×220 mm – ff. 85
EN, *LI*, *Mech.* (ff. 69–78, with diagrams), excerpt from Iamblichus, *VP*.
Copyist: Georgius Tzangaropulos
Provenance: probably Crete
Lit.: Stornajolo (1895), 48–49; Harlfinger (1971), 266–268.

Bibliography

Andrist, Patrick. 2007. *Les manuscrits grecs conservés à la Bibliothèque de la Bourgeoisie de Berne – Burgerbibliothek Bern: catalogue et histoire de la collection*. Zurich: Urs Graf.

Bandini, Angelo Maria. 1768. *Catalogus codicum manuscriptorum Bibliothecae Mediceae Laurentianae*, vol. 2. Florence: Typis Regiis.

Bekker, Immanuel (ed.). 1831. *Aristotelis Opera II*. Berlin: Reimer.

Berger, Friederike. 2005. *Die Textgeschichte der Historia animalium des Aristoteles*. Wiesbaden: Reichert.

Bloch, David. 2008. The text of Aristotle's *De Sensu* and *De Memoria*. *RHT* 3: 1–58.

Bottecchia, Maria Elisabetta (ed.). 1982. *Aristotele: ΜΗΧΑΝΙΚΑ, Tradizione manoscritta, testo critico, scolii*. Padua: Antenore.

Burnikel, Walter. 1974. *Textgeschichtliche Untersuchungen zu neun Opuscula Theophrasts*. Wiesbaden: Franz Steiner Verlag.

Canart, Paul. 1977–1979. Démétrius Damilas, *alias* le "librarius Florentinus". *RSBN* N.S.14-16: 281–347. Rome.

Capocci, Valentino. 1958. *Codices Barberiniani Graeci*. Tomus 1: Codices 1–163. Rome.

Cirillo, Salvadore. 1826. *Codices Graeci manuscripti Regiae bibliothecae Borbonicae*. Naples: Regia Typogr.

Diels, Hermann. 1905. *Die Handschriften der antiken Ärzte I*. Berlin: Verl. der Königl. Akad. der Wiss.

Escobar, Angel. 1990. *Die Textgeschichte der aristotelischen Schrift Περὶ ἐνυπνίων: Ein Beitrag zur Überlieferungsgeschichte der Parva Naturalia*. Berlin.

Faye, Christopher U. and William H. Bond. 1962. *Supplement to the census of Medieval and Renaissance manuscripts in the United States and Canada*. New York: The Bibliographical Society of America.

Formentin, Mariarosa. 1997. Codici greci di medicina nella Biblioteca Nazionale Vittorio Emanuele III di Napoli: le vie di acquisizione. In *Lingue tecniche del Greco e del Latino II*, ed. S. Sconocchia, 207–216. Bologna: Patron.

Forshall, Josiah. 1840. *Catalogue of manuscripts in the British Museum*, vol.1, pt.2: The Burney manuscripts. London: The Trustees.

Gamillscheg, Ernst and Herbert Hunger et al. 1981–1997. *Repertorium der griechischen Kopisten: 800–1600*. 3 Bände. Vienna: Verl. der Österreichischen Akademie der Wissenschaften.

Hadot, Ilsetraut. 1978. La tradition manuscrite du commentaire de Simplicius sur le Manuel d'Épictète. *RHT* 8: 1–108.

Harlfinger, Dieter. 1971. *Die Textgeschichte der pseudo-aristotelischen Schrift ΠΕΡΙ ΑΤΟΜΩΝ ΓΡΑΜΜΩΝ: Ein kodikologisch-kulturgeschichtlicher Beitrag zur Klärung der Überlieferungsverhältnisse im Corpus Aristotelicum*. Amsterdam: Hakkert.

Harlfinger, Dieter. 1980. Einige Grundzüge der Aristoteles-Überlieferung. In *Griechische Kodikologie und Textüberlieferung*, ed. D. Harlfinger, 447–483. Darmstadt: Wissenschaftliche Buchgesellschaft.

Hunger, Herbert. 1961. *Katalog der griechischen Handschriften der Österreichischen Nationalbibliothek*. Vienna: Prachner; Hollinek.

Lilla, Salvator. 1985. *Codices Vaticani Graeci*. Codices 2162–2254. Rome.

Martini, Emidio and Dominicus Bassi. 1906. *Catalogus codicum Graecorum Bibliothecae Ambrosianae*. Milan: Hoepli.

Mercati, Giovanni and Pio Franchi de' Cavalieri. 1923. *Biblioteca Apostolica Vaticana: Codices Vaticani Graeci*. Codices 1–329. Rome.

Mioni, Elpidio. 1972. *Bibliothecae Divi Marci Venetiarum codices Graeci manuscripti*, Classis II, codices 121–198; classes III, IV, V; Indices. Rome: Ist. Poligrafico e Zecca dello Stato.

Mioni, Elpidio. 1981. *Bibliothecae Divi Marci Venetiarum codices Graeci manuscripti*, Codices 1–299. Rome: Ist. Poligrafico e Zecca dello Stato.

Mondrain, Brigitte. 2007. Les écritures dans les manuscrits byzantins du XIVe siècle. *RSBN* N.S.44: 157–196.

Moraux, Paul. 1976. *Aristoteles Graecus: Die griechischen Manuskripte des Aristoteles*. 1.Band: Alexandrien – London. Berlin: De Gruyter.

Napolitano, Felicia, Maria Luisa Nardelli and Luigi Tartaglia. 1977. *Manoscritti greci non compresi in cataloghi a stampa*. Naples: Arte Tipografica.

Nussbaum, Martha Craven. 1976. The text of Aristotle's *De Motu Animalium*. *HSCP* 80: 111–159.

Omont, Henri. 1888. *Inventaire sommaire des manuscrits grecs de la Bibliothèque Nationale*. Seconde partie/troisième partie. Paris: Picard.

Pingree, David (ed.). 1968. *Albumasar: De revolutionibus nativitatum*. Leipzig: Teubner.

Rashed, Marwan. 2001. *Die Überlieferungsgeschichte der aristotelischen Schrift De generatione et corruptione*. Wiesbaden: Reichert.

Rose, Paul Lawrence and Stillman Drake. 1971. The pseudo-Aristotelian *Questions of Mechanics* in Renaissance culture. *Studies in the Renaissance* 18: 65–104.

Roselli, Amneris (ed.). 1992. *Aristotele: De spiritu*. Pisa: ETS Editrice.

Shailor, Barbara A. 1984. *Catalogue of Medieval and Renaissance manuscripts in the Beinecke Rare Book and Manuscript Library, Yale University*. Vol. I: codices 1–250. Binghamton: Center for Medieval & Early Renaissance Studies.

Siwek, Paul. 1961. *Les manuscrits grecs des Parva naturalia d'Aristote*. Rome: Desclée.

Stevenson, Enrico. 1885. *Codices manuscripti Palatini Graeci Bibliothecae Vaticanae*. Rome: Ex Typographeo Vaticano.

Stevenson, Enrico. 1888. *Codices manuscripti Graeci Reginae Suecorum et Pii PP. II Bibliothecae Vaticanae*. Rome: Ex Typographeo Vaticano.

Stornajolo, Cosimo. 1895. *Codices urbinates Graeci Bibliothecae Vaticanae*. Rome: Ex Typographeo Vaticano.

Todd, Robert B. (ed.). 1990. *Cleomedis Caelestia*. Leipzig: Teubner.

Tziatzi-Papagianni, Maria. 1994. *Die Sprüche der sieben Weisen: Zwei byzantinische Sammlungen; Einleitung, Text, Testimonien und Kommentar*. Stuttgart: Teubner.

Vendruscolo, Fabio. 1996. Manoscritti greci copiati dall'umanista e filosofo Niccolò Leonico Tomeo. In *ΟΔΟΙ ΔΙΖΗΣΙΟΣ. Le vie della ricerca. Studi in onori di Francesco Adorno*, ed. M.S. Funghi, 543–555. Florence: Olschki.

Vogel, Marie and Victor Gardthausen. 1909. *Die griechischen Schreiber des Mittelalters und der Renaissance*. Leipzig: Harrassowitz.

Vogt, Sabine. 1999. *Opuscula VI: Physiognomonica*. Munich: Oldenbourg Akademieverlag.

Wartelle, André. 1963. *Inventaire des manuscrits grecs d'Aristote et de ses commentateurs: contribution à l'histoire du texte d'Aristote*. Paris: Les Belles Lettres.

Zorzi, Marino. 1993. *Collezioni veneziane di codici greci: dalle raccolte della Biblioteca Nazionale Marciana*. Venice: Il Cardo.

Chapter 3
The Textual Transmission

My analysis of the transmission of the Aristotelian *Mechanics* is based on collations of all extant manuscripts.[1] With a length of about 11 Bekker pages, the *Mechanics* is one of the shorter treatises in the *Corpus Aristotelicum*, thus enabling me to make full collations of all of the manuscripts. As I will point out at the end of this chapter, all critical editions of the text contain some major shortcomings. We have three editions of the *Mechanics* from the nineteenth century, namely those by Johannes van Cappelle (1812), Immanuel Bekker (1831) and Otto Apelt (1888); more recently, an edition was produced in the twentieth century by Maria Elisabetta Bottecchia (1982). Bottecchia's edition is a clear improvement on the previous editions in the extent of its research. Whereas the other editors of the *Mechanics* taken together consulted a total of nine manuscripts, Bottecchia considered nearly all of the extant manuscript material for her critical edition of the text. When I began my project, I did not expect to find significant new results which would necessitate a completely new critical edition of the text.

However, as I compared Bottecchia's analysis of the manuscript tradition with my own examination of the text and collations of the manuscripts, it became clear that her edition contains numerous errors. First, Bottecchia establishes several affiliations between manuscripts that are impossible from a chronological point of view. She argues for instance that the manuscript V^2 derives from W^a. This is incompatible with strong evidence concerning the chronology of the manuscripts: there are convincing reasons to think that V^2 was composed in the first half of the fourteenth century and W^a more than a century later, in about 1465.[2] Another example can be found in Bottecchia's making codex U^m dependent on W^a. Whereas in the first chapter she agrees with the generally acknowledged dating of U^m to before the year 1446, and of W^a to about 1465,[3] she still believes that the manuscript U^m was copied

[1] Results from my analysis of the Greek manuscript tradition of the *Mechanics* have already been published in van Leeuwen (2013).

[2] Hadot (1978), 89, dates V^2 between 1317 and 1338; see Harlfinger (1971), 266 on the date of W^a.

[3] See Bottecchia (1982), 21 and 28.

J. van Leeuwen, *The Aristotelian Mechanics*, Boston Studies in the Philosophy and History of Science 316, DOI 10.1007/978-3-319-25925-3_3

from W^a.[4] Second, and even more important, is Bottecchia's failure to recognize the influence exerted on the textual tradition by Georgius Pachymeres, parts of whose Byzantine paraphrase of the *Mechanics* were incorporated into several manuscripts of the text. Bottecchia, for example, regarded *Mut.* 76 (M^u), which is an apograph of Pachymeres' paraphrase, as an authentic Aristotelian manuscript and included readings of this paraphrase in her critical edition of the text.[5]

A new examination of the manuscripts of the *Mechanics* is therefore necessary, and will yield a different perspective on the textual transmission of the treatise. I have made complete collations of all 31 manuscripts in which the *Mechanics* has been handed down to us. Three of these manuscripts, H^d, N^b and N^c, were collated by me for the first time. Different readings led me to divide the writings into three manuscript families: family *a*, which contains two of the oldest codices of the treatise, H^a and L; family *b*, of which P, M^v and D^1 are the most important representatives, and family *c*, which includes W^a and V^1. Apart from these families, there is a group of contaminated manuscripts which include aspects of different manuscript families. The Byzantine paraphrase of the *Mechanics* contained in Pachymeres' compendium of Aristotle's philosophy offers an almost verbatim quotation of the Aristotelian text. Therefore, it can be considered as a part of the textual tradition of the *Mechanics*. Every now and then, Pachymeres makes some smaller emendations or adds an illustrative remark to the manuscript, giving his text an independent position among the other manuscripts. His paraphrase derives from the same hyparchetype as family *a*; it is, however, recorded in the *stemma codicum* as a different branch, because through Pachymeres' emendations and additions paraphrastic elements came to be incorporated into the text. Pachymeres' paraphrase of the *Mechanics* is important in constituting the text, since it was written at the same time or even shortly before the authentic Aristotelian manuscripts were copied.

The present chapter offers my analysis of the single manuscripts and manuscript families, including the Byzantine paraphrase by Pachymeres. Furthermore, both the Latin and Arabic tradition of the *Mechanics* will shortly be discussed, as well as the different critical editions of the treatise. Finally, the results of my analysis of the manuscript tradition will be presented in a *stemma codicum* which illustrates the affiliations of the manuscripts.

---

[4] See Bottecchia (1982), 63.

[5] This has already been noticed by Hilgers (1992), 63.

3.1 Manuscript Families

3.1.1 Family a

The two manuscripts in this family, H^a and L, are among the oldest preserved manuscripts of the *Mechanics*, and derive independently of each other from the hyparchetype α. They share readings which separate them from the other families, for example at:

847b25	διεστήκατον *a*	διέστηκεν *cett.*
848b24	πλευρῶν *a*	πλειόνων *cett.*
850a32	κινεῖν *a*	κινῆσαι *cett.*
851b18	ὁ τῆς ἁμάξης τροχὸς *a*	ὁ τροχὸς ὁ τῆς ἁμάξης *cett.*
853a25	ὄντι τῷ μόχλῳ *a*	ὄντι *cett.*
853a27	ἐφ' ᾧ ΔΕΖΗ *a*	ΔΕΗΖ *cett.*
853b32	πρόσκειται *a*	ἔγκειται *cett.*
854b20	τὴν ἐλάττω διάμετρον *a*	τὴν διάμετρον τὴν ἐλάττω *cett.*
854b27	πλευράν *a*	πλείω *cett.*
855a13	ὀξείας ὥσπερ συμβαίνει *a*	ὀξείας συμβαίνει *cett.*
856a24	κωλύει *a*	κυλίει *cett.*
856b28	ΒΓ *a*	ΑΓ *cett.*
857b12	οὕτως *a*	ἐλάττων *cett.*
857b38	ἀνίσταται ἀναστῆναι *a*	ἀναστῆναι *cett.*
858b11	καὶ τότε μένει *a*	καὶ τιθέαμεν εἰ *cett.*

Further variants of family *a* can be found at 848a6, 848a11, 848b16, 849b20, 850a1, 850a24, 850a27, 850a31, 851a7, 851a13, 851a28, 852a6, 853a30, 854a11, 854a14, 854b5, 854b11, 854b20, 855a3, 855a4, 855b3, 855b24, 856a34, 857a2, 857b5 and 858b24.

The dating of codex H^a has been a subject of much controversy over the past decades (Fig. 3.1). Whereas earlier scholars dated H^a to anywhere from the eleventh to the fifteenth century, Dieter Harlfinger was able to determine the date more precisely as falling between 1270 and 1370.[6] Marwan Rashed settled the controversy about H^a definitively, fixing the date between 1290 and 1303.[7] He identified one of the scribes who worked on H^a as the copyist of another manuscript, *Ambr.* G51 sup., and argued convincingly that parts of H^a were likely copied from *Ambr*. The *terminus ante quem* of *Ambr.* is 1303, the year in which Conradus Beginus acquired this manuscript and took it along to Kaffa. Therewith, H^a can be dated more precisely,

[6] See Harlfinger (1971), 168. Without giving any reasons based on the textual evidence, Bottecchia (1982), 17 rejects the dating by Harlfinger and proposes a date for H^a in the twelfth century: "È più probabile che esso sia stato scritto nel secolo XII, come si desume dall' analisi calligrafica."

[7] See Rashed (2001), 250.

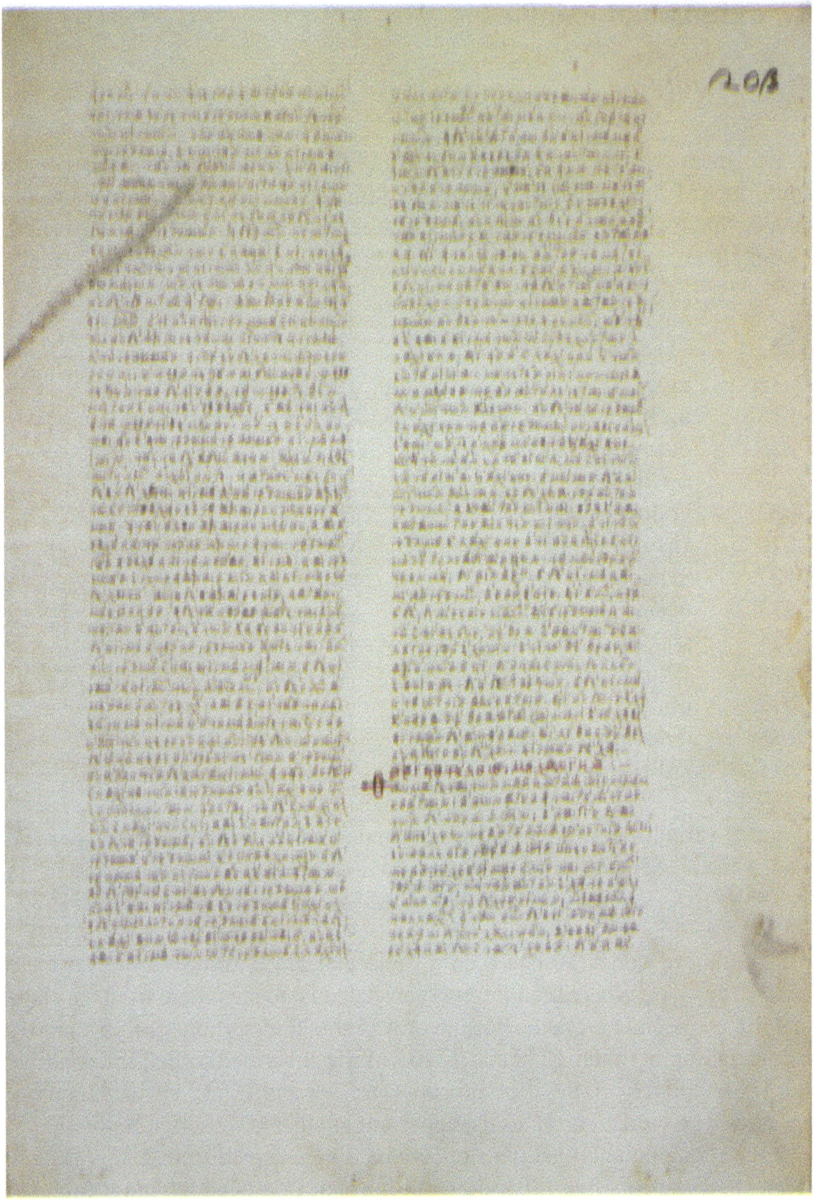

Fig. 3.1 Cod. Marc.Gr. Z.214 (H^a), f.203r. Title page of the *Mechanics*. Su concessione del Ministero dei Beni e delle Attività Culturali e del Turismo – Biblioteca Nazionale Marciana. Divieto di riproduzione

namely after the making of *Ambr.* and before Conradus' acquisition; these facts speak for a date in the final years of the thirteenth century, possibly around 1300.

H^a, which was written in double columns throughout, has influenced a group of manuscripts that emanates from O^a. Since O^a was copied from a manuscript in family *b* and contains the variants of family *a* only after emendation, this group of manuscripts will be presented in a separate Sect. 3.1.3.

The second manuscript of family *a*, L, was copied by an anonymous scribe at the beginning of the fourteenth century (Fig. 3.2).[8] For the *Mechanics*, no direct descendants of this codex are known. Both H^a and L derive independently of each other from the hyparchetype α. On account of a number of separative errors, a dependence in either direction between H^a and L can be ruled out, for instance:

848b23	φανερὸν add. μὲν H^a	φανερὸν L + α
850a25	γὰρ H^a	δὲ L + α
851a13	om. τὴν H^a	τὴν L + α
852b16–17	/	om. αἱ τῶν μειζόνων κύκλων ἢ αἱ τῶν ἐλαττόνων · ὑπὸ τῆς αὐτῆς γὰρ ἰσχύος L
852b22	περὶ H^a + α	παρὰ L
853a7	ἑκατὸν H^a + α	ἐκ τῶν L
853a12–14	om. μᾶλλον · ὅσῳ γὰρ ἂν πλέον ἀπέχῃ τοῦ ὑπομοχλίου, τοσούτῳ ἀνάγκη κάμπτεσθαι H^a	/
853a34	ἑκατέρου H^a	ἐκ θατέρου L + α
854a22	ἁρμόττει H^a	περιαρμόττει L + α
855b34	κινεῖ H^a + α	κινῆσαι L
858a12	κινούμενον H^a + α	κείμενος L

Further separative errors can be found at 847b25, 849b20, 852b39, 853b9, 853b14, 854a3, 854a13, 854b3, 854b10, 855b38, 856a26, 856b3, 856b14, 857a20, 857b14 and 858b26. The same relation between manuscripts H^a and L, as independent textual witnesses to a lost exemplar, was assessed by Harlfinger for the treatise *On Indivisible Lines*.[9] Harlfinger further confirms that this relationship holds for *On Colours* and the *Mechanics*, and Siwek states for the *Parva Naturalia* that both manuscripts can be traced back to the same exemplar.[10] It is therefore very likely that these treatises were all copied into H^a and L from the same lost exemplar, especially since they are contained in both H^a and L in the very same order. This sequence starts with the *Parva Naturalia* and is followed by the treatises *On Colours*,

[8] See Harlfinger (1971), 165–166.

[9] See Harlfinger (1971), 159–160.

[10] See Siwek (1961), 42–46.

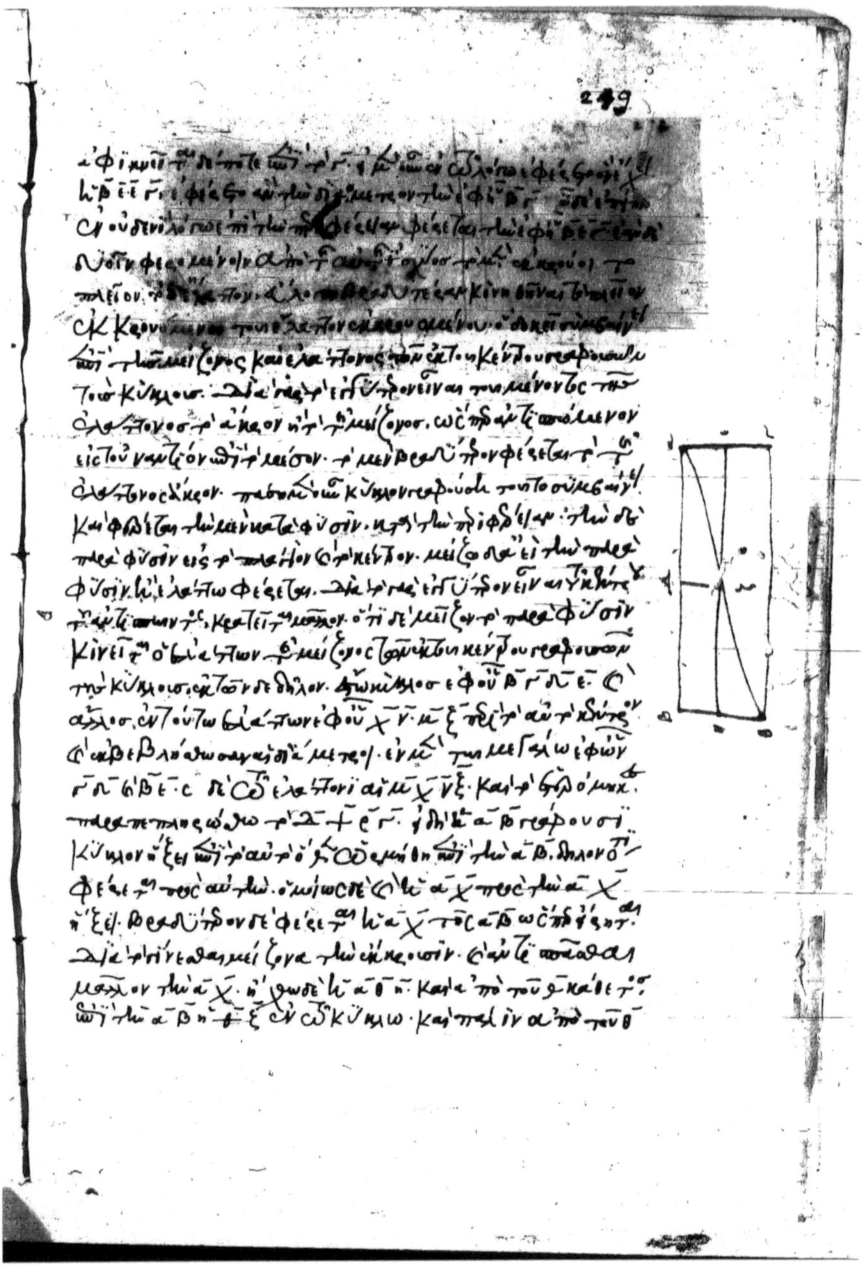

Fig. 3.2 Codex Vat.gr. 253 (L), f.249r. Diagram of the parallelogram of motions to problem 1 of the *Mechanics*. © [2015] Biblioteca Apostolica Vaticana. By permission of Biblioteca Apostolica Vaticana, with all rights reserved

On Indivisible Lines, *Mechanics* and *On Breath*; from this we can plausibly infer that the *Mechanics* was contained in codex α after *On Indivisible Lines*, but before *On Breath*. Like both of its copies H^a and L, codex α was probably written in Constantinople. There is no certainty about the date of codex α[11]; the dating of the oldest extant manuscript H^a provides at least a *terminus ante quem* of 1300.

Since H^a and L can be traced independently of each other back to the hyparchetype α, the readings of both manuscripts are important in constituting the text. Notwithstanding her errors in dating and collation of H^a and L,[12] Bottecchia was the first editor to recognize the equal importance of these two manuscripts. In her analysis of the manuscripts, she regards H^a and L as independent witnesses derived from a common source, and for this reason she consulted both manuscripts for her critical edition of the *Mechanics*.

3.1.2 Family b

Family *b* consists of the following manuscripts: P, B^a, B^e, L^v, M^v and D^1. P is the oldest and best-known manuscript of this family (Fig. 3.3). The value of codex P as a textual witness has been questioned in relation to various treatises from the *Corpus Aristotelicum*.[13] P shows many traces of contamination and a greater part of the peculiar readings in this manuscript turn out to be plain errors. Nevertheless, scholars generally agree that P contains readings which constitute independent and relevant textual evidence. For the *Mechanics*, there are agreements between P and family *c*,[14] but on account of certain relevant textual variants P can be traced back to

[11] See Harlfinger (1971), 162–163.

[12] See n.6. Because of Bottecchia's incorrect early dating of H^a to the twelfth century, she can exclude the possibility that H^a was copied from the younger codex L. In order to show an independence of both manuscripts, she thus only needs to prove that L is not a copy of H^a. She does so on p. 40 by giving some examples in which L contains a reading that is more complete than the variant in H^a. Although most of the examples stated here are correct, Bottecchia forgets to mention those cases where L is less complete than H^a, e.g. lines in 852b16-17 that are only omitted in L. To sum up, Bottecchia reaches the correct conclusion that H^a and L are independent witnesses derived from a common source. However, she draws this conclusion based on an erroneous examination of the codicological evidence.

[13] See e.g. Harlfinger (1971), 247–261 on *LI*; Nussbaum (1976), 128–131 on *MA*; and Bloch (2008), 51–56 on *Sens.* and *Mem.*

[14] See e.g. at 848a6, 850a24, 850a31, 851a13, 852b26, 854b27, 855a3, 856a24, 856a34, 857b12 and 858b11.

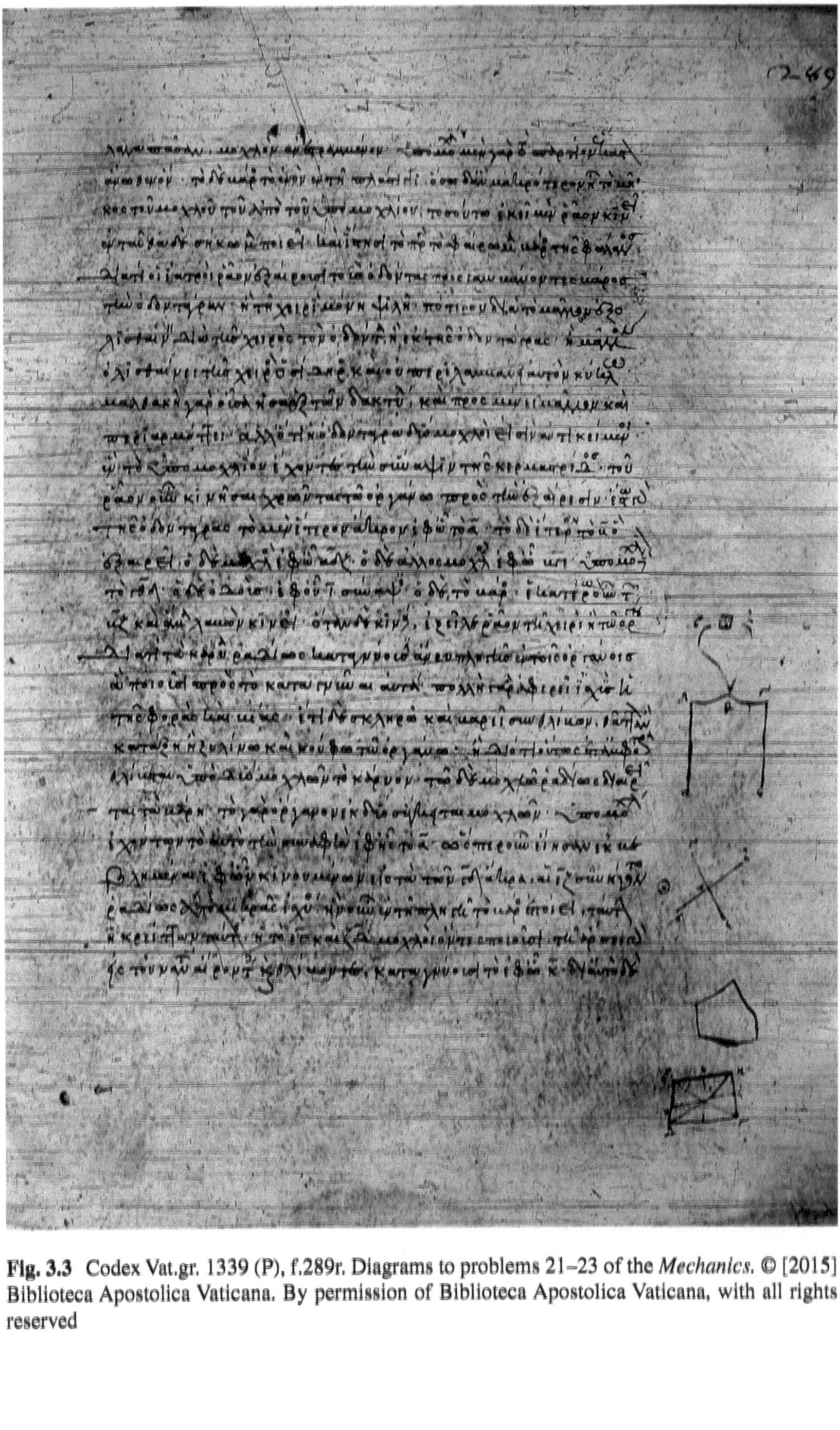

Fig. 3.3 Codex Vat.gr. 1339 (P), f.289r. Diagrams to problems 21–23 of the *Mechanics*.

its own hyparchetype β. Some characteristic readings of P, which are also present in manuscripts B[a], B[e] and L[v],[15] can be found at:

847b19	σὺν ἀλλήλοις *b*	μετ' ἀλλήλων *cett.*
848a23	ὑπεναντίους *b*	ὑπεναντίως *cett.*
848b33	προειρημένα πρότερον *b*	προειρημένα *cett.*
849b2	om. τῷ μείζονι *b*	/
849b18	κατὰ φορὰν *b*	κατὰ φύσιν *a*; τὸ κατὰ φύσιν *c*
852a9	αὐτῶν *b*	αὐτὸν *cett.*
852a17	χωρίῳ *b*	χρόνῳ *cett.*
854a34	ἀφερεῖ *b*	ἀφαιρεῖται *cett.*
854b27	AE add. XAB *b*	AE *cett.*
854b29	πεπληρώσθω *b*	παραπεπληρώσθω *cett.*
855b26	μὴ *b*	μὲν *cett.*
855b27	μείζω *b*	μείζονι *cett.*
855b39	αὐτὴν *b*	αὐτοῦ *cett.*
856b13	ἴσαι εἰσίν *b*	εἰσίν ἴσαι *cett.*
857a38	καθιέναι *b*	καθεικέναι *cett.*
857b11	μόνῳ *b*	μέσῳ *cett.*

Codex P was written by a copyist by the name of Joseph some time in the second half of the fourteenth century.[16] Most editions of the *Mechanics* acknowledge the importance of P as a direct descendant of the hyparchetype β. Both Bekker and Apelt adopted many readings of P, and Bottecchia too included various readings of this codex in her edition of the treatise. Although P contains valuable variants, in a new critical edition of the text one should carefully consider whether a peculiar reading in P constitutes independent textual evidence. Excepting instances of contamination from family *c*, other variants in P might well be errors due to a corrupted source, as I will explain below. Therefore, the editor should decide in each case whether a peculiar reading is the result of contamination, constitutes an independent variant, or is simply an error.

Codex B[a] was written by Johannes Scutariotes in the middle of the fifteenth century. It is an accurate copy of P that contains only few own errors, e.g. at 848b19 ἔχει B[a]: ἔχειν P; 849a7 φερομένην B[a]: φερομένοιν P; 850a32 om. τὸ; 851a6 κινεῖσαι B[a]: κινῆσαι P; 856a21 om. ὁ B[a]; 857b15 om. γὰρ B[a]; 858a23 φαίνεται B[a]: φέρεται P.

[15] Manuscripts M[v] and D[l], which contain only part of the readings of family *b*, will be discussed below.

[16] See Vogel and Gardthausen (1909), 221. Vogel was the first to decipher the name Ἰωσήφ as the copyist of P, although the name was hardly recognizable. Harlfinger (1971), 253 confirmed this assumption by finding another manuscript *Vind.Hist.gr.* 16 by the hand of the same scribe. This enabled him to dispel the controversy surrounding the dating of P and to propose a date in the second half of the fourteenth century.

Two further manuscripts in family *b*, which share part of the readings of P, are M^v and D^1. Both of these manuscripts were copied in Constantinople in the middle of the fifteenth century.[17] Several copyists collaborated on D^1. The final pages of the *Mechanics* were written by Matthaios Kamariotes, by whose hand M^v was also copied.

Although M^v and D^1 have many similarities with P, it is not possible that these manuscripts are dependent on P, since P contains many peculiar readings, for example at:

848a3	om. κύκλον P	/
852a23	κινῆται P	κινεῖται M^v D^1
853b7	ὀλίγες P	ὀλίγαις M^v D^1
854b27	AE add. XAB P	AE M^v D^1
855a27	φερομένου P	φερομένη M^v D^1
856b20	om. ἔχει P	/
857b6	ἐστίν P	ἂν M^v D^1
857b11	μόνῳ P	μέσῳ M^v D^1

Moreover, M^v and D^1 contain further readings that coincide with variants from family *c*,[18] as, for example, at:

850a35	ἀντὶ σπαρτίου γίνεται *c*+M^v D^1	ἐστι τὸ σπαρτίον *a*: ἐστι τί σπαρτίου/-τίον γίνεται *b*
854a20	μαλακὴ *c*+M^v D^1	μαλθακὴ *cett.*
857a2	τῇ *c*+M^v D^1	τῷ *cett.*
858b9	ἐλαττονα *c*+M^v D^1	ἐλάττω *cett.*
858b25	τοῖς *c*+M^v D^1	τοίνυν *cett.*

The readings from family *c* show a proximity to manuscript W^a: for example at 852b22, 855a5, 855b26 and 855b37. Finally, both manuscripts share numerous significant errors which are not shared by P, as, for example, at 850a19 PT M^v D^1: P P; 851b4 πορρωτέρω M^v D^1: πορρώτερον P; 853b7 πολλαί γίνονται διαφοραί M^v D^1: πολλὴ γίνεται διαφορά P; 853b14 μέγαν πέλεκυν M^v D^1: πέλεκυν μέγαν P; 855a19 φέρεσθαι M^v D^1: φέρεται P; 856b1 [lac.] om. ποδῶν M^v D^1; 857a29 γὰρ ὁ μὲν M^v D^1: ὁ μὲν om. γὰρ P.

Certain passages show that the copyists of P, M^v and D^1 had similar difficulties in reading their source: 856b2 [lac.] om. τὸ P: [lac.] om. ἢ τὸ μὲν μέγεθος M^v D^1; 856b4 οὔτ [lac.] om. ω διπλασι όπλευροι P: οὕτως [lac.] om. διπλασιό πλευροι M^v D^1. At the end of D^1 Kamariotes even mentions the corruption of his source at f.353v:

---

[17] See Harlfinger (1971), 250.

[18] See Bottecchia (1982), 80–82. Bottecchia mentions the contaminated character of D^1, but does not notice its close connection with M^v. She lists M^v among the manuscripts collated by her, but apart from a description in the first chapter, this manuscript does not appear in her classification and, in her view, it is therefore not clear to which family it belongs.

σημειωτέον ὅτι πάνυ διεφθορὸς ἦν τὸ ἀντίγραφον. The exemplar of M^{v} and D^{1} must be a corrected or contaminated codex, since many readings from family *c* are incorporated in these manuscripts. Similar problems the scribes of all three manuscripts had with their corrupt source, suggest that M^{v} and D^{1} were copied from the same hyparchetype as P, but one that had undergone some emendations in the century between the composition of P, on the one hand, and of M^{v} and D^{1}, on the other. Since D^{1} cannot be the exemplar of M^{v} and vice versa,[19] both manuscripts must be direct copies of the hyparchetype β, although not completely independent of each other. The fact that the hyparchetype was in such poor condition makes it reasonable to assume that Kamariotes, and the copyists working for him, constantly referred to M^{v} while copying D^{1}. This would also explain the prevalence of shared errors in both manuscripts.

A highly interesting and influential family *b* manuscript, which can be traced back to B^{a}, is B^{e}.[20] The copyist of this manuscript is Niccolò Leonico Tomeo, an Italian humanist and professor of Aristotelian philosophy in Padua.[21] Leonico Tomeo owned this manuscript and used it for his 1525 Latin translation of the *Mechanics*.[22] My collations of B^{e} confirm Martin Sicherl's statement that this manuscript, together with some influences from the paraphrase by Pachymeres, provided the basis for the first printed edition of the *Mechanics* by Aldus Manutius in Venice.[23]

B^{e} contains many emendations made by the same hand both in the margins and within the text. Some of these emendations agree with variant readings contained in the paraphrase by Pachymeres, for example at 850a13 add. ἔσται, 851a27 add. δὲ, 852a17 add. χρόνῳ, 853a25 add. τῷ μόχλῷ, 855b36 add. καθ' αὑτὸ and 856a9 add. μικρὸς. Others, however, have no connection with any extant manuscript and are probably based on Leonico Tomeo's own interpretation of the text. Interestingly, some of the emendations in B^{e} were adopted in the first printed edition by Manutius, and have found their way via the Aldine even into modern editions of the treatise. Furthermore, in his edition of the treatise, Bekker included many other variants offered by Leonico Tomeo which are not contained in the Aldine: for example, at

[19] Dependence of M^{v} on D^{1} can be ruled out by reason of omissions in D^{1}, e.g. at 847a27, 848a16, 849b5, 851b38, 853a22 and 854a19, by reason of more complete readings in M^{v} at 849b17, 853b10 and 854a24, and of individual errors in D^{1} at 851a25, 851b8, 851b30, 852b1, 854b13 and 856b14. Dependence of D^{1} on M^{v} seems unlikely, since in D^{1} Kamariotes had other difficulties in reading his source than he had in M^{v}: at 849a8 [lac.] om. εὔλογον D^{1}: εὔλογον M^{v}; 849b17 ἔσται [lac.] om. ἦν D^{1}: ἔσται δὲ A om. ἦν M^{v}; 857a4 αἱ [lac.] πλευραὶ D^{1}: αἱ πλευραὶ M^{v}.

[20] Codex *Ambr.* P80 sup. might form an additional connecting link between P, B^{a} and B^{e}. Burnikel (1974), 72 states that *Ambr.* P80 sup. is mutilated and contains in its present form only the *Opuscula* of Theophrastus. It likely contained also the Aristotelian texts that were transmitted together with Theophrastus in B^{a} and B^{e}, including the *Mechanics*, as Sicherl (1997), 92 notes. In that case, in order to explain the shared errors of B^{a} and B^{e} against P (e.g. at 858a23 φαίνεται B^{a} and B^{e}: φέρεται P), *Ambr.* P80 sup. would occupy a position in the *stemma* as a copy of P and the exemplar of the independent copies B^{a} and B^{e}.

[21] See inter alia Vendruscolo (1996), 549–550 and Andrist (2007), 188–196.

[22] Chapter 5 treats in more detail Leonico Tomeo's Latin translation of the text, the *Mechanical Questions*. This work is especially interesting regarding the diagrams contained in it, which are influenced by the diagrams in B^{e}.

[23] See Sicherl (1997), 96.

850b25 ἧ B^e: εἰ *cett.*; 851b9 ἀντισπᾶν B^e: ἀντισπᾶ *cett.*; 855a38 ἡ δὲ τοῦ ἐλάττονος ἐλάττων B^e: om. *cett.*; 856a29 μικρὸς B^e: μέγας *cett.*; 857a18 τὸ μέσον B^e: om. *cett.*: ἔστω μέσον Bekker; 858b29 δὲ B^e: om. *cett.*; 858b30 εἰς B^e: om. *cett.* As he made his edition of the *Mechanics*, Bekker probably had access to Leonico Tomeo's translation and commentary on the treatise, and thus included Leonico Tomeo's emendations in his edition.[24] An important goal for a new critical edition of the *Mechanics* is to distinguish Leonico Tomeo's work on the treatise from the authentic Aristotelian manuscript readings.

The manuscript L^v is a precise copy of B^e, before Leonico Tomeo added any of the emendations in this codex. In her division of the manuscripts, Bottecchia does not note that L^v descends from B^e. The differences between L^v and B^e listed by her[25] all concern corrections in codex B^e; in every case the first reading in B^e perfectly coincides with the text of L^v.

The hyparchetype of family *b* was probably copied in Constantinople, given that two of its immediate descendants, M^v and D^1, were produced here as well. It was very likely a corrupted manuscript, since the scribes of P, M^v and D^1 had equal difficulties in deciphering their source. As we can see from the sequence of works contained in three of its copies, codex β probably contained a collection of the works of Aristotle: the treatises on animals, the *Parva Naturalia*, as well as *On Colours*, *On Indivisible Lines*, *Mechanics* and *On Breath*. The origins of both families *a* and *b* can thus be traced back to Constantinople. Manuscript B^a, which was copied in Florence in the middle of the fifteenth century, brought the text of the *Mechanics* to Italy, where it served as the source of a group of manuscripts that will be discussed in the next section.

3.1.3 *O^a and Its Descendants*

This group of manuscripts combines readings from two manuscript families. The primary readings in these manuscripts come from family *b*, whereas variants from family *a* are only contained in secondary instances. The central member of this group, O^a, stems from manuscript B^a and was subsequently emended with the help of codex H^a. Shortly after the emendations from family *a* were added, O^a provided the main source for the other manuscripts in this group, namely Q, H^d and V^{ph}.

O^a was copied in 1445 by Johannes Scutariotes on behalf of cardinal Bessarion. This is the same scribe who was also responsible for copying codex B^a. Although B^a deviates only slightly from P, it is most likely that O^a was copied from B^a and not from P, since it contains all errors peculiar to B^a: 849a7, 850a32, 851a6, 851b30,

[24] Apelt (1888), VI, assumes that Bekker included some of Leonico Tomeo's emendations in his edition; this has now been confirmed by my collations. It might also be possible that Bekker did not consult Leonico Tomeo's commentary himself, but borrowed Leonico Tomeo's emendations from the edition by van Cappelle. Van Cappelle included many emendations by Leonico Tomeo in his critical apparatus.

[25] See Bottecchia (1982), 56–58.

853a24, 856a21, 856b12 and 857b15.[26] Thus, we are able to date B^a more precisely than Bottecchia, who dates the manuscript to between 1442 and 1459.[27] B^a must predate codex O^a and for that reason must have been written before 1445, at least that part of the codex that contains the *Mechanics*.

Many variants in O^a have been changed from B^a's reading of the text into the version as contained in H^a, as can be seen at 850a35 ἐστι τὸ σπαρτίον ~~γίνεται~~; 850b7–8 ~~τὸ δὲ ἐφ' ᾧ τὸ Δ~~; 850b17 add. τὸ πλεῖστον; 852b26 ἔνθα μὲν ~~γὰρ~~; 853a25 ὄντι add. τῷ μόχλῳ; 857b6 add. τὸν κάδον; 858a1 ~~ἴσας~~ εὐθείας, and 858b11 καὶ ~~τιθέαμεν εἰ~~ add. τότε μένει. In some cases, extant variants of both B^a and H^a exist next to each other: at 852a5 ἔχοντες/ἔχοντα; 853b32 ἔγκειται/πρόσκειται; 854a34 ἀφερεῖ/ἀφαιρεῖται; 854b27 πλείω/πλευράν; 855a3 πλευρὰ/πολλαπλασία; 855b3 γίνονται/γίνεται; 855b27 μείζω/μείζονι, and 857b18 ὑποκειμένου/ἐπικειμένου. In Bottecchia's arrangement of the manuscript families, O^a assumes a much more important position than it actually has. Bottecchia recognizes the close connection of O^a to H^a and P, but at the same time she establishes the direct derivation of O^a from the archetype. As my collations have proven, O^a was not copied from codex α, since all primary readings in this manuscript coincide with B^a and all secondary readings can be explained by means of H^a.[28]

O^a with emendations provided the basis for the other manuscripts in this group: Q, H^d and V^{ph}. Codex Q was also copied on behalf of cardinal Bessarion by Johannes Rhosos in 1457. It can be excluded as a possible exemplar of the other two manuscripts, because of some peculiar errors, e.g. at 851a18 πλείῳ, 855a31 μεγέθει, 855b16 ἔλθοι, 858b26 ἀδύναται, and some omissions that only occur in Q: 851b39-40, 852a2-3, 855b30-31 and 856b25-26.[29]

Manuscript H^d, which I have collated for the first time, shares nearly all of its peculiar readings with V^{ph}. However, no parent-child relationship exists between these manuscripts; V^{ph} cannot be a copy of H^d for chronological reasons, and H^d cannot have been copied from V^{ph} because there are some passages that are omitted only in V^{ph}, for example, lines 850b14-15, 853a15, 854b27 and 856a23. Nor is it possible that these manuscripts are independent copies of O^a, since they share many peculiar errors that are absent from O^a. Furthermore, in every case where O^a provides two different readings, H^d and V^{ph} share one and the same reading: 852a5 ἔχοντα; 853b32 πρόσκειται; 854a34 ἀφαιρεῖ; 854b27 πλείω; 855b3 γίνεται; 855b27

[26] Dependence of B^a on O^a before latter received emendation can be ruled on the basis of the omission of 858b25-26 in O^a.

[27] See Bottecchia (1982), 22.

[28] See Bottecchia (1982), 54. Bottecchia argues for this double derivation from P and the archetype by means of the following example, where she claims that the scribe was uncertain how to choose between the readings in his sources and in the end combined both: Bottecchia reads at 849b12 ἐν τ[lac.] ἐφ' οὗ X σημεῖον O^a: ἐν ἐφ' οὗ X σημεῖον H^a: ἐν τῷ ἐφ' οὗ X σημεῖον P. As my collations have proven, the readings in O^a and P are not deviant at all, since both read: ἐν τ[lac.] ἐφ' οὗ X σημεῖον. Therefore it is not necessary to assume a further dependence apart from that of P on the archetype.

[29] See Bottecchia (1982), 46. Here Bottecchia cites some small variants contained in both Q and V^{ph}, on the basis of which she concludes that a direct dependence of V^{ph} upon Q is evident. The omission of several passages in Q alone already shows that such a relation must be dismissed.

μείζονι; 857b18 ἐπικειμένου. It is very unlikely that two copyists working independently of each other made the same decision every time they were confronted with two different readings. Therefore, it is natural to conclude that H^d and V^{ph} are dependent on the same lost manuscript, which in turn stems from manuscript O^a.

3.1.4 Family *c*

Family *c*, which can be traced back to its own hyparchetype γ, is somewhat younger than the other manuscript families. Codex γ probably originated in Italy, since all of its copies are found in and around Italy. It was not copied as a collection of the works of Aristotle, but either had a hybrid character, like many of its copies, or contained the text of the *Mechanics* only. Family *c* has the following members: W^a, U^m, V^1, M^1, F^2, N^b, q, N^h, F^1, Par and V^2.[30] This family is distinguished, amongst other things, by the presence of scholia, both within the text and in the margins.[31] The manuscripts in family *c* are also characterized by the omission of various longer passages and have characteristic readings which are not preserved in the other manuscript families, for example at:

848a5	καὶ εἰς τὸν ὄπισθεν *c*	καὶ τὸν ὄπισθεν *cett.*
848b10	λίγῳ/ὀλίγῳ *c*	λόγῳ *cett.*
850a35	ἀντὶ σπαρτίου γίνεται *c*	ἐστι τὸ σπαρτίον *a*: ἐστι τί σπαρτίου/-τίον γίνεται *b*
850b18	ἀμφοτέροις *c*	ἀμφότερα/ἀμφότερον *cett.*
852b17	θᾶττον *c*	μεῖζον *cett.*
854a20	μαλακὴ *c*	μαλθακὴ *cett.*
854b9	ἐστιν μὲν οὖν τὸ *c*	ἐστιν οὖν τὸ μὲν *cett.*
854b14	εὕρεσις *c*	ἄρσις *cett.*
855b16	om. περιφέρεια *c*	περιφέρεια *cett.*
856a1	ἡ A *c*	ὁ A *cett.*
857a12	ἐπειδὴ διότι *c*	αἴτιον δὲ διότι *a*: αἴτιον δὲ ὅτι *b*
857b13	τῶν ξύλων *c*	τὸ ξύλον *cett.*
858b9	ἐλάττονα *c*	ἐλάττω *cett.*
858b25	τοῖς *c*	τοίνυν *cett.*

[30] Codex V^2 contains abstracts from the *Mechanics* and is on the basis of its many modifications and strong paraphrastic character presented in Sect. 3.1.5 on contaminated and incomplete manuscripts.

[31] Bottecchia (1982), 145–165, was the first editor to include the scholia to the *Mechanics* in her edition. However, it is important to notice that not only the scholia from family *c* are contained in this edition, but also the glosses added by Pachymeres (from the codex M^u), which should be distinguished from the "real" scholia.

V^1 and W^a emanate independently of each other from the hyparchetype γ. Manuscript V^1, written by Demetrius Branas around 1430, is the older of the two.[32] In her assessment of the manuscripts, Bottecchia adjusts the dating of V^1 by several decades, because she believes that this manuscript is a copy of W^a (circa 1465).[33] She points to some similar readings in both manuscripts; however, she does not argue that the relationship of W^a and V^1 can only be that of father and child. The examples given by Bottecchia do not exclude a reverse dependence of W^a upon V^1, or that both are independent copies from a common source. V^1 does not contain all the scholia from family *c*, but only those at 848a32 and 852a23. In the tradition of the *Mechanics*, it has one direct descendant, M^1. Codex M^1 was copied by two hands: Manuel, a pupil of Konstantin Laskaris wrote the first part, and Laskaris himself the second part, to which the *Mechanics* belongs. For *On Indivisible Lines* and *History of Animals*, codex V^1 formed the exemplar of M^1 as well.[34] The other treatises contained in M^1 were copied from H^a, or, in the case of *On Generation and Corruption*, can be traced back to a now lost manuscript.[35] Apart from some longer omissions (e.g. at 847a27–28, 847b25–26, 852a2–3 and 852a10–11), M^1 is an accurate copy of V^1 and contains few errors of its own. Laskaris was, however, not a very talented draftsman, as a couple of diagrams in the margins of M^1 show.

W^a was written some 35 years after V^1 was composed. It was copied by Georgius Tzangaropulos in an accurate hand. All the scholia of family *c* were added in the margins of the text by a different hand, probably in the sixteenth century.[36] Both V^1 and W^a have several longer omissions, although of different passages. Since all manuscripts in family *c* are characterized by such omissions, it is reasonable to assume that the hyparchetype γ was already corrupted, at least to some degree, and may have contained different stages of emendation.

Of the other manuscripts in family *c*, nearly all contain the full range of scholia, and all share a great number of longer omissions, as for example at 851a20–22, 855a30–32, 855b6–10, 856a27–29, 857a30–32 and 857b30–32. On the basis of further omissions and peculiar errors, these manuscripts can be split into two groups. U^m, q, N^b and F^2 form the first group and are closely connected. U^m and q are the oldest of these manuscripts and it is possible that they are the exemplars of the remaining manuscripts in this group. U^m was copied in Crete by Giovanni Simeonachis on behalf of Marco Lippomano. The *terminus ante quem* for this manuscript is 1446, the year in which Lippomano died.[37] Manuscript q can, on the basis

[32] See Harlfinger (1971), 266, on the date of V^1. The full name of the copyist remained unknown until recently. As Harlfinger informed me in October 2009 he was able to decipher the name of the copyist from another manuscript *Berol.Phill.* 1582 by the same hand.

[33] See Bottecchia (1982), 28.

[34] See Harlfinger (1971), 285–290 on the relation between V^1 and M^1 for *LI*, and Berger (2005), 147–151 on *HA*.

[35] See Rashed (2001), 69.

[36] See Stornajolo (1895), 49.

[37] See Mioni (1972), 246–247 and Zorzi (1993), 40. As mentioned at the beginning of this chapter, Bottecchia (1982), 21 acknowledges a dating of U^m before 1446, but still believes that this codex

of watermarks, be dated with certainty to the first half of the fifteenth century, and some parts of the codex can even be ascribed to the first quarter of that century, about 1420.[38] Whereas the catalogues agree on a date for N^b in the fourteenth century,[39] Harlfinger has established that the codex is of a later date in the second half of the fifteenth century, possibly even the beginning of the sixteenth century.[40] Hence the possibility of N^b being the original of the other manuscripts in this group can be discarded. Because of peculiar omissions, especially the omission of lines 847a18 and 855a12 in U^m, and of 848a4-5, 855a35-36 and 856a16 in q, it is impossible to ascertain whether one of the manuscripts descends from the other. Therefore, it is likely that U^m, N^b and q stem independently of each other from the same lost manuscript. F^2 contains all the peculiar omissions and errors of U^m, and must therefore be a copy of this manuscript.

The second group, consisting of N^h, F^1 and Par, shares four further omissions, namely of lines 852b23-24, 854b20, 854b26 and 856b25-26. There is a close connection between N^h and Par, which can be seen in the corresponding errors, for example, at 851a5, 851a9, 851a24, 851b33, 852a2 and 852a9. There are, however, further peculiar errors in N^h that are not contained in Par, from which it follows that this manuscript cannot descend from N^h. The Par codex underwent elaborate emendation after completion; all omissions were completed and many variants were added or overwritten, once the first reading had been erased. Par was emended with a manuscript containing the paraphrase of Pachymeres, as is shown by the remarks added at 847b26 and 848a37, and by the variants at 850a35, 850b37, 851a20, 851b33, 852a5, 852a12, 852a18, 853a10, 853a25, 853a27, 853b6, 853b32, 854b1, 854b20, 855a13, 855b39-856a1, 856a35, 856b21, 856b32, 857a24, 857b2, 857b6, 857b38, 858b9, 858b24 and 858b25. Finally, manuscript F^1 is a copy of N^h.

Family *c* offers some valuable textual variants which are not contained in the other manuscript families, as for example at:

849a36	ὀρθὰς *c*	ὀρθὴ/ὀρ[lac.] *cett.*
849b18	τὸ κατὰ φύσιν *c*	κατὰ φύσιν *a*: κατὰ φορὰν *b*
850a35	ἀντὶ σπαρτίου γίνεται *c*	ἐστι τὸ σπαρτίον *a*: ἐστι τί σπαρτίου/-τίον γίνεται *b*
852b17	θᾶττον *c*	μεῖζον *cett.*
857a28	ἀπὸ τῆς *c*	τῆς *a*: [lac.] ἀπὸ τῆς *b*

is a copy from the manuscript W^a, written around 1465. At p.63–64 she comes to the following conclusion: "È più facile, dunque, supporre che sia U^m a copiare da W^a, anziché W^a da U^m, anche se U^m dovette colmare con l'aiuto di altri codici le due omissioni di W^a."

[38] See Tziatzi-Papagianni (1994), 346.

[39] See Diels (1905), 38; Napolitano et al. (1977), 26–27; Formentin (1997), 213–214.

[40] As Harlfinger informed me in a conversation in May 2010.

Even more important for a new critical edition of the *Mechanics* is the position of family *c*, where readings from this family agree with family *a* against family *b*. As I mentioned above, the hyparchetype of family *b* was a corrupted and contaminated manuscript, and the editor should be very careful in adopting readings from this family. Given the agreement between family *c* and *a*, he or she might reconsider the following variants:

851b1	δεῖ κινεῖν *b*	κινεῖν δεῖ *cett.*
853b2	ἕλξει *b*	om. *cett.*
854a17	τῇ χειρὶ μόνῃ ψιλῇ *b*	ψιλῇ τῇ χειρὶ μόνῃ *cett.*
856b13	ἴσαι εἰσίν *b*	εἰσίν ἴσαι *cett.*

Whereas the nineteenth-century editors of the treatise consulted manuscripts W^a and Par, family *c* was neglected in Bottecchia's edition. In her division of the families, these manuscripts all depend on H^a and cannot be traced back to a separate hyparchetype. Such an attribution is completely unfounded and exceedingly unlikely, since it does not explain any of the readings characteristic of family *c*. In order to support her thesis that W^a was copied from H^a, Bottecchia cites four examples in which W^a and H^a agree against L.[41] Two of these examples are concerned with passages that are highly problematical in all manuscripts, and do not offer strong evidence for her thesis, since we have seen that scribes make different decisions when confronted with corrupted passages. Another example from 850b1 concerns an error in L, where L's reading differs from all other manuscripts (not just H^a and W^a). Finally, the omission of lines 853a12–14 in H^a and W^a presents a case of homeoteleuton. This error does not point to a relation between the two manuscripts, but rather shows that two copyists made a similar mistake independently of one another. In reading and copying a manuscript, it frequently happens that scribes unintentionally skip a line (or two) in the exemplar, and consequently continue their copy with the same word found one or few lines below. The fact that this omission is not present in the other manuscripts of family *c*, such as V^1 or U^m, shows that it is rather an individual scribal error. Moreover, since the manuscripts in family *c* do not contain any of the peculiar readings from family *a* (and vice versa), it is only reasonable to assume that family *c* derives from an independent source.

Furthermore, Bottecchia makes several errors concerning the chronology of the manuscripts. For example, as I mentioned above, she derives codex V^1 from W^a. She agrees in dating W^a to around 1465, but she rejects Harlfinger's dating of V^1, since it does not support her hypothesis: "In base ad un confronto con altro manoscritto sembra ad Harlfinger di poterne fissare la data al 1430 circa; a me sembra che il manoscritto sia posteriore di qualche decennio: tale ipotesi è confortata dai dati della collazione e classificazione."[42] Bottecchia does not give any codicological reasons for adjusting the date of V^1; for her it is sufficient to say that her classification of W^a as an antigraph of V^1 speaks against dating V^1 to around 1430. Future editors

[41] See Bottecchia (1982), 68.

[42] Bottecchia (1982), 28.

of the *Mechanics* should especially consult W^a and V^1, since both manuscripts directly stem from the hyparchetype γ. On the other hand, editors should be very hesitant to adopt readings from Par, which was examined by van Cappelle and Apelt. Because of the fact that Par contains many emendations by Pachymeres, this codex is only of minor significance to the textual transmission of the treatise.

3.1.5 Contaminated and Incomplete Manuscripts

Finally, we must mention a group of manuscripts which are of no great value for ascertaining Aristotle's text. Some of these manuscripts contain a mixture of readings from different manuscript families, as, for example, N^a; others contain only extracts of the *Mechanics*, for example V^2 and B^u.

Codex V^2, which was written by several hands, originates from the fourteenth century, at some point between 1317 and 1338.[43] It contains excerpts from the *Mechanics*, in which whole problems have been left out. Thus, there are many lacunae: at 848b9–850a29, 851b6–851b14, 851b40–852a8, 852a23–852a37, 852b11–852b21, 853a5–853b13, 853b22–857a4, 857a22–857b20, 857b26–857b38 and 858a1–858a12. Nevertheless, it is possible to determine which family V^2 is affiliated with. The first scholium from family *c*, at 848a32, is present in this manuscript, and there are further variants that show a proximity to family *c*: at 850a35, 850b30, 857a12, 858a14, 858b9 and 858b25. This codex also includes modifications and expressions that cannot be found in any other manuscript, and are probably inventions of the scribe: for example, at 850b4 ἔχει V^2: γράφει *cett.*; 850b25 μέρος αὐτῆς V^2: μέρος *cett.*; 852b3 λίθον V^2: τὸ βέλος *cett.*; 852b9 θᾶττον καθὼς προείρηται V^2: θᾶττον *cett.*; 857b23 διότι V^2: πότερον ὅτι *cett.*; 858a19 καὶ τοῦθ' πάλιν ἕτερον V^2: καὶ τοῦθ' ἕτερον *cett.*; 858b10 φερόμενον ὥστε ἐν δυσὶ κύκλοις εἶναι V^2: φερόμενον *cett.* V^2 does not derive from any of the extant manuscripts in family *c*, since it is considerably older than the other manuscripts contained in this family.[44] It might be a direct copy of the hyparchetype γ, if we assume that this hyparchetype was composed in the first quarter of the fourteenth century.[45] Nevertheless, given its individual character, V^2 is not a very valuable witness of the textual transmission, since it is remote from the authentic Aristotelian text.

Codex N^a was copied between 1465 and 1472 on behalf of cardinal Bessarion. It is the work of four different copyists, of which Harlfinger was able to identify Demetrius Triboles for the first part.[46] For the part that contains the *Mechanics*,

[43] See n.2 on the dating of V^2.

[44] As I mentioned at the beginning of this chapter, in Bottecchia's assessment of the manuscripts, V^2 is a copy of W^a. This attribution speaks against the chronology of the manuscripts. At p.28 Bottecchia dates V^2 to the fifteenth century, without referring to a source or providing grounds for this late dating.

[45] Harlfinger (1971), 265 gives a *terminus post quem* for this manuscript of between 1300 and 1325.

[46] See Harlfinger (1971), 310.

he recognized the hand of his so-called Anonymous KB. Bottecchia believes to have found in codex O^a the exemplar of N^a. This, she asserts, is visible from the almost complete accordance of the readings in both manuscripts.[47] A first glance comparison of the first pages of these manuscripts, however, shows that there can be no such relationship, as is clear from, for example, the variants at 847b19, 847b28, 848a2 and 848b8. In fact, N^a is a highly contaminated manuscript containing aspects of family *a* (for example at 848b24, 856a34, 856b28 and 857a12), family *b* (at 855b26, 857a38) and family *c* (at 850a35, 855b16). In addition, it has many variants of its own: for example, at 851a33 μέσον τι τοῦ κινουμένου N^a: τι τοῦ κινουμένου μέσον *cett.*; 851b9 μὲν οὖν N^a: μὲν *cett.*; 852a16 ὡσαύτως N^a: ὁμοίως *cett.*; 853b32 ἐπίκειται N^a: πρόσκειται *a*: ἔγκειται *cett.*; 857a5 μικρὰ/ἀπὸ μίκρου N^a: μακρὰ/ἀπὸ μακροῦ *cett.* Neither these readings from different manuscript families nor the manuscript's own variants were added or produced by emendation at a later stage, but are perfectly integrated in N^a; therefore, it is likely that this manuscript stems from a source that was already contaminated.[48]

Manuscript P^t, which was consulted by van Cappelle for his edition of the *Mechanics*, belongs to a different tradition and is transmitted together with astronomical works. It was wrongly attributed to the fifteenth century by Omont,[49] but Harlfinger was able to date it on the basis of the watermarks to the fourteenth century, more precisely to between 1370 and 1380. With this dating, Pingree's attribution to Isidore of Kiev (fifteenth century) in his edition of Albumasar is no longer possible,[50] and Harlfinger assumes that is was copied by the Byzantine astronomer Johannes Ambramios. Of the two codices that are known to have been copied by Ambramios, *Marc.gr.* V.13 (dated 1376) and *Laur.Plut.* 28.16 (dated 1381/1382),[51] the lettering of the first is somewhat smaller than that of P^t, but the handwriting of the second is very similar to our codex. Mondrain, however, proposes another copyist for P^t: Isaac Argyros, a fourteenth century mathematician and astronomer, and the student of Nicephorus Gregoras.[52] The circle around Gregoras could very well have been responsible for a manuscript like P^t, which contains almost exclusively astronomical texts. Although there are some similarities between the handwriting of Argyros in *Marc.gr.* 155 and codex P^t,[53] I am not convinced that these manuscripts were copied by the same hand. More samples of the styles of both Ambramios and Argyros need to be examined, in order to solve the problem of which of these scribes was responsible for copying P^t. Codex P^t contains a revised version of the *Mechanics*.

[47] See Bottecchia (1982), 45.

[48] See Harlfinger (1971), 309–310 where he mentions the highly contaminated character of N^a in the case of the treatise *LI*.

[49] See Omont (1888), 274. Bottecchia (1982), 27 adopted this date by Omont and attributed P^t to the fifteenth century.

[50] See Pingree (1968), XI.

[51] See Turyn (1972), 241–244 on *Marc.gr.* V.13 and 245–248 on *Laur.Plut.* 28.16. For the hand of Ambramios see plates 201 and 203. See further Vogel and Gardthausen (1909), 167.

[52] See Mondrain (2007), 167.

[53] See Mondrain (2007), Figs. 1 and 2.

This is already evident from the title, which is rather paraphrastic: αἰτιολογία τῆς τῶν μηχανικῶν ἐνεργίας ὥς τινες λέγουσιν 'Αριστοτέλους. The peculiar readings, lacunae (at 849b7–852a14 and 855a27–858b31) and paraphrastic character of P^t, make it impossible to ascertain its dependence, although there are some similarities with family *a*: for example at 848b24, 853a27, 853b32, 855a3 and 855a13. Bottecchia's claim that this manuscript descends from N^a must be rejected for chronological reasons.[54]

Manuscripts P^4 and P^8 both contain important variants from family *c*, but cannot be definitely attributed to this family.[55] P^4 contains the greater part of the readings from family *c* (at 850a35, 852b17, 854a20, 854a36, 854b9, 854b14, 855b16, 856a1, 857a12, 857b13 and 858b9), but none of the scholia or omissions that are particular to this family. The fact that there are no emendations or additions in P^4 itself implies that it is based on some earlier, non-extant manuscript, which had already been corrected by means of a manuscript from another family.

The other Parisian manuscript, P^8, contains not only the majority of the readings from family *c* (at 848b10, 850a35, 850b18, 852b17, 854a20, 854b9, 855b16, 856a1, 857a12, 858b9 and 858b25), but also the scholia from this family and its main omissions, which were written in the margins. However, P^8 fits not unproblematically into this family, since some readings from family *a* are present too. These readings are for one part added at a later stage, and suggest an emendation with manuscript L. For example, at 847b25, a scholium contained exclusively in L has been added in the margins; at 848a11 μαθημάτων has been added; lines 850b7–8 are still absent after completing 850b6–9; at 853a27 <u>ἐφ' ᾧ</u> ΔΕΗΖ has been added in the margin; at 858b11 τότε μένει was probably written over an erased τιθέαμεν εἰ. For another part, these variants from family *a* do not appear to be later additions: for example 848b24 πλευρῶν and 850a1 μέρει. The exemplar of P^8 might have been an already contaminated manuscript. Furthermore, P^8 contains many peculiar errors and paraphrastic passages, and is therefore not of particular interest for the reconstruction of the text of the *Mechanics*.

The seventeenth-century manuscript B^u contains compendia of many Aristotelian works. For the *Mechanics*, it contains abstracts of most of the problems; some of these problems were, however, reduced to just a few lines. For that reason, it is difficult to detect an affiliation between B^u and one of the manuscript families. The following variants at the beginning of B^u show a proximity to family *a*: 847a16 πολλαχῶς; 847b19 μετ' ἀλλήλων; and 848b24 πλευρῶν.

---

[54] See Bottecchia (1982), 84.

[55] From my analysis of both manuscripts, it turns out that Bottecchia's certain attribution of P^4 as a copy of M^1, and P^8 as a copy of U^m, is highly problematic and unlikely.

3.2 The Paraphrase by Pachymeres

The next group of manuscripts, which directly descends from the hyparchetype α, does not contain authentic Aristotelian writings, but the Byzantine paraphrase of Aristotle's works by Georgius Pachymeres. This paraphrase influenced the first printed edition of the treatise by Aldus Manutius and thereby also manuscripts V^3, N^c and u, which are copies of Manutius' edition.

Georgius Pachymeres (1242 – circa 1310) was a Byzantine philosopher and the leading historian and scholar of his time.[56] He was one of the most productive Byzantine commentators and played an important role in the interpretation of Platonic and Aristotelian philosophy. Around 1300, he wrote a compendium of Aristotle's philosophy in 12 books, the *Philosophia.* Almost every book contains an epitome of one Aristotelian text, except for e.g. the final book, which contains the shorter treatises *On Colours*, *On Indivisible Lines* and *Mechanics*. For most of the treatises Pachymeres' text is shorter than Aristotle's original.[57] Again, exceptions can be found in book 12, for the text of *On Indivisible Lines*,[58] and, as we will see, for the *Mechanics*. Pachymeres' method is to produce a running text in which literal quotations and paraphrased passages, together with Pachymeres' own comments, examples and references form a coherent whole. The text of the *Mechanics*, however, is less a paraphrase than an almost verbatim quotation of the Aristotelian text. For this reason, it is a complex task to distinguish Pachymeres' writings from the authentic Aristotelian text, with the result that manuscripts containing Pachymeres' paraphrase have often been considered as Aristotelian writings. In Wartelle's *Inventaire* many apographs of Pachymeres are still listed among the Aristotelian manuscripts,[59] and, as mentioned above, Bottecchia did not recognize M^u as a manuscript by Pachymeres and included many paraphrastic elements in her edition. The paraphrase, however, is of great importance for the transmission of the *Mechanics*, since it was written at the same time or even shortly before the authentic manuscripts were copied. We even possess two autographs of Pachymeres' *Philosophia*, *Berol.Ham.* 512 ($Pach^b$) and *Par.gr.* 1930 ($Pach^p$), which contain the *Mechanics*.[60] $Pach^p$ is posterior to $Pach^b$ and is a copy of this manuscript.

[56] See Zografidis (2011), 394.

[57] See Pappa (2002), 29 on the *Metaph*.

[58] See Harlfinger (1971), 349.

[59] Wartelle (1963); see no. 1038, where he lists M^u as a manuscript containing the Aristotelian *Mechanics*.

[60] These autographs were identified by Harlfinger (1971), 357–358.

Pach^b Berol.Ham. 512 – 14th century, beginning – paper – 308×220 mm – ff. 234
Prooimion; Book I: *Org.*; Book II: *Ph.*; Book III: *Cael.*;
Book IV: *GC*; Book V: *Mete.*; Book VI: *PA*, *IA*; Book VII: *de An.*;
Book VIII: *PN*, *MA*; Book IX: *GA*; Book X: *Metaph.*; Book XI: *EN*;
Book XII: *Col.*, *LI*, Mech. (ff. 224–234, with diagrams).
Copyist: Georgius Pachymeres
Provenance: probably Constantinople
Lit.: Harlfinger (1971), 345–360; Wilson (1983), 241–242.

The pages near the beginning and end of a manuscript had to endure the most wear and tear, and these are the ones that are often mutilated. This is also the case in Pach^b, where the text of Pachymeres breaks off at 857b14 with φερόντων and is continued with a supplement by a later hand. The copyist of these supplementary pages was identified as Bartolomeo Zanetti,[61] who was working in Venice when he replaced the damaged pages of the manuscript in about 1530. Zanetti did not have another Pachymeres manuscript at his disposal to use in restoring the last pages of the *Mechanics*. My collations have proven that the variants in Pach^b from 857b14 to the end of the text perfectly agree with readings from the Aldine: for example, at 857b15 om. γὰρ/ἀπέχη, 857b16 τὸ, 857b29 ὀρθὸν, 857b33 κνῆμαι, 857b38 ἅμα ἀνίστηται ἀναστῆναι, 858a19 τοῦτο, 858a24 ῥιπτόμενα, 858a29 ἀντερείδει, 858a31 μηθὲν, 858b5 φέρεται, 858b9 αὐτὸν, 858b14 δινομένου, 858b21 ὅσω, 858b24 ὑπολείπεσθαι τὸ πᾶν, 858b29 om. δὲ and 858b30 om. εἰς.[62] Printed a couple of years earlier in Venice, the Aldine should have been easily accessible to Zanetti.

The other autograph of Pachymeres, Pach^p, which is the work of different copyists, contains the text of the paraphrase for the whole *Mechanics*. The first pages of the *Mechanics* are by the hand of Pachymeres himself. At 849a31, however, a change of copyist takes place, and the text is completed by an assistant of Pachymeres.

The manuscripts that contain the writings of Pachymeres share many of the readings characteristic of family *a*, as, for example, at:

847b25	διεστήκατον *a*+Pach.	διέστηκεν *cett.*
850a32	κινεῖν *a*+Pach.	κινῆσαι *cett.*
850b7–8	om. ὑπομόχλιον ἐφ’ ᾧ τὸ Ε, τὸ δὲ ἐφ’ ᾧ τὸ Δ *a*+Pach.	/
851b18	ὁ τῆς ἁμάξης τροχὸς *a*+Pach.	ὁ τροχὸς ὁ τῆς ἁμάξης *cett.*

[61] See Harlfinger (1971), 358 n.1.

[62] For this reason Pach^b is a not a new Aristotelian manuscript as Hilgers (1992), 62–64, assumes. Hilgers noticed for the text passage 857b14–858b31 a proximity to the manuscripts L and Q. Apart from some small similarities between these manuscripts, there are too many peculiar readings in both L and Q, which make such an affiliation impossible. Peculiar errors in L can be found at 857b18, 857b33, 858a1, 858a12, 858a15, 858a16 and 858b25. And in Q: at 857b18, 858a1, 858a15, 858a16, 858a23, 858b11, 858b25 and 858b26. Furthermore, L and Q do not share the variants of Pach^b at 857b15, 857b33, 858b5 and 858b11.

853a25	ὄντι τῷ μόχλῷ *a* + Pach.	ὄντι *cett.*
853a27	ἐφ' ᾧ ΔΕΖΗ *a* + Pach.	ΔΕΗΖ *cett.*
854b20	τὴν ἐλάττω διάμετρον *a* + Pach.	τὴν διάμετρον τὴν ἐλάττω *cett.*
855a13	ὀξείας ὥσπερ συμβαίνει *a* + Pach.	ὀξείας συμβαίνει *cett.*
856a24	κωλύει *a* + Pach.	κυλίει *cett.*
856b28	ΒΓ *a* + Pach.	ΑΓ *cett.*
857b38	ἀνίσταται ἀναστῆναι *a* + Pach.	ἀναστῆναι *cett.*

Furthermore, there are many peculiar readings not contained in any of the other manuscript families:

849a4	ἡ ΒΔ πρὸς τὴν ΔΓ Pach.	ἡ ΒΕ ΕΓ *cett.*
850a39	βάρη δὲ δύο Pach.	δύο δὲ βάρη *cett.*
850b4	μεῖζον Pach.	μείζονα *cett.*
851a27	δὲ οὗ τὸ Δ Pach.	οὗ τὸ Δ *cett.*
851b36	ἐλάττους Pach.	ἐλάττονας *cett.*
852b32	ῥᾷον Pach.	θᾶττον *cett.*
854a30	μᾶλλον Pach.	ῥᾷον *cett.*
854b20	φορᾶς Pach.	πλευρᾶς *cett.*
857a10	φέρεται Pach.	αἴρεται *cett.*
857a29	τὸ ξύλον μακρότερον Pach.	μακρότερον τὸ ξύλον *cett.*
858b14	δυναμένου Pach.	δινουμένου *cett.*

Pachymeres repeatedly adds illustrative remarks and phrases to the text,[63] as, for example, in the introduction to the *Mechanics*, where the properties of the circle are explained. It is said by the author that a point along the radius of a rotating circle moves faster the further it is from the centre. At 848a17 Pachymeres explains: ὡς πολλάκις καὶ ἡμῖν ἐλέχθη, ἐν τῷ περὶ οὐρανοῦ καὶ ἐν ἄλλοις· ὅτι θάττονα εἰς κίνησιν τὰ πορρώτερα τοὺ μέσου.

Because it is often difficult to distinguish Pachymeres' paraphrase of the *Mechanics* from the authentic Aristotelian manuscripts, the first printed edition by

[63] It should be noted that codex L contains at 847b26 after ἀλλήλων a scholium in the text that is similar in part to a remark in the text of Pachymeres. It is possible that it was added by Pachymeres to his exemplar and copied from here by the scribe of L; since this is the only case in the whole text of the *Mechanics*, we cannot say anything definite on possible interventions by Pachymeres in his exemplar. Something similar happens in some variants, where H^{a} agrees with manuscripts from other families in a minor error, whereas L and Pachymeres contain the correct variant, for example, at 851b19 μόνον L/Pach.: μένον H^{a}; 853a34 ἐκ θατέρου L/Pach.: ἑκατέρου H^{a}; 856b36 τρυπήματα L/Pach.: τρυπήματι H^{a}. This is possibly a further indication that Pachymeres might have been responsible for making some smaller adjustments in his exemplar. In this case, his corrections influenced only codex L, since Ha had probably already been copied from the hyparchetype when Pachymeres accessed this manuscript.

Manutius, and, via the Aldine, all modern editions, contain variants from his paraphrase: for example at 849a4 ἡ ΒΔ πρὸς τὴν ΔΓ Pach./Aldine: ἡ ΒΕ ΕΓ codd.; 853a10 ὁ ἡ χεὶρ αἴρει Pach./Aldine: om. codd.; 855b39–856a1 ἀλλὰ τὴν τοῦ κινοῦντος Pach./Aldine: om. codd.; 856a4 ὑπὸ τοῦ ἐλάττονος Pach./Aldine: om. codd. One of the fundamental tasks for a new edition of the treatise will be to remove all such traces of Pachymeres from the text of the *Mechanics*.

Whereas the manuscript B^e provided the basis for Manutius' edition, Pachymeres' paraphrase also exerted an influence on the Aldine. Some of the variants offered by Pachymeres found their way into the Aldine through the manuscript B^e; other variants were included directly from Pachymeres in the first printed edition of the *Mechanics*. Bottecchia, however, believes she has identified the exemplar of the Aldine in codex V^3.[64] As Sicherl pointed out, the relationship between these manuscripts is exactly the contrary: V^3 is a copy of the Aldine.[65] My collations of both manuscripts have confirmed this assumption. The copyist of V^3 is Demetrius Damilas, an important and prolific copyist, who is identical with Harlfinger's "librarius Florentinus".[66] Damilas worked at the end of the fifteenth century in Florence together with the Byzantine humanist Demetrius Chalkondyles, but he was still active at the beginning of the sixteenth century. Between 1490 and 1504 his name appears eight times in the borrowing register of the Vatican Library. Furthermore, in February 1506 Pope Julius II sent a *Motu proprio* in favour of Damilas.[67] Damilas was assigned to copy a certain number of diverse texts for the pontifical library. Thus, a dating of V^3 towards the beginning of the sixteenth century is likely,[68] and allows V^3 to descend from the Aldine.

Of the two remaining manuscripts in this group, the u codex is a copy of the Aldine, whereas N^c derives from V^3.[69] Although the Aldine and V^3 are to a large extent identical, there are some minor errors where V^3 and N^c agree against the Aldine: for example, at 850a29, 851b31, 852a1, 852a9, 852b2, 852b12 and 857a29.

---

[64] See Bottecchia (1976), 389–392. Bottecchia here lists the text passages where V^3 and the Aldine agree against the other manuscripts. In most cases manuscript B^e shares the reading of V^3/Aldine, and where it has a different reading, the paraphrase by Pachymeres provides the variant of V^3/ Aldine, see, for example at 854b20, 856a4, 857a24, 857b12, 857b32 and 857b38. Since Bottecchia did not recognize the paraphrase of Pachymeres, she does not give a satisfying explanation for those "peculiar" readings in V^3, where V^3 has a reading different from all other manuscripts. They are definitely not a result of a derivation from the manuscripts L and P as presented in her stemma at p.392. Bottecchia probably supposes a different hyparchetype for V^3, as she does in her edition of the *Mechanics*. However, as my collations have proven, these "peculiar" readings in V^3 do not stem from a new hyparchetype, but are variants by Pachymeres.

[65] See Sicherl (1997), 95–96.

[66] See Canart (1977–1979), 282.

[67] See Canart (1977–1979), 281.

[68] Capocci (1958), 24, also dates it to the beginning of the sixteenth century.

[69] That V^3 descends from N^c can be ruled out on the basis of omissions in N^c, as e.g. at 847b23–24 and 849b5. The arrangement of the title page of N^c is a perfect copy of V^3.

3.3 The Latin Tradition

Different studies have shown that the Aristotelian *Mechanics* influenced the mechanical tradition of the *scientia de ponderibus* in the Middle Ages in general, and the school of Jordanus de Nemore in particular.[70] Because of connections between Jordanus' work and the *Mechanics* on the one hand, and the medieval interest in mechanical texts, as is evident from the comprehensive translation movement in the twelfth and thirteenth centuries on the other, it is likely that a Latin translation of the *Mechanics* was made in this period. However, to this day there is no trace of such a Latin translation. We only know of a reference to the *Mechanics* by Emperor Frederick II in his *De arte venandi cum avibus*.[71] This work belongs to the later years of Frederick's reign and can be dated to between 1244 and 1250. In a chapter on the wings of birds, Frederick mentions a principle that we know from the introduction of the *Mechanics*, namely that a point on the radius of a circle that is further away from the centre moves faster than a point that is closer by (848a15–17). He refers to the *Mechanics* by the Latin title *Liber de ingeniis levandi pondera* and attributes it to Aristotle:

> When in motion a wing describes an arc of a circle; all the largest flight feathers follow the same movement and describe shorter or longer arcs depending on whether they are near to or remote from the body of the bird, and these arcs are all parallel with each other. Those feathers that are farthest from the body of the bird and describe in flight the greatest arcs have the greatest power to lift and carry the bird forward. About this Aristotle says in his book on the art of raising weights that the greater the orbit, the greater the power to lift a weight.[72]

Further, Clagett mentions one other allusion to the *Mechanics* in the fifteenth century.[73] A license for the export of books from Bologna in the year 1413, includes the title *Repertorium super mechanica Aristotilis*. The existence of a repertorium suggests a prior translation, which might possibly be found in the *Liber mechanicorum* that is listed among a group of books at Padua in 1401, as Clagett notes.[74] Although the references to the *Mechanics* in the Middle Ages are very meagre, there is still reason to hope that a Latin translation of the treatise will one day be found.

---

[70] See e.g. Moody and Clagett (1952), amongst others 124–125, 147, 170 and Crombie (1967), 114–120.

[71] See Haskins (1927), 316–317.

[72] Transl. Wood and Fyfe (1943), 70. See Haskins (1927), 316 n.113 for the Latin text.

[73] See Clagett (1959), 71.

[74] See Clagett (1959), 71–72.

3.4 The Arabic Tradition

The evidence for a transmission of the *Mechanics* in the Arabic world is conclusive. As early as 1908, Thomas Ibel identified a passage in al-Khazini's *Kitab Mizan al-Hikma* (Book of the Balance of Wisdom) as a partial Arabic version of the *Mechanics*.[75] However, this discovery went largely unnoticed,[76] until Mohammed Abattouy in recent years published several articles on the Arabic tradition of the Aristotelian *Mechanics*.[77]

This abridged Arabic version of the *Mechanics*, titled *Nutaf min al-hiyal* (Elements/Extracts of Mechanics), was included in the twelfth century by al-Khazini in the fifth book of his *Kitab Mizan al-Hikma*.[78] The title *Nutaf min al-hiyal* itself indicates that it is an excerpt from a longer text, which might even point to the existence of a complete Arabic translation of the *Mechanics*. The text of the *Nutaf min al-hiyal* is attributed to Aristotle and begins with the sentence: "Aristotle said." It contains an abridged version of the introduction and the first problem of the *Mechanics* on the question of why larger balances are more accurate than smaller ones. The introduction to the text in Arabic presents a reliable abstract of the Greek. In it the remarkable properties of the circle are described, as well as the importance of the circle as the underlying principle of all mechanical phenomena. In contrast, the text of the first problem in the *Nutaf min al-hiyal* is very short and leaves out the whole geometrical proof that is essential to the author of the *Mechanics*.[79] It focuses on the question of the precision of balances of different sizes, which is answered by the fact that the point on the radius that is further from the centre travels further than the closer point, although they are moved by the same force. Therefore, in the case of a longer balance beam, the motion will be stronger than in a smaller balance. When a very small weight is used, a smaller balance does not have any inclination, whereas the motion is clearly visible in a larger balance.

The fact that the geometrical analysis of motion in the first problem is not included in the Arabic version of the *Nutaf min al-hiyal* can be explained by the fact that al-Khazini's interests lay elsewhere. The text of the *Nutaf min al-hiyal* is included by al-Khazini after a long technical discussion on balance equilibrium.[80]

[75] See Ibel (1908), 123–125.

[76] See e.g. Peters (1968), 61. Peters notes that the *Mechanics* was possibly employed, though not cited by name, in the *Mizan al-Hikma* of al-Khazini. He further states that there is no distinct trace of a translation in either the Arabic or Arabic-Latin manuscript tradition.

[77] See most importantly Abattouy (2001a, b).

[78] See Abattouy (2001a), 96.

[79] I do not agree with Abattouy (2001a), 122, that the text of the *Nutaf min al-hiyal* comprises all important aspects of the Greek text. The geometrical analysis of motion cannot be regarded as a mere digression, since it offers the author's justification of the central argument in the text that the longer radius moves more quickly than the smaller one. Abattouy's conclusion that the Arabic version presents the authentic text, whereas the Greek contains many later comments and additions, including the geometrical proof, should be rejected on these grounds.

[80] See Abattouy (2001a), 104.

He considers three cases involving the equilibrium of the balance, which depend on the position of the axis with respect to the beam's centre of gravity.[81] Two of these cases, namely when the axis is above or below the centre of the beam, correspond to the question on the suspension of balances in the second problem of the Aristotelian *Mechanics*. The third case, where the axis coincides with the beam's centre of gravity, was not considered by Aristotle. Abattouy mentions the close connection between al-Khazini's discussion of the equilibrium of the balance and the second problem of the *Mechanics*.[82] Although two of the cases correlate with the Aristotelian question on the suspension of balances, al-Khazini does not necessarily rely here on the text of the *Mechanics*. In his discussion of balance equilibrium, he takes a rather different approach to our author in considering the centre of gravity of the balance. The concept of centre of gravity was not yet known to the author of the *Mechanics*; this thus shows that al-Khazini was working within a later tradition in which mechanical notions were further developed. The fragment of the *Nutaf min al-hiyal* is included directly after this discussion on balance equilibrium in the *Kitab Mizan al-Hikma*, probably, as Abattouy argues, in order to supply a theoretical framework for it.

Interestingly, the texts of Ibel and Abattouy explaining al-Khazini's balance equilibrium question contain two similar diagrams.[83] The first diagram shows the balance beam released from its needle, whereas the second diagram presents the beam with needle. In both texts, it remains unclear whether these diagrams are also contained (in this form) in the Arabic manuscripts of the *Kitab Mizan al-Hikma*.[84] The existence of diagrams in the Arabic manuscripts might have important implications for my research on the diagrams in the Greek manuscripts of the *Mechanics*. If there are any connections between the figures in the Arabic manuscripts and the Greek diagrams, they can be traced back to an early source and are likely to have their origins in antiquity. However, if the diagrams in Ibel and Abattouy represent the ones in the manuscripts, the Arabic figures are probably unrelated to the Greek diagrams. Each of the diagrams represents all three cases (where the axis either coincides with, is below, or is above the centre of gravity) in one diagram, whereas the Greek manuscripts contain two diagrams, one for each case. Neither do

[81] See Abattouy (2001a), 106.

[82] See Abattouy (2001a), 106–107.

[83] See Ibel (1908), 120–121 and Abattouy (2001b), 192 and 194. The same diagrams, now with Arabic labelling as well, can be found in Khanikoff's partial publication and translation of the *Kitab Mizan al-Hikma*; for these see Khanikoff (1860), 89 and 92.

[84] Unfortunately, I have not yet been able to consult either of the manuscripts myself. As Abattouy (2001a), 108 mentions, two complete copies of the *Kitab Mizan al-Hikma* are preserved: one in Saint Petersburg (Khanikoff MS 117), and the other in Hyderabad (MS Riyāḍī 125). Apart from the diagrams preceding the *Nutaf min al-hiyal*, I do not consider it likely that the text of the *Nutaf min al-hiyal* itself contains any diagrams in the Arabic manuscripts. Since the geometrical analysis of motion was left out of the Arabic version, the text does not contain any letter labels that refer to a corresponding diagram.

I notice any similarities between the sets of letters that are used in the geometrical constructions of the Arabic and Greek texts.[85]

Besides the partial Arabic version of the *Mechanics* included in al-Khazini's *Kitab Mizan al-Hikma*, there are several references to the Aristotelian *Mechanics* in the Arabic world. The earliest reference is found in a chronicle from the middle of the eleventh century, where the *Book of Geometrical Mechanics* is mentioned among Aristotle's works.[86] Further similar references are found in Arabic works, which are discussed in Abattouy.[87] They show us that the text of the *Nutaf min al-hiyal* was not an isolated phenomenon in the Arabic world, but that the Aristotelian *Mechanics* was more widely known.

3.5 Critical Editions

3.5.1 The Editio Princeps

The *editio princeps* of the *Mechanics* is the complete edition of Aristotle, printed by Aldus Manutius in Venice (1495–1498). We do not possess the printer's copy of the *Mechanics*, but, as I mentioned above, the edition is based upon manuscript B^e, together with influences from the Byzantine paraphrase by Pachymeres.

Many readings in the Aldine agree with the first, non-emended reading of B^e:

847b19	σὺν ἀλλήλοις Aldine/B^e	μετ' ἀλλήλων B^e corr./Pach.
850a22	ἡμίσεως Aldine/B^e	ἡμίσεος Pach.
850a27	τοῦ Aldine/B^e	τῷ B^e corr./Pach.
850a31	προσλαμβάνοντες Aldine/B^e	προσλαβόντες Pach.
852a17	χωρίῳ Aldine/B^e	χρόνῳ B^e corr./Pach.
854b30–31	om. ὥστε τὸ A ἐπὶ τῆς πλευρᾶς ἐνήνεκται τῆς AE Aldine/B^e	/
856a26	αὐτὴν Aldine/B^e	αὐτοῦ B^e corr./Pach.

Several other variants coincide with codex B^e after emendation, as, for example, at 850b17 μέσῳ B^e: μέσῃ Aldine/B^e corr.; 851a27 οὗ τὸ Δ B^e: add. δὲ οὗ τὸ Δ Aldine/B^e corr.; 853a25 ὄντι B^e: ὄντι add. τῷ μόχλῳ Aldine/B^e corr. The paraphrase by Pachymeres also exerted a direct influence on the Aldine, as is evident from the following readings, which are not contained in B^e:

---

[85] Abattouy (2000), 10–11 notes that the same set of letters is used in the second problem of the *Mechanics* as well as in the Arabic version. Some letters coincide, which is self-evident when many of them are used in a geometrical construction, but apart from that I cannot find any similarities.

[86] See Abattouy (2001a), 101–102.

[87] See Abattouy (2001a), 101–103.

852a5	ἔχοντα Aldine/Pach.	ἔχοντες B^e
852b2	τὸ βέλος Aldine/Pach.	τὰ βέλη B^e
853a10	ὃ ἡ χεὶρ αἴρει Aldine/Pach.	om. B^e
853b32	πρόσκειται Aldine/Pach.	ἔγκειται B^e
854b20	τὴν ἐλάττω διάμετρον Aldine/Pach.	τὴν διάμετρον τὴν ἐλάττω B^e
855a13	ὀξείας add. ὥσπερ Aldine/Pach.	ὀξείας B^e
856a4	ὑπὸ τοῦ ἐλάττονος Aldine/Pach.	om. B^e
858b24	ὑπολείπεσθαι add. τὸ πᾶν Aldine/Pach.	ὑπολείπεσθαι πᾶν B^e

The *editio princeps* is contained in my *stemma codicum*, since it was the source for some early sixteenth-century manuscripts: u, V^3 and N^c, which is a copy of V^3, as I explained in Sect. 3.2.

3.5.2 Later Editions

The nineteenth century brought us three main editions of the *Mechanics*: by Johannes van Cappelle (1812), Immanuel Bekker (1831) and Otto Apelt (1888). These editions are all likewise unsatisfactory, in that they are too much affected by Pachymeres' paraphrase. In addition, these editors considered only a small part of the textual tradition; all together they consulted a total of nine manuscripts. Van Cappelle's edition contains the Greek text together with comments and a Latin translation. In his introduction, van Cappelle mentions having collated three manuscripts of the *Mechanics*, namely L^v, Par and P^t. The Parisian manuscripts, Par and P^t, both include readings that are of little or no value for the textual tradition. Par departs in places from the authentic Aristotelian text, as a result of influence from the paraphrase of Pachymeres. And P^t, despite being an early codex from the fourteenth century, has a strong paraphrastic character and contains many peculiar readings and lacunae.

In his complete edition of Aristotle, Bekker cites manuscripts P and W^a as his main sources for the *Mechanics*. These manuscripts, which represent different branches of the textual tradition, contain valuable readings, and, when analysed together with L and H^a, cover the most important textual variants. However, apart from numerous errors in collation, many of Bekker's readings are printed without any justification, not being contained in any of the manuscripts cited by him, for example, at 855a38 ἡ δὲ τοῦ ἐλάττονος ἐλάττων Bekker/B^e corr.: om. codd.; 856a29 μικρὸς Bekker/B^e corr.: μέγας codd.; 857a5 ἀπ' ἄκρου Bekker: ἀπὸ μακροῦ codd. While collating the manuscripts, I discovered many agreements between Bekker's edition and variants included in B^e, which was written by Niccolò Leonico Tomeo. In 1525, Leonico published a Latin translation of the *Mechanics* with commentary. Bekker most likely consulted this commentary in preparing his edition of the treatise, and tacitly included some of Leonico Tomeo's emendations in his text.[88]

Apelt's edition contains the Greek text of the *Mechanics* and five further smaller works by Aristotle. Apelt did not examine any new manuscripts, but relied on the manuscripts consulted by Bekker, P and W^a, and also on Par, previously analysed by van Cappelle. In his foreword, Apelt praises the edition by van Cappelle and the "excellent" codex Par. He holds that Par offers many good readings of corrupt passages and fills the lacunae of other manuscripts.[89] Although this is true to some extent,[90] we should be careful in accepting the variants from Par, since many of them were borrowed from the paraphrase by Pachymeres and do not contain the authentic text. Given that Apelt did not add any new manuscripts to the ones collated by his predecessors, his edition does not make a significant contribution to our understanding of the textual tradition.

The Loeb Classical Library edition of Walter S. Hett (1936) relies upon the Bekker text, except in a few cases where the edition of Apelt has been employed. Hett himself has also made some smaller emendations.

For her 1982 edition of the *Mechanics*, Maria Elisabetta Bottecchia examined nearly all the manuscripts of the text. Her analysis, however, did not bring forth a reliable critical edition of the text. As I have pointed out at various places in this chapter, Bottecchia has made some serious errors in the dating of the codices as she establishes dependences between manuscripts that speak against their chronology. Furthermore, her edition contains many traces from the paraphrase by Pachymeres, either directly through codex M^u which contains a copy of Pachymeres' text, or indirectly through consulting several manuscripts that are strongly influenced by the Byzantine paraphrase. Bottecchia divided the manuscripts into four different families, namely *a*, *b*, *c* and *d*. Her text relies primarily upon manuscripts in her families *a* and *b*. H^a, L and O^a, all in family *a*, together with P and V^3, both belonging to family *b*, are the main sources for her edition of the text. Codex V^3 is of especial importance for Bottecchia, since she believes that this manuscript constitutes independent textual evidence. She finds many variants that are uniquely present in V^3 and adopts them in her critical edition: e.g. at 849a4, 853a10, 853b3, 853b23 and 856a4. These readings, however, cannot be derived from an independent source, but are Pachymeres' emendations of the text. Manuscript V^3 is a copy of the first printed edition of the text by Manutius and for that reason contains many elements from the paraphrase by Pachymeres.[91] Thus, V^3 is not valuable for the reconstruction of the *Mechanics* and should not be considered in a new critical edition of the text. In Bottecchia's analysis of the manuscripts, families *c* and *d* cannot be traced back to

---

[88] See n.24 on the influence of Leonico Tomeo's *Mechanical Questions* on Bekker's edition.

[89] See Apelt (1888), VI–VII.

[90] See e.g. Ch. 4 n.62 for the readings in Par at 855b8 and 855b22, and Ch. 5 n.19 for 849b12-14.

[91] See Sect. 3.2 on the relationship between Pachymeres' paraphrase and the Aldine.

their own hyparchetype, but are dependent upon codex H^{a}. For this reason, she did not consult these families for her edition, and overlooked some important variants offered by these manuscripts. In the year 2000 Bottecchia produced an Italian translation of the *Mechanics* along with a Greek text. This Greek text differs in certain aspects from her 1982 edition.[92] It remains, however, unclear whether these variants are the result of a re-assessment of the manuscript material or derive from other texts of the *Mechanics*. Unfortunately, Bottecchia's text lacks a critical apparatus, nor is any information on the variants contained in the commentary section of this work. A comparison between these variants and the manuscript material points to the influence of the manuscript Par (one of the less important manuscripts in Bottecchia's 1982 edition of the text), which is visible from some peculiar variants, for instance at 851a22 and 857a17. However, text passages at 849b12 or 851a27, where Bottecchia's readings perfectly agree with van Cappelle, reveal that Bottecchia probably did not re-examine Par herself, but rather adopted these readings from Par via van Cappelle's edition.

In recent years, two editions have been produced, by Albert Presas I Puig and Joan Vaqué Jordi (2006), and Maria Fernanda Ferrini (2010). These editors did not examine the manuscript material themselves, but reprinted the Greek text from existing editions. Presas I Puig's edition contains the Greek text by Bottecchia together with a Catalan translation. He also relies on the edition by van Cappelle, since many variants from codex Par, and conjectures by van Cappelle arc included in his text and critical apparatus. Ferrini reproduces the Bekker text; her edition does not contain a critical apparatus; instead she includes some critical notes at the end on those passages where her text differs from Bekker's edition. It is unclear where these emendations come from, but she seems to be especially influenced by the edition by van Cappelle.

3.6 Results

3.6.1 Stemma Codicum

Figure 3.4 illustrates the analysis of the manuscripts presented in this chapter:

[92] Variants can be found at 849a23, 849a27, 849a31, 849b3, 849b7, 849b12, 849b14, 849b16, 850a27, 851a22, 851a24, 851a27, 851a36, 851a39, 853a30, 854b23, 854b36, 855a6, 855a7, 855b9, 855b16, 855b23, 855b28, 855b37, 855b39, 856b9, 856b15, 857a10, 857a11, 857a17, 857a26, 857b7, 858a2, 858a30 and 858b2.

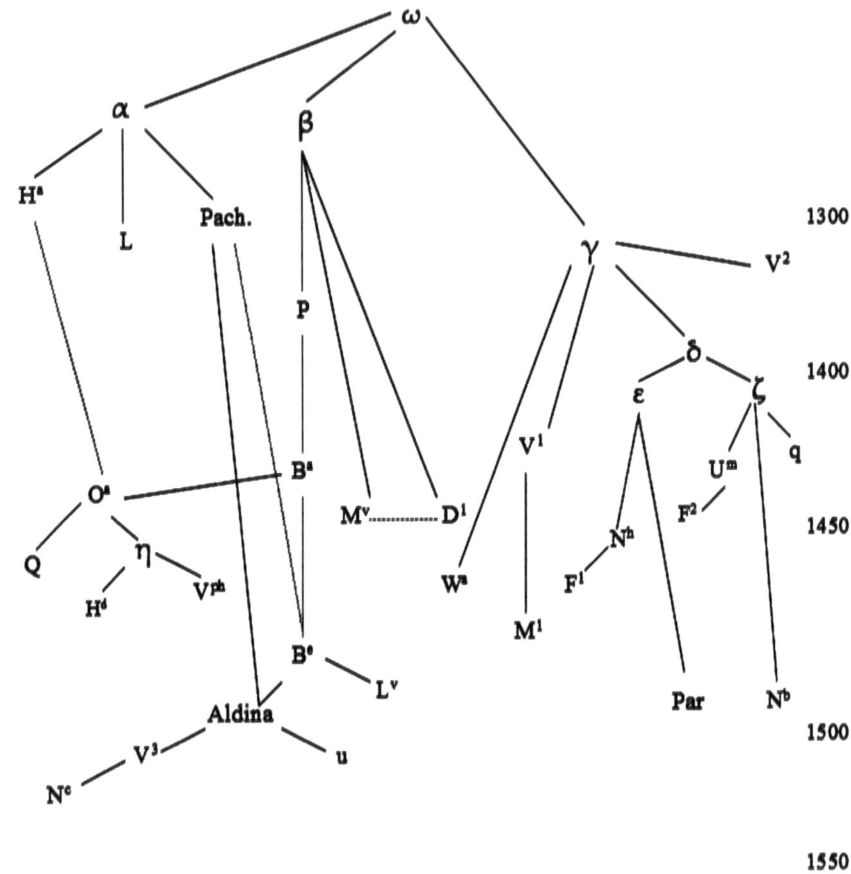

Contaminated manuscripts: B^u, N^a, P^t, P^4 and P^s.

Fig. 3.4 *Stemma codicum* of all manuscripts. *Black lines* in the figure show a dependence of one manuscript on another; *dashed lines* show that one manuscript contains emendations from another

3.6.2 *On a New Text*

Assessing the earliest phases of the tradition is more complicated for the *Mechanics* than for most other texts in the *Corpus Aristotelicum*. Whereas in general for Aristotelian treatises, at least one manuscript from the ninth or tenth century has been preserved, in the case of the *Mechanics* we have to content ourselves with the fact that the oldest manuscripts date from the beginning of the fourteenth century. Consequently, the tradition of the *Mechanics* cannot be traced all the way back to a heavily reworked exemplar in majuscule script from the seventh or eighth century, unlike the transmission of the *Physics*, as attested by Hermann Diels in his authoritative paper on that treatise.[93] What do we know about the earliest tradition and the archetype? Both hyparchetypes of families *a* and *b* were probably copied in Constantinople as collections of the works of Aristotle. There is no certainty about the dates of codices α and β; manuscripts H^a and P at least determine the *terminus ante quem* of these manuscripts. However, it seems likely that both hyparchetypes were copied somewhere in the twelfth century before the sack of Constantinople in 1204. The hyparchetype of family *c* is somewhat more recent and can possibly be dated in the first quarter of the fourteenth century. This codex γ has its origin in the Italian world. We further know that it did not contain the collected works of Aristotle, but was rather a hybrid, or contained the *Mechanics* as a single text. If we count one generation back from the oldest hyparchetypes to the archetype, we can infer that the archetype of the *Mechanics* was a manuscript written in minuscule somewhere between the ninth and twelfth centuries, most likely in the tenth century. It was probably composed in Constantinople and contained a collection of the works of Aristotle.

In a new critical edition of the *Mechanics*, the manuscripts of all three families should be analysed, with particular attention given to those manuscripts that can be traced back to a hyparchetype, namely H^a, L, P, M^v, W^a and V^1. Family *b*, however, has a somewhat different status compared to the other manuscript families. Since the manuscripts in this family have a deteriorated and contaminated ancestor, the editor is not always in a position to recover the authentic reading, when two families agree against the third. It is entirely possible that families *b* and *c* agree, whereas family *a* contains the correct variant, e.g. at 848b24 πλευρῶν *a*: πλειόνων *cett.*; 854b27 πλευράν *a*: πλείω *cett.*; 855b24 μένειν *a*: μὲν *cett.* Although family *b* is contaminated, it looks back to a different tradition and certainly contains some valuable readings, for instance at 852a9 αὐτῶν *b*: αὐτὸν *cett.*; 853b2 ἕλξει *b*: om. ἕλξει *cett.*; 854a12 τοῦ ἀπὸ *b*: τὸ ἀπὸ *cett.*; 855b26 μὴ *b*: μὲν *cett.*; 857a12 αἴτιον δὲ ὅτι *b*: αἴτιον δὲ διότι/ἐπειδὴ διότι *cett.*; 857b13 ὑπομόχλιον *b*: ὑπομοχλίων *cett.* However, the editor should consider carefully whether a peculiar variant in P or M^v contains an authentic reading, a contamination, or an error.

An analysis of all three families would lead to a clear improvement upon the previous editions of the treatise, of which Bekker's edition is still the most complete; manuscripts from family *b* and *c* provided the main sources for his text, while manuscripts from the oldest family, family *a*, are accorded only secondary importance. Although Bottecchia studied nearly the complete manuscript material, she neglected to consult family *c* for her edition, and thereby overlooked some important readings. She did not regard, for example, the variant from family *c* at 850a35 ἀντὶ σπαρτίου γίνεται. Nevertheless, she included some other readings from family *c* in her edition, as for example at 849a36 ὀρθᾶς; 849b18 τὸ κατὰ; 852b17 θᾶττον, although she did not recognize these variants as belonging to family *c*, and ascribed them to Bekker or manuscript N^a.

[93] See Diels (1969), 203.

The future editor might further reconsider the importance of family *a*. Bottecchia mostly favoured the readings from manuscript V^3, and included the variants from family *a* in her apparatus. There are, however, certain passages in the *Mechanics* where this family alone might contain the correct reading, for example, at 858a1 εὐθείας *a*: ἴσας *cett.*

Although a new examination of all three families will most likely favour certain variants different from the ones favoured in previous editions of the treatise, a far more important task for a new edition is to remove all foreign influences from the text of the *Mechanics*. The Byzantine paraphrase by Georgius Pachymeres, and to some extent the variants offered by the humanist Niccolò Leonico Tomeo, have influenced all previous editions of the treatise, including the latest edition by Bottecchia. These paraphrastic traces should be individually evaluated, but must be separated from the authentic Aristotelian text.

Bibliography

Abattouy, Mohammed. 2000. *Nutaf Min Al-Hiyal: An Arabic partial version of pseudo-Aristotle's Mechanica Problemata.* Preprint 153. Berlin: Max Planck Institute for the History of Science.

Abattouy, Mohammed. 2001a. Nutaf Min Al-Hiyal: A partial Arabic version of pseudo-Aristotle's "Problemata Mechanica". *Early Science and Medicine* 6(2): 96–122.

Abattouy, Mohammed. 2001b. Greek mechanics in Arabic context: Thābit ibn Qurra, al Isfizārī and the Arabic traditions of Aristotelian and Euclidean mechanics. *Science in Context* 14: 179–247.

Andrist, Patrick. 2007. *Les manuscrits grecs conservés à la Bibliothèque de la Bourgeoisie de Berne – Burgerbibliothek Bern: catalogue et histoire de la collection.* Zurich: Urs Graf.

Apelt, Otto (ed.). 1888. *Aristotelis quae feruntur De Plantis, De Mirabilibus Auscultationibus, Mechanica, De Lineis Insecabilibus, Ventorum Situs et Nomina, De Melisso Xenophane Gorgia.* Leipzig: Teubner.

Bekker, Immanuel (ed.). 1831. *Aristotelis Opera II.* Berlin: Reimer.

Berger, Friederike. 2005. *Die Textgeschichte der Historia animalium des Aristoteles.* Wiesbaden: Reichert.

Bloch, David. 2008. The text of Aristotle's *De Sensu* and *De Memoria. RHT* 3: 1–58.

Bottecchia, Maria Elisabetta (ed.). 1982. *Aristotele: MHXANIKA, Tradizione manoscritta, testo critico, scolii.* Padua: Antenore.

Burnikel, Walter. 1974. *Textgeschichtliche Untersuchungen zu neun Opuscula Theophrasts.* Wiesbaden: Franz Steiner Verlag.

Canart, Paul. 1977–1979. Démétrius Damilas, *alias* le "librarius Florentinus". *RSBN* N.S.14-16: 281–347. Rome.

Capocci, Valentino. 1958. *Codices Barberiniani Graeci.* Tomus 1: Codices 1–163. Rome.

Clagett, Marshall. 1959. *The science of mechanics in the Middle Ages.* Madison: University of Wisconsin Press.

Crombie, Alistair C. 1967. *Medieval and early modern science: Volume 1: Science in the Middle Ages: V-XIII centuries.* Cambridge, MA: Harvard University Press.

Diels, Hermann. 1905. *Die Handschriften der antiken Ärzte I.* Berlin: Verl. der Königl. Akad. der Wiss.

Diels, Hermann. 1969. Zur Textgeschichte der Aristotelischen Physik. In *Kleine Schriften zur Geschichte der antiken Philosophie*, ed. W. Burkert, 199–238. Hildesheim: Olms.

Ferrini, Maria Fernanda (ed.). 2010. *Aristotele: Meccanica*. Milan: Bompiani.

Formentin, Mariarosa. 1997. Codici greci di medicina nella Biblioteca Nazionale Vittorio Emanuele III di Napoli: le vie di acquisizione. In *Lingue tecniche del Greco e del Latino II*, ed. S. Sconocchia, 207–216. Bologna: Patron.

Hadot, Ilsetraut. 1978. La tradition manuscrite du commentaire de Simplicius sur le Manuel d'Épictète. *RHT* 8: 1–108.

Harlfinger, Dieter. 1971. *Die Textgeschichte der pseudo-aristotelischen Schrift ΠΕΡΙ ΑΤΟΜΩΝ ΓΡΑΜΜΩΝ: Ein kodikologisch-kulturgeschichtlicher Beitrag zur Klärung der Überlieferungsverhältnisse im Corpus Aristotelicum*. Amsterdam: Hakkert.

Haskins, Charles Homer. 1927. *Studies in the history of mediaeval science*, 2nd ed. Cambridge, MA: Harvard University Press.

Hett, Walter S. (ed.). 1936. *Aristotle: Minor works*. Cambridge, MA: Harvard University Press.

Hilgers, Robert. 1992. Eine neue Aristoteles-Handschrift in Berlin. *Codices Manuscripti* 16: 62–64.

Ibel, Thomas. 1908. *Die Wage im Altertum und Mittelalter*. Erlangen: Junge.

Khanikoff, N. 1860. Analysis and extracts of the *Book of the Balance of Wisdom*, an Arabic work on the water-balance, written by Al-Khâzinî in the twelfth century. *Journal of the American Oriental Society* 6: 1–128.

Leonico Tomeo, Niccolò. 1525. *Nicolai Leonici Thomaei Opuscula nuper in lucem aedita quorum nomina proxima habentur pagella*. Venice: Bernardino Vitali.

Manutius, Aldus (ed.). 1495–1498. *Opera Aristotelis*. Venice.

Mioni, Elpidio. 1972. *Bibliothecae Divi Marci Venetiarum codices Graeci manuscripti*, Classis II, codices 121–198; classes III, IV, V; Indices. Rome: Ist. Poligrafico e Zecca dello Stato.

Mondrain, Brigitte. 2007. Les écritures dans les manuscrits byzantins du XIVe siècle. *RSBN* N.S.44: 157–196.

Moody, Ernest A. and Marshall Clagett. 1952. *The medieval science of weights: Treatises ascribed to Euclid, Archimedes, Thabit ibn Qurra, Jordanus de Nemore and Blasius of Parma*. Madison: University of Wisconsin Press.

Napolitano, Felicia, Maria Luisa Nardelli and Luigi Tartaglia. 1977. *Manoscritti greci non compresi in cataloghi a stampa*. Naples: Arte Tipografica.

Nussbaum, Martha Craven. 1976. The text of Aristotle's *De Motu Animalium*. *HSCP* 80: 111–159.

Omont, Henri. 1888. *Inventaire sommaire des manuscrits grecs de la Bibliothèque Nationale*, Seconde partie / troisième partie. Paris: Picard.

Pappa, Eleni (ed.). 2002. *Georgius Pachymeres, Philosophia. Buch 10. Kommentar zur Metaphysik des Aristoteles: Editio princeps. Einleitung, Text, Indices*. Athens: Akademia Athenon.

Peters, Francis E. 1968. *Aristotle and the Arabs: The Aristotelian tradition in Islam*. New York: New York University Press.

Pingree, David (ed.). 1968. *Albumasar: De revolutionibus nativitatum*. Leipzig: Teubner.

Presas i Puig, Albert and Joan Vaqué Jordi (eds.). 2006. *Aristòtil: Questions mecàniques*. Barcelona: Fundació Bernat Metge.

Rashed, Marwan. 2001. *Die Überlieferungsgeschichte der aristotelischen Schrift De generatione et corruptione*. Wiesbaden: Reichert.

Sicherl, Martin. 1997. *Griechische Erstausgaben des Aldus Manutius: Druckvorlagen, Stellenwert, kultureller Hintergrund*. Paderborn: Schöningh.

Siwek, Paul. 1961. *Les manuscrits grecs des Parva naturalia d'Aristote*. Rome: Desclée.

Stornajolo, Cosimo. 1895. *Codices urbinates Graeci Bibliothecae Vaticanae*. Rome: Ex Typographeo Vaticano.

Turyn, Alexander. 1972. *Dated Greek manuscripts of the thirteenth and fourteenth centuries in the libraries of Italy*. Urbana: University of Illinois Press.

Tziatzi-Papagianni, Maria. 1994. *Die Sprüche der sieben Weisen: Zwei byzantinische Sammlungen; Einleitung, Text, Testimonien und Kommentar*. Stuttgart: Teubner.

Van Cappelle, Johannes Petrus (ed.). 1812. *Aristotelis Quaestiones Mechanicae*. Amsterdam: Den Hengst.

Van Leeuwen, Joyce. 2013. The text of the Aristotelian *Mechanics*. *Classical Quarterly* 63(1): 183–198.

Vendruscolo, Fabio. 1996. Manoscritti greci copiati dall'umanista e filosofo Niccolò Leonico Tomeo. In *ΟΔΟΙ ΔΙΖΗΣΙΟΣ. Le vie della ricerca. Studi in onori di Francesco Adorno*, ed. M.S. Funghi, 543–555. Florence: Olschki.

Vogel, Marie and Victor Gardthausen. 1909. *Die griechischen Schreiber des Mittelalters und der Renaissance*. Leipzig: Harrassowitz.

Wartelle, André. 1963. *Inventaire des manuscrits grecs d'Aristote et de ses commentateurs: Contribution a l'histoire du texte d'Aristote*. Paris: Les Belles Lettres.

Wilson, Nigel G. 1983. *Scholars of Byzantium*. London: Duckworth.

Wood, Casey A. and F. Marjorie Fyfe. 1943. *The Art of Falconry by Frederick II of Hohenstaufen*. Stanford: Stanford University Press.

Zografidis, George. 2011. George Pachymeres. In *Encyclopedia of medieval philosophy: Philosophy between 500 and 1500*, vol. 2, ed. Henrik Lagerlund, 394–397. Dordrecht: Springer.

Zorzi, Marino. 1993. *Collezioni veneziane di codici greci: dalle raccolte della Biblioteca Nazionale Marciana*. Venice: Il Cardo.

Chapter 4
Editing Diagrams

Critical studies of diagrams from manuscript traditions exhibit many similarities with the branch of textual criticism. As in the case of texts, errors are the most important guide in determining relationships between manuscript diagrams. Generally, we can state that the diagram that is the least corrupted, is most likely to be the oldest diagram. Each subsequent copy contains additional errors in comparison with its exemplar. When the diagrams in two codices share the exact same set of errors and alterations, it is much more probable that these manuscripts are connected with each other than that two scribes made similar changes independently. Despite these coherences between textual criticism and diagram criticism, there are also important differences between the two, by reason of which the findings for texts cannot be automatically applied to diagrams. The copyist of the text and the diagrams might not have been the same person, which could result in separate textual and diagrammatic traditions, or diagrams could have been added many centuries later to a text that was not originally illustrated. Therefore, it cannot be simply assumed that the *stemma codicum* of the text is identical to that of the diagrams. They should be worked out separately and only compared afterwards, as Kurt Weitzmann recommends.[1] From such isolated analyses of text and diagram traditions in the *Mechanics*, it turns out that the two *stemmata* are in fact the same; thus, results from the one support and strengthen the other. In this case, the diagrams become a powerful tool to verify the *stemma codicum* of the text and to confirm authentic readings from the archetype.

The diagrams in the manuscripts of the Aristotelian *Mechanics* have not been subject to critical examination, although they form an important part of the tradition of the treatise. Contrary to the opinion of certain scholars who still believe that the manuscripts were transmitted without any diagrams,[2] more than half of all manuscripts contain diagrams. Such misconceptions were facilitated by the fact that none of the editors of the *Mechanics* refer to existing diagrams in manuscripts. Instead,

[1] See Weitzmann (1970), 189.

[2] See e.g. Drachmann (1963), 13 and Vilain (2008), 153–154.

J. van Leeuwen, *The Aristotelian Mechanics*, Boston Studies in the Philosophy and History of Science 316, DOI 10.1007/978-3-319-25925-3_4

they include diagrams of their own that interpret the mechanical principles in modern terms. Whereas all variants of the text are noted in the critical apparatus, in the case of the diagrams the reader is not informed whether a given diagram is also contained in the manuscripts or what it looks like. This situation is characteristic of the diagrams in many ancient scientific works. Even though Otto Neugebauer called for the critical assessment of diagrams in 1975,[3] it was not until recently that scholars began to take into account the manuscript diagrams in their editions of classical texts. The situation mostly changed for Greek mathematical texts, which is due especially to the publications of Reviel Netz and Ken Saito on the relations between the argument of the text and the corresponding diagrams.[4]

The diagrams in the Aristotelian *Mechanics* are not as complex as those in ancient mathematics, and, unlike the latter, can, in most cases, be constructed on the basis of the text. However, the diagrams were envisioned in the argument of the text and are crucial to our understanding of the mechanical principles in question. The author of the *Mechanics* employs a technique known from ancient mathematics, namely the practice of using letter labels to refer to an accompanying diagram. His constant use of letter labels makes it difficult to understand the text without diagrams. It is hard to see how editors could have examined the text of the *Mechanics* without consulting the diagrams in the manuscripts. Critical examination of the diagrams is therefore necessary. In this chapter, the first critical edition of the diagrams will be produced. To begin with, however, I will describe the transmission of the diagrams and explore some of their important characteristics, such as the use of letter labels, their dependence on the text and their position.

In this context, the question of the authenticity of the manuscript diagrams needs to be addressed. The main argument against studying the figures in the codices is the claim that these figures are not authentic.[5] I will show that the diagrams in their present form are later Byzantine additions that are most probably not identical with the original diagrams. However, the fact that the diagrams cannot be traced back to the author's lifetime, by no means implies that they have no value for further research. Many scientific texts as we possess them now are the result of a creative tradition, having been studied and reworked during the centuries following their composition. This makes the study of these texts an important part of the study of reception and an invaluable aid to understanding scientific texts as an evolving form. Even though the diagrams in the manuscripts of the *Mechanics* are not likely to be authentic, they reflect certain practices from antiquity. At the same time, they illustrate later Byzantine practice, and thus shed light on the transfer of mechanical knowledge during this period.

---

[3] See Neugebauer (1975), 751–755. There are some notable exceptions of scholars who made critical studies of diagrams; see e.g. Jones (1986).

[4] See e.g. Netz (1999) and Saito (2006).

[5] E.g. Krafft (1970), 28–30 argues against the authenticity of the diagrams in the *Mechanics*. He further makes the strong assumption that diagrams were never part of the Aristotelian corpus.

4.1 Transmission of Diagrams

4.1.1 Manuscript Families

Of the 31 manuscripts in which the text of the Aristotelian *Mechanics* has been handed down to us, 17 contain diagrams. Some of these are among the oldest preserved manuscripts of the treatise, dating from the beginning of the fourteenth century, as, for example, codex L in family *a*. All of the following manuscripts contain diagrams:

B^e Bern. 402, fifteenth century, before 1497.
D^1 Berol.Phill. 1507, fifteenth century, circa 1455.
F^1 Laur.Acq. 65, fifteenth century, 1450–1475.
F^2 Laur.Plut. 28.45, fifteenth century, 1445.
L Vat.gr. 253, fourteenth century, beginning.
M^1 Matr. 4563, fifteenth century, 1470.
M^v Mosq. 453, fifteenth century, circa 1450.
N^b Nap. IIIAA14bis, fifteenth century, 1450–1500.
N^h New Haven Phill. 7488, fifteenth century, middle.
P Vat.gr. 1339, fourteenth century, 1350–1400.
Par Par.gr. 2115, fifteenth-sixteenth century.
P^s Par.Suppl.gr. 541, fifteenth century.
P^t Par.gr. 2507, fourteenth century, 1370–1380.
q Urb.gr. 76, fifteenth century, 1400–1450.
U^m Marc.gr. IV.57, fifteenth century, before 1446.
V^1 Vat.gr. 905, fifteenth century, circa 1430.
W^a Urb.gr. 44, fifteenth century, circa 1465.

The diagrams in four of these manuscripts, B^e, Par, P^s and P^t, represent a departure from the tradition. Certain properties of the diagrams hint at an influence from one of the manuscript families, but others are unrelated to the extant diagrams and appear to be the scribe's personal interpretations of the mechanical problems. Manuscript B^e, as presented in the previous chapter, was copied and owned by the Italian humanist Niccolò Leonico Tomeo. This codex contains later emendations and diagrams in the margins in Leonico Tomeo's hand.[6] Leonico Tomeo made these figures independently of the manuscript tradition; a comparison with his Latin translation of and commentary on the *Mechanics*, published in 1525, shows that these diagrams influenced the figures in his Latin translation.[7] Leonico Tomeo's diagrams will be analysed in chapter 6, as they enable us to observe the diagrammatic practices in the process from handwritten manuscript to printed edition, as

[6] We know that these diagrams and emendations by Leonico Tomeo are of a later date, since the manuscript L^v, which is an accurate copy of B^e, does not contain any of these diagrams or corrections. Therefore, L^v was copied before the diagrams were included in B^e.

[7] See Andrist (2007), 15.

well as to investigate the different functions and conception of diagrams in the early modern period. The Par codex underwent elaborate emendation after completion. The figures in Par were also added at this later stage, and they show no connection with the diagrams contained in the exemplar of Par. Both P^s and P^t are contaminated manuscripts that cannot be attributed to a certain manuscript family. Many diagrams in these codices appear to be unrelated to the ones in the extant manuscripts; the diagrams contained in P^t in particular offer very different interpretations of the text. Due to their independent status in the manuscript tradition, these diagrams do not contribute to the recovery of the diagrams from the archetype. For that reason, they are not considered in my critical edition of the diagrams. However, a separate Sect. 4.5 will take account of their broader significance for our knowledge of the reception of the *Mechanics*, as these diagrams constitute individual commentaries on the text.

The diagrams contained in the other manuscripts of the *Mechanics* were all copied from the same source that is identical with the textual archetype, a Byzantine manuscript most likely composed in the tenth century. This is by no means standard in the transmission of ancient texts, as we often see that text and diagram *stemmata* do not coincide. It could, for example, occur that two illustrators included figures independently of one another, so that there are two *stemmata* for the figures, while we have a single *stemma* for the text.[8] We may also find divergences in the *stemma* of the diagrams, for example, when text and diagrams were not copied at the same time and a later illustrator was working from a different exemplar than the codex used for copying the text. In working with diagram traditions, one should always remain attentive to the fact that diagrams may have undergone alterations. Sometimes the number of figures is extended or figures are enriched with decorative elements over time. This is, however, much more likely to occur in those cases where diagrams are less necessary to the text, for example, in literary or liturgical texts, rather than in scientific texts where the reasoning depends on diagrams clearly defined by the text. In the case of the *Mechanics*, we notice that the diagrams are very similar to each other, and that, apart from some new errors in every copy, they are faithful replicas of their exemplars. Certain diagrammatic properties present in all manuscript diagrams, that could not possibly have been made by each of the individual scribes, are further evidence of a common source. Thus, it is very likely that the diagrams derive from a common archetype, and it is possible to reconstruct the earliest figures from the archetype with reasonable certainty. My examination of the manuscript diagrams has shown that the *stemma codicum* I have already established for the text is similar for the diagrams. Figure 4.1 illustrates the relations between the codices that contain diagrams.

The codices that contain diagrams are again divided into three different families, *a*, *b* and *c*. These families are dependent on the manuscripts α, β and γ respectively, which in turn can be traced back to an archetype from about the tenth century. In order to reconstruct the diagrams in the archetype, it was first necessary to reproduce

[8] See Weitzmann (1970), 189.

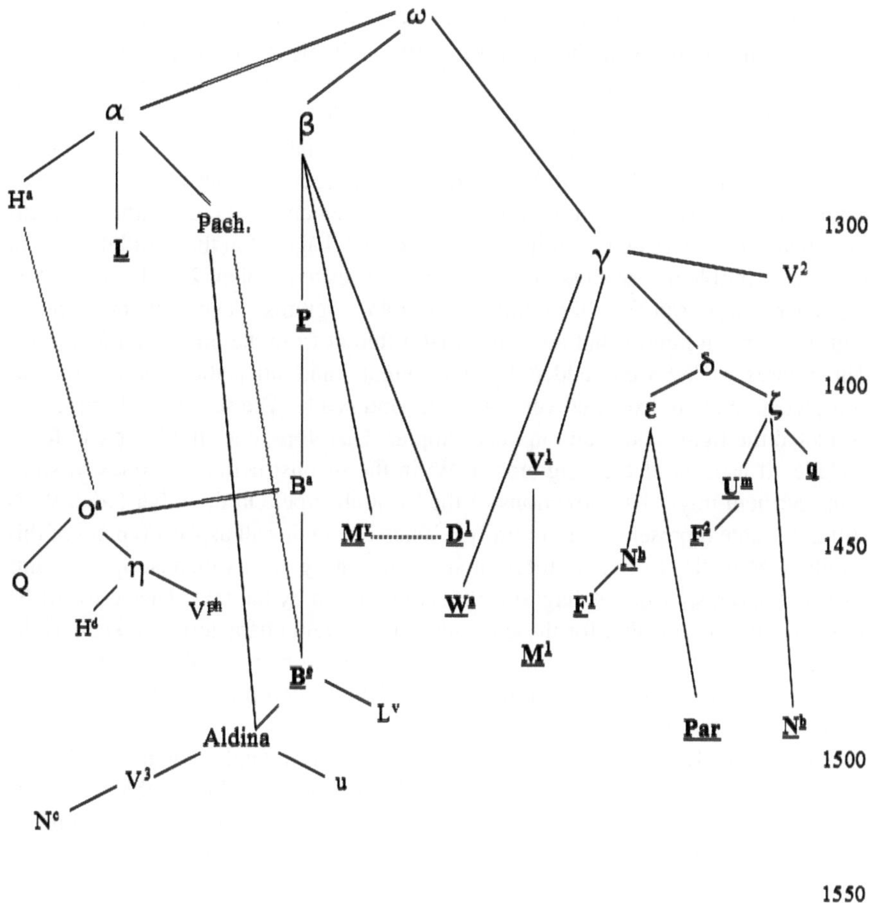

Contaminated manuscripts: B^u, N^a, P^t, P^4 and P^s.

Fig. 4.1 *Stemma codicum* of the manuscripts that contain diagrams. *Black lines* in the figure show a dependence of one manuscript on another; *dashed lines* show that one manuscript contains emendations from another; *bold* and *underscored sigla* show those manuscripts containing diagrams

all of the extant figures, and then to reconstruct the figures from the lost manuscripts: α, β, γ, δ, ε and ζ.

In the critical assessment of diagrams we can apply methods already known from textual criticism. For example, when we find a similar set of alterations in diagrams, it is much more likely that these diagrams derive from a common source than that single scribes were responsible for these changes. Errors are again an important means in tracing relationships between the manuscript diagrams. For each manuscript family we find certain errors, omissions and additions that are characteristic of that family. In family *a*, for instance, several diagrams are missing, such as the diagram for problem 5 of the text. We notice that the copyists in family *b* had similar difficulties in deciphering their source. In the analysis of the manuscripts we have already seen that the hyparchetype β was corrupted; this is reflected in the diagrams in P, M^v and D^1. For example, the diagram to circular motion in the first problem is highly flawed in P, whereas the scribe of M^v, after drawing the circles, did not even attempt to complete the labels. In other diagrams the labelling in P is incorrect and incomplete, while the figures in M^v and D^1 are very precise; in these cases, Matthaios Kamariotes, the scribe of both codices, probably emended the diagrams from his exemplar. Characteristic of family *c* are the second diagonal in the parallelogram of motions to problem 1, or three diagrams of balances to problem 2 of the text, while the other manuscript families contain only two diagrams. Several manuscripts in family *c* contain figures that are somewhat different from the others in this family. The figures in W^a were added by a different hand after the manuscript was completed,[9] and are executed very precisely and neatly. The scribe of W^a probably corrected the figures he found in his exemplar. Therefore, one should be careful to include all aspects of the diagrams in W^a in the reconstruction of codex γ, since some of them might be corrections of the hyparchetype. On the other hand, W^a is still a valuable representative of family *c*, since it contains all aspects typical of this family. Codex N^b also has a different status in family *c*; this manuscript contains variant diagrams, or omits diagrams that are required by the text. The scribe of N^b was probably responsible for these changes, for certain characteristics, such as the second *alpha* contained in the diagram to the third problem, or the absence of the label Θ from the centre of the rhombus in problem 23, confirm an attribution of this manuscript to family *c*.

The fact that L is the only manuscript with diagrams in family *a* makes it difficult to reconstruct the diagrams from codex α. Sometimes it is impossible to decide whether a certain error was already contained in the hyparchetype, or was actually a new error by the scribe of L. The paraphrase by Georgius Pachymeres, which derives from the same hyparchetype as L, might be useful in those cases to confirm a certain diagrammatical variant. Both autographs of Pachymeres' *Philosophia*, $Pach^b$ and $Pach^p$, contain diagrams.[10] The diagrams in these manuscripts are not, however, considered in the critical edition, since they partially depart from codex α and represent Pachymeres' own interpretation of the mechanical principles. For instance, the diagram of the parallelogram of motions in problem 1 of the text differs from the other manuscript figures. Pachymeres depicts the motion of an object with the ratio of AB to AΓ, with points other than Δ and E that are mentioned in the text, and that are represented in all other manuscript figures. Besides, in Pachymeres' figure of a series of equal circles in the introduction to the treatise, his circles are actually touching one another, unlike in other manuscripts. Finally, the labelling in the diagrams in Pachymeres' paraphrase is more complete and correct than in the other manuscripts, which indicates that he deliberately deviated from his source.

---

[9] See Stornajolo (1895), 49.

[10] See Sect. 3.2 on the Byzantine paraphrase of the *Mechanics* by Pachymeres, including descriptions of the autographs *Berol.Ham.* 512 ($Pach^b$) and *Par.gr.* 1930 ($Pach^p$).

Certain characteristics of the diagrams in L and $Pach^b$ point to a common source. The same figures are missing from these codices, namely to problems 1, 5, 22 and 24. If the diagrams contained in the manuscripts by Pachymeres stood apart from the tradition, and Pachymeres had drawn his own figures independently of those in codex α, he would also have included diagrams to these problems.[11] The figure to problem 30, however, poses a puzzle in this context. The diagram for this problem is absent from L and $Pach^b$, but present in $Pach^p$. In the previous chapter, I explained that the final pages of $Pach^b$ were mutilated and replaced by Bartolomeo Zanetti in the first half of the sixteenth century.[12] It is very likely that the authentic pages in $Pach^b$ contained the figure for problem 30, which was copied into $Pach^p$ before the final pages were damaged. The question about the provenance of this diagram, however, is not yet solved. Either it was contained in codex α and the scribe of L forgot to copy it, or it was added by Pachymeres on his own account.

Despite the similarities between the diagrams in L, and those in $Pach^b$ and $Pach^p$, we can exclude the possibility that Pachymeres was responsible for former. If Pachymeres had added or emended diagrams in codex α, L would contain at least part of these emendations, instead of being consistent with the errors in the other manuscript families. For that reason, L is important in the reconstruction of the diagrams in the archetype. Since the diagrams in Pachymeres derive from the same codex as L, these figures can sometimes be deployed in the critical edition to confirm a certain diagrammatical variant in L.

The fact that the *stemma codicum* I have already established for the text is similar for the diagrams provides us with a very useful tool for verifying the *stemma*. Diagrammatical errors, omissions or additions that occur in only a few manuscripts allow us to confirm previously stated affiliations among codices. This applies equally to the attribution of a manuscript to a certain family and to dependences between single manuscripts. Moreover, as will be described in the next chapter, the study of the manuscript diagrams enables us to discover many authentic readings of the text.

4.1.2 Correspondence of Diagrams

The manuscript families and single manuscripts differ in the number of diagrams they contain. Some manuscripts contain variant diagrams to a certain problem, or diagrams that are not referred to in the text, or leave out a diagram that is required by the text. The text of the *Mechanics* comprises 35 different problems, preceded by an introduction in which the circle is presented as the underlying principle of all mechanical phenomena: all principles concerning the balance can be referred to the

[11] This assumption is further reinforced by the fact that Pachymeres' diagrams contain errors similar to those in the other manuscripts. For instance, all versions of the figure of a rhombus in problem 23 of the text contain the label *iota*, which is not mentioned in the text. It seems too much of a coincidence for Pachymeres to have added the very same letter to his figure completely independently of the tradition.

[12] See Harlfinger (1971), 358 n.1.

Table 4.1 Number of diagrams per problem in each codex

Problem ↓ Manuscript →	L	P	M^v	D^1	W^a	V^1	M^1	N^h	F^l	U^m	F^2	N^b	q
Introduction: peculiarities of the circle	1	3	0	–	3	3	3	4	4	3	3	1	3
1: rectilinear and circular motion	2	5	3	–	4	4	3	4	4	4	4	3	4
2: balance supported from above and below	2	2	2	–	3	0	–	3	3	3	2	2	3
3: working of the lever	1	1	1	–	1	0	–	1	1	1	1	1	1
5: working of the rudder	0	1	1	–	1	0	–	1	1	1	1	1	1
17: working of the wedge	1	1	1	1	1	1	–	1	1	0	0	0	0
21: working of the dentist's forceps	1	1	1	1	1	1	–	1	1	1	1	1	1
22: working of the nutcracker	0	1	1	1	1	1	–	1	1	1	1	1	1
23: rectilinear motion in a rhombus	1	3	1	1	1	1	–	2	2	1	1	1	2
24: paths of different concentric circles	1	2	2	2	2	0	–	2	2	3	2	2	3
25: different cording of beds	3	4	2	2	3	3	–	3	3	3	3	1	3
26/27: carrying and raising of long timbers	1	1	1	1	1	1	–	1	1	1	1	1	1
30: how men rise from a seated position	0	1	1	1	1	1	–	1	1	1	1	1	1

circle, those concerning the lever, in turn, to the balance, whereas all other mechanical principles are derived from the lever. After this introduction, separate mechanical problems are addressed, each having a similar structure starting with a question of the type διὰ τί, "why is it that", for example, "Why is it that great weights and bodies are split with a wedge, which is only small, and why is strong pressure applied with it?" (problem 17). This question is referred to one of the mechanical principles, in the case of the wedge to two levers working in opposite directions. The explanation of the principle is either purely text-based, or makes reference to a diagram. Table 4.1 lists the number of diagrams per problem in each codex.

The table shows that sometimes a diagram is missing from a group of manuscripts, for example, the figure of the wedge in problem 17 is absent from U^m, F^2, N^b and q. These manuscripts are all dependent on codex ζ, from which the figure was likely missing too. Furthermore, the manuscripts D^1 and M^1 both contain figures for only part of the text of the *Mechanics*. D^1 was composed collaboratively by different copyists, and we notice that the second part of the *Mechanics* was written by a different hand. Whereas the first scribe copied only the text, the second also included the diagrams he found in his exemplar. We can only guess the reason why the scribe of M^1 only copied the first diagrams from his source V^1. These few diagrams reveal, however, that the scribe of M^1 was a very poor draftsman. He had the greatest difficulties in drawing a series of equal circles from the introduction of the text and probably gave up after his first attempts (Fig. 4.2).

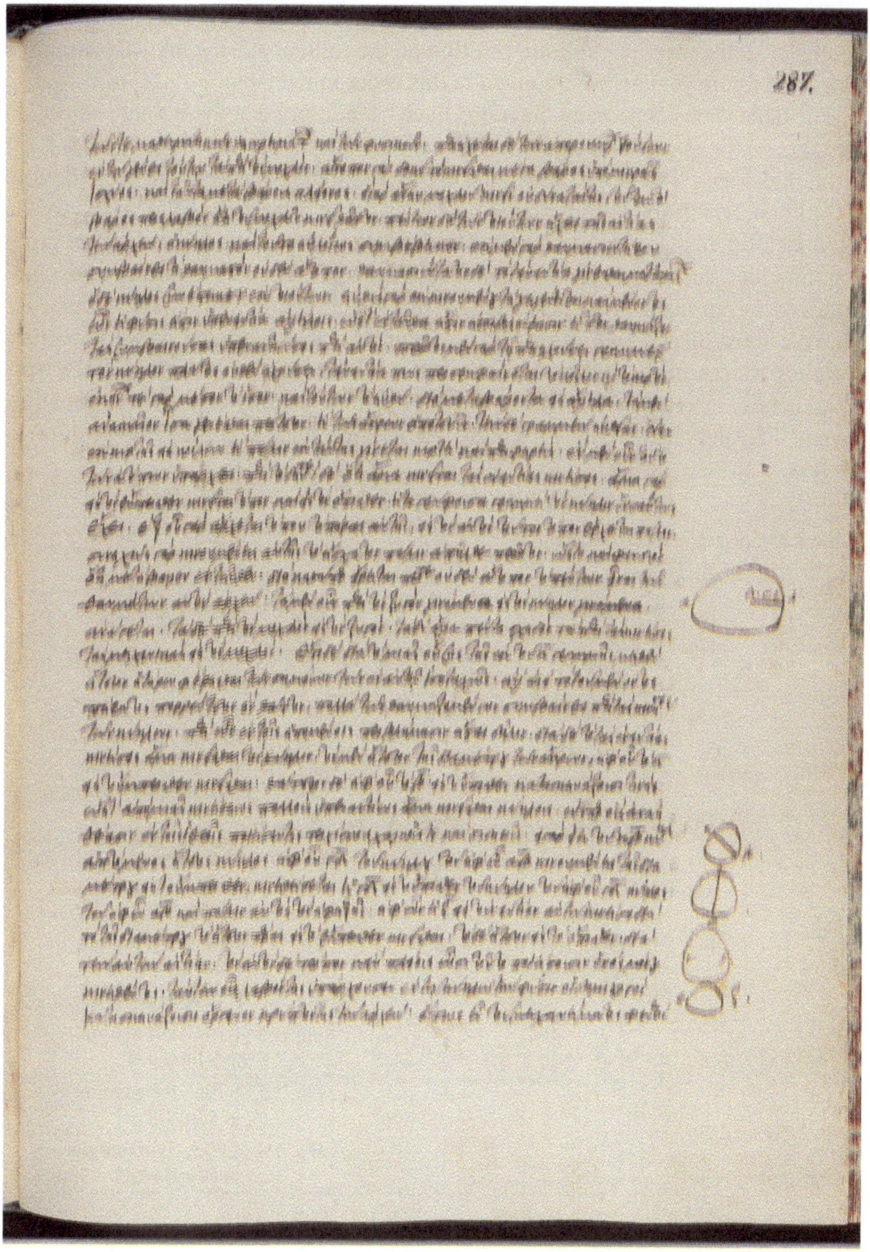

Fig. 4.2 Madrid, Biblioteca Nacional, Ms. Matr. 4563 (M[1]), f.287r. Diagram of a series of equal circles to the introduction of the *Mechanics*

Table 4.2 compares references to diagrams in the text with the corresponding figures in all manuscripts: folium, position on the folium, and the textual passage to which the figure is attached are given in this order in the table.

The table shows that many diagrams have been inserted at the wrong position in the text, as, for example, the position of the diagram for problem 3 in codices P, N^h and F^1 illustrates. This might be related to the positioning of a diagram at the bottom of a folium, which sometimes makes it impossible to insert a diagram at exactly the right position. In those cases where we find diagrams to a completely different problem, however, such an argument cannot apply; codices L and P especially are characterized by an inadequate placement of the diagrams. In some manuscripts, for example those that derive from codex ζ, namely U^m, F^2, N^b and q, we recognize certain patterns in the position of the figures. The figures in these manuscripts are generally inserted at the same correct or incorrect position in the text. From this, we can infer that the diagrams in codex ζ probably appeared at the same position in the text.

Table 4.2 Correspondence of diagrams

Description →	Introduction: 847a11-848a37		1: 848b1-850a2	
	Fig. I.2: 848a19-25		Fig. 1.2: 848b13-23	
Manuscript ↓	Fig. I.3: 848a26-34		Fig. 1.3: 849a2-6	
			Fig. 1.4: 849a21-b19	
L	Fig. I.3	f.248r, bottom, at 848b12	Fig. 1.2	f.249r, right margin, turned, at 849a14-24
			Fig. 1.4	f.249v, bottom left, indentation, at 849b20
P	Fig. I.1	f.282r, right margin, at 847b20-25	Fig. 1.1	f.284v, left margin, at 850a10
	Fig. I.2	f.282v, left margin, at 848a13-15	Fig. 1.2	f.283v, left margin, at 849a23-32, 2 figures
	Fig. I.3	f.283r, right margin, at 848b27-849a5	Fig. 1.3	f.283v, left margin, at 849a15
			Fig. 1.4	f.284v, bottom, at 850a35
M^v			Fig. 1.2	f.70v^1, left margin, at 848b29-31
			Fig. 1.3	f.70v^1, left margin, at 849a3-5
			Fig. 1.4	f.70r^2, right margin, at 849a24-28
D^1				
W^a	Fig. I.1	f.69v, left margin, at 848a5	Fig. 1.1	f.69v, left margin, at 848b10
	Fig. I.2	f.69v, left margin, at 848a23	Fig. 1.2	f.69v, bottom, at 848b11
	Fig. I.3	f.69v, left margin, at 848a28-b8	Fig. 1.3	f.69v, bottom, at 848b11
			Fig. 1.4	f.70v, left margin, at 849a20-30

(continued)

Table 4.2 (continued)

Description →	Introduction: 847a11-848a37		1: 848b1-850a2	
	Fig. I.2: 848a19-25		Fig. 1.2: 848b13-23	
Manuscript ↓	Fig. I.3: 848a26-34		Fig. 1.3: 849a2-6	
			Fig. 1.4: 849a21-b19	
V^1	Fig. I.1	f.36v, left margin, at 848a9-12	Fig. 1.1	f.37r, bottom, at 848b20
	Fig. I.2	f.36v, left margin, at 848a23-25	Fig. 1.2	f.37r, bottom, at 848b20
	Fig. I.3	f.36v, left margin, at 848a25-30 and f.37r, top, at 848a30, this figure is split up in two parts	Fig. 1.3	f.37r, bottom, at 848b20
			Fig. 1.4	f.38r, right margin, partly in text, at 849a20-25
M^1	Fig. I.1	f.287r, right margin, at 848a8-12	Fig. 1.1	f.287v, bottom, at 849a18
	Fig. I.2	f.287r, right margin, at 848a25-26	Fig. 1.2	f.287v, bottom, at 849a18
	Fig. I.3	f.287r, right margin, at 848a27-35	Fig. 1.3	f.287v, bottom, at 849a18
N^h	Fig. I.1	f.2v, bottom, at 848a10 (at f.2 r, bottom, at 847b26 just a circle)	Fig. 1.1	f.4r, bottom, at 848b6
	Fig. I.2	f.3 v, left margin at 848a22-23	Fig. 1.2	f.4v, bottom, at 848b19
	Fig. I.3	f.3v, left margin at 848a26-33	Fig. 1.3	f.5v, bottom, at 849a6
			Fig. 1.4	f.6v, bottom, at 849a31
F^1	Fig. I.1	f.1v, left margin, at 848a6-10 (f.1v, left margin at 847b25-848a2 just a circle)	Fig. 1.1	f.2r, right margin at 848b4-8
	Fig. I.2	f.1v, left margin, at 848a19-22	Fig. 1.2	f.2r, bottom, at 848b16
	Fig. I.3	f.2r, right margin, at 848a24-35	Fig. 1.3	f.2v, left margin, at 848b36-849a6
			Fig. 1.4	f.3r, bottom, at 849b3
U^m	Fig. I.1	f.1v, left margin, at 847b23-25	Fig. 1.1	f.3r, bottom, at 848b32
	Fig. I.2	f.1v, left margin, at 848a1-4	Fig. 1.2	f.3r, bottom, at 848b32
	Fig. I.3	f.2r, right margin, at 848a7-17	Fig. 1.3	f.3r, bottom, at 848b32
			Fig. 1.4	f.4r, left, integrated in text, at 849a20-25
F^2	Fig. I.1	f.4v, left margin, at 847b23-24	Fig. 1.1	f.7v, bottom, at 849a1
	Fig. I.2	f.4v, left margin, at 848a1-3	Fig. 1.2	f.7v, bottom, at 849a1
	Fig. I.3	f.2r, right margin, at 848a7-8 and f.2v, left margin, at 848a12-17, this figure is split up in two parts	Fig. 1.3	f.7v, bottom, at 849a1
			Fig. 1.4	f.8r, left, integrated in text, at 849a21-23
N^b	Fig. I.3	f.49v, bottom, at 848a36	Fig. 1.1	f.49r, bottom, at 848a15
			Fig. 1.3	f.51r, bottom, at 849a22
			Fig. 1.4	f.51v, bottom, at 849a35
q	Fig. I.1	f.80v, left margin, at 847b22-23	Fig. 1.1	f.81v, bottom, at 848b28
	Fig. I.2	f.80v, left margin, at 848a1-3	Fig. 1.2	f.81v, bottom, at 848b28
	Fig. I.3	f.80v, left margin, at 848a7-13	Fig. 1.3	f.81v, bottom, at 848b28
			Fig. 1.4	f.82v, left margin, at 849a20-25

(continued)

Table 4.2 (continued)

Description →	2: 850a3-29		3: 850a30-b9	
	Fig. 2.1/2.2: 850a11-19		Fig. 3: 850b6-9	
Manuscript ↓	Fig. 2.3: 850a24-29			
L	Fig. 2.1	f.250v, left margin, at 850a13-18	Fig. 3	f.251r, right margin, at 850b8-15
	Fig. 2.3	f.250v, left margin, at 850a30-35		
P	Fig. 2.1	f.284v, bottom, at 850a35	Fig. 3	f.285r, right margin, at 850b25-30
	Fig. 2.3	f.285r, right margin, at 850b15		
M^v	Fig. 2.1	f.70v², left margin, at 850a8-13	Fig. 3	f.71r, right margin, at 850b4-10
	Fig. 2.3	f.70v², left margin, at 850a24-28		
D^l				
W^a	Fig. 2.1	f.70v, bottom left, at 849b29	Fig. 3	f.71v, left margin, at 850b8-15
	Fig. 2.2	f.71r, right margin, at 850a15-20		
	Fig. 2.3	f.71r, right margin, at 850a25-32		
V^l				
M^l				
N^h	Fig. 2.1	f.8v, bottom, at 850a3	Fig. 3	f.10v, bottom, at 850b22
	Fig. 2.2	f.9r, bottom, at 850a17		
	Fig. 2.3	f.9v, bottom, at 850a32		
F^l	Fig. 2.1	f.3v, bottom, at 849b38	Fig. 3	f.4v, left margin, at 850b10-20
	Fig. 2.2	f.4r, right margin, at 850a15-24		
	Fig. 2.3	f.4r, bottom, at 850a36		
U^m	Fig. 2.1	f.5r, bottom, at 849b38	Fig. 3	f.6v, left margin, at 850b7-11
	Fig. 2.2	f.5v, bottom, at 850a22		
	Fig. 2.3	f.6r, right margin, at 850a25		
F^2	Fig. 2.2	f.12v, bottom, at 850a23	Fig. 3	f.14r, right margin, at 850b3-9
	Fig. 2.3	f.13r, right margin, at 850a24		
N^b	Fig. 2.2	f.53v, left margin, at 850a13-16	Fig. 3	f.54r, right margin, at 850b7-11
	Fig. 2.3	f.53v, left margin, at 850a24-27		
q	Fig. 2.1	f.84r, top, at 849b35	Fig. 3	f.85r, right margin, at 850b3-14
	Fig. 2.2	f.84r, bottom, at 850a18		
	Fig. 2.3	f.84v, left margin, at 850a22-25		

(continued)

Table 4.2 (continued)

Description →	5: 850b28-851a37		17: 853a19-31	
Manuscript ↓	Fig. 5: 851a17-27		Fig. 17: 853a27-31	
L			Fig. 17	f.255r, right margin, at 853a31-34
P	Fig. 5	f.285v, left margin, at 851a15	Fig. 17	f.288r, right margin, at 853b10
M^{v}	Fig. 5	f.71v, left margin, at 851a16-20	Fig. 17	f.73v, left margin, at 853a26-29
D^{1}			Fig. 17	f.351r, right margin, at 853a26-31
W^{a}	Fig. 5	f.71v, left margin, at 851a1	Fig. 17	f.73v, left margin, at 853a25
V^{1}			Fig. 17	f.44r, top, at 853a26
M^{1}				
N^{h}	Fig. 5	f.11v, left margin, at 850b37-851a2	Fig. 17	f.18v, bottom, at 853a31
F^{1}	Fig. 5	f.4v, bottom, at 850b32	Fig. 17	f.8r, bottom, at 853b12
U^{m}	Fig. 5	f.7v, bottom, at 851a31		
F^{2}	Fig. 5	f.17v, bottom, at 851a33		
N^{b}	Fig. 5	f.55v, left margin, at 851a23-25		
q	Fig. 5	f.85v, left margin, at 851a9-12		

(continued)

Table 4.2 (continued)

Description →	21: 854a16-31		22: 854a32-b15	
Manuscript ↓	Fig. 21: 854a25-31		Fig. 22: 854a38-b15	
L	Fig. 21	f.256v, left margin, at 854a22-28		
P	Fig. 21	f.289r, right margin, at 854a30-37	Fig. 22	f.289r, right margin, at 854b1-5
M^{v}	Fig. 21	f.74v, left margin, at 854a20-28	Fig. 22	f.74v, left margin, at 854a36-39
D^{1}	Fig. 21	f.351v, left margin, at 854a15-28	Fig. 22	f.351v, left margin, at 854a38-854b2
W^{a}	Fig. 21	f.74v, left margin, at 854a30-35	Fig. 22	f.74v, left margin, at 854a30-35
V^{1}	Fig. 21	f.45r, right margin, at 854a25-30	Fig. 22	f.45v, top, at 854a35
M^{1}				
N^{h}	Fig. 21	f.22v, left margin, at 854a28-30	Fig. 22	f.22v, left margin, at 854a33-35
F^{1}	Fig. 21	f.9r, bottom, at 854a23	Fig. 22	f.9v, left margin, at 854a33-37
U^{m}	Fig. 21	f.14v, left margin, at 854a25-30	Fig. 22	f.14v, bottom, at 854b2
F^{2}	Fig. 21	f.33v, left margin, at 854a23-25	Fig. 22	f.34v, left margin, at 854b2-4
N^{b}	Fig. 21	f.63r, right margin, at 854a28-32	Fig. 22	f.63r, right margin, at 854b3-5
q	Fig. 21	f.91v, left margin, at 854a23-26	Fig. 22	f.92r, right margin, at 854a35-854b1

(continued)

Table 4.2 (continued)

Description →	23: 854b16-855a27		24: 855a28-856a38	
	Fig. 23.1: 854b22-855a7		Fig. 24.1: 855b5-23	
	Fig. 23.2: 855a19-27		Fig. 24.2: 856a1-16	
Manuscript ↓	(reference to the same diagram)			
L	Fig. 23.1	f.257r, right margin, at 854b25-30	Fig. 24.1	f.258v, left margin, at 855b12-17
P	Fig. 23.1	f.289r, bottom right, at 854b5, 2 figures (1 try-out)	Fig. 24.1	f.290r, right margin, at 855a16-22
	Fig. 23.2	f.290r, right margin, at 855a7-9	Fig. 24.2	f.290v, left margin, at 855b29
M^v	Fig. 23.1	f.75r, right margin, at 854b21-26	Fig. 24.1	f.75v, left margin, at 855b1-5
			Fig. 24.2	f.76r, right margin, at 856a1
D^1	Fig. 23.1	f.351v, left margin, at 854b21-27	Fig. 24.1	f.352r, right margin, at 855a38-855b7
			Fig. 24.2	f.352r, right margin, at 856a1
W^a	Fig. 23.1	f.75r, right margin, at 854b15-20	Fig. 24.1	f.75v, bottom, at 855b30
			Fig. 24.2	f.76r, right margin, at 855b37
V^1	Fig. 23.1	f.45v, left margin, at 854b10-15		
M^1				
N^h	Fig. 23.1	f.23v, left margin, at 854b16-21	Fig. 24.1	f.27r, bottom, at 855b30
	Fig. 23.2	f.25r, bottom, at 855a20	Fig. 24.2	f.27v, bottom, at 856a5
F^1	Fig. 23.1	f.9v, bottom, at 854b14	Fig. 24.1	f.11r, bottom, at 855b23
	Fig. 23.2	f.10v, bottom, at 855a30	Fig. 24.2	f.11v, left margin, at 856a5
U^m	Fig. 23.1	f.16r, right margin, at 855a13-16	Fig. 24.1	f.17r, bottom, at 855b15 (at f.17v, left margin, at 855b16-19 just a circle)
			Fig. 24.2	f.18v, bottom, at 856a23
F^2	Fig. 23.1	f.37v, bottom, at 855a17	Fig. 24.1	f.39v, bottom, at 855b16
			Fig. 24.2	f.42v, bottom, at 856a23
N^b	Fig. 23.1	f.64r, right margin, at 854b26-27	Fig. 24.1	f.65v, bottom, at 855b22
			Fig. 24.2	f.66r, bottom, at 856a6
q	Fig. 23.1	f.92v, left margin, at 854b20-22	Fig. 24.1	f.93v, left margin, at 855b5-11 and f.94r, right margin, at 855b17-20, 2 identical figures
	Fig. 23.2	f.93r, right margin at 855a16-19	Fig. 24.2	f.94v, bottom, at 856a8

(continued)

Table 4.2 (continued)

Description →	25: 856a39-857a4		26/27: 857a5-21/ 857a22-33	
	Fig. 25.1: 856b11-38		Fig. 26: 857a18-21	
Manuscript ↓	Fig. 25.3: 856b38-857a4		Fig. 27: 857a29-33	
L	Fig. 25.1	f.260r, right, partly text integrated, at 856b28-32	Fig. 26/27	f.261r, right margin, at 857a35
	Fig. 25.2	f.260v, left margin, at 857a1-3		
	Fig. 25.3	f.260v, left margin, at 857a17-21		
P	Fig. 25.1	f.291v, bottom left, at 856b20	Fig. 26/27	f.293r, right margin, at 857b20
	Fig. 25.2	f.292v, left margin, at 857a25		
	Fig. 25.3	f.292v, left margin, at 857a30-857b5 2 figures (1 try-out)		
M^v	Fig. 25.1	f.76v, left margin, at 856b13-16	Fig. 26/27	f.77r, right margin, at 857a18-19
	Fig. 25.3	f.77r, right margin, at 856b35-38		
D^1	Fig. 25.1	f.352v, left margin, at 856b9-13	Fig. 26/27	f.353r, right margin, at 857a16-19
	Fig. 25.3	f.352v, left margin, at 856b35-857a2		
W^a	Fig. 25.1	f.76r, bottom right, at 856a37	Fig. 26/27	f.77r, right margin, at 857a29
	Fig. 25.2	f.76v, left margin, at 856b10		
	Fig. 25.3	f.76v, left margin, at 856b15		
V^1	Fig. 25.1	f.48r, bottom, at 856b10	Fig. 26/27	f.49v, left margin, at 857a35
	Fig. 25.2	f.48v, left margin, at 856b28-30		
	Fig. 25.3	f.48v, left margin, at 856b30-35		
M^1				
N^h	Fig. 25.1	f.29v, bottom, at 856b10	Fig. 26/27	f.31v, bottom, at 857a35
	Fig. 25.2	f.30v, left margin, at 856b31-34		
	Fig. 25.3	f.30v, bottom, at 857a1		
F^1	Fig. 25.1	f.12v, bottom, at 856b30	Fig. 26/27	f.13v, left margin, at 857a34-37
	Fig. 25.2	f.13r, right margin, at 856b32-34		
	Fig. 25.3	f.13r, right margin, at 857a2-5		
U^m	Fig. 25.1	f.19v, bottom, at 856b14	Fig. 26/27	f.21r, right margin, at 857a16-19
	Fig. 25.2	f.20v, left margin, at 856b34-37		
	Fig. 25.3	f.20v, left margin, at 857a1-4		
F^2	Fig. 25.1	f.45r, bottom, at 856b20	Fig. 26/27	f.47r, right margin, at 857a14-16
	Fig. 25.2	f.46,r right margin, at 856b33-35		
	Fig. 25.3	f.46v, left margin, at 857a1-3		
N^b	Fig. 25.1	f.67v, top, at 856b9	Fig. 26/27	f.68v, left margin, at 857a29-31
q	Fig. 25.1	f.96r, top, at 856b10	Fig. 26/27	f.97r, right margin, at 857a27-30
	Fig. 25.2	f.96v, left margin, at 856b35-37		
	Fig. 25.3	f.96v, left margin, at 857a1-5		

(continued)

Table 4.2 (continued)

Description →	30: 857b21-858a2	
Manuscript ↓	Fig. 30: 857b32-858a2	
L		
P	Fig. 30	f.293r, right margin, at 858a10
M^v	Fig. 30	f.77v, left margin, at 857b26-31
D^1	Fig. 30	f.353r, right margin, at 857b23-30
W^a	Fig. 30	f.77v, left margin, at 857b37-858a7
V^1	Fig. 30	f.50r, right margin, at 858a3-10
M^1		
N^h	Fig. 30	f.33r, bottom, at 858a7
F^1	Fig. 30	f.14r, right margin, at 858a2-8
U^m	Fig. 30	f.22v, left margin, at 857b34-858a2
F^2	Fig. 30	f.50r, right margin, at 857b35-38
N^b	Fig. 30	f.69r, right margin, at 857b26-36
q	Fig. 30	f.98r, right margin, at 857b33-858a5

The majority of the figures are placed in the margins of the text, and only a very few are integrated into the text. Sometimes a similar diagram is contained on different folia of a codex. In problem 23, for instance, we find two references in the text to the same diagram. Some of the copyists consequently included this diagram twice in their manuscripts. Another reason for including the same diagram twice might be that scribes wanted to assist their readers in navigating the rather lengthy descriptions by saving them the trouble of scrolling backwards and forwards while reading.

4.2 Practices of Diagrams

4.2.1 Terminology

The fact that the extant works of Aristotle have been handed down to us as lecture notes that were not meant for publication, raises some questions about the form of illustrations in them. From the famous passage in Plato's *Meno* (82b–85b), in which Socrates explains the doubling of the area of a square by drawing figures in the sand, we know that illustrations were a fundamental part of teaching activities in antiquity. But, it is not immediately clear whether and to what extent Aristotle himself included diagrams in his texts or explained his teaching material by illustrating it in front of his students. We may, for example, look at the diagrams contained in the manuscripts of Aristotle's logical treatises. Although it certainly makes sense to represent syllogisms by means of diagrams, it is not, as Alfred Stückelberger argues, required for an understanding of the text.[13] The presence of these diagrams, therefore, cannot necessarily be traced back to Aristotle himself and later scribes or commentators may have been responsible for adding these figures to the manuscripts. Rather different is the situation in the mathematical descriptions contained in works such as the *Meteorology* and *On the Heavens*, in which the extensive use of letter labels in the text refers to accompanying diagrams. These complicated descriptions cannot be understood without a corresponding diagram. The presence of diagrams integrated into the textual space in some of the oldest Aristotelian manuscripts, such as *Vind.Phil.gr.* 100 or *Par.gr.* 1853, shows that these diagrams most likely go back to some point in late antiquity, maybe even to Aristotle himself.

Clarification through diagrams plays an important role in the *Corpus Aristotelicum*. In fact, Aristotle gives us the earliest evidence for the lettered diagram, as Netz points out.[14] Aristotle employs lettered diagrams in many different contexts, for instance in his logical treatises, biological works and geometrical proofs. The terminology he uses in referring to diagrams is diverse, as the *Meteorology* effectively illustrates. In his discussion of the Milky Way, Aristotle refers to a *hupographē* (346a31-32): "The circle and the constellations in it may be seen in the diagram (ὑπογραφῆς)."[15] From this passage, it remains unclear precisely what kind of illustration the *hupographē* refers to. However, we find another reference to the same term in line 363a25-26 of the *Meteorology*, this time in a discussion of the position of the winds: "What we say about their position must be followed with the help of the diagram (ὑπογραφῆς)."[16] In the following description of the diagram, Aristotle employs letter labels and connects these with the different winds, e.g. Zephyrus blows from A (363b12). In spite of his use of letter labels and other geometrical terminology in this context, as Liba Taub argues, the description is not purely mathematical, that is, the text is not about mathematical objects.[17] These usages of *hupographē* in the *Meteorology* show that the meaning of the term is distinct from a mathematical diagram. An analogous term used by Aristotle to refer to a non-mathematical diagram is *diagraphē*. It is, for example, used in the *History of Animals* in lines 525a7-9: "For the disposition of these parts I must refer to my anatomical diagrams (διαγραφῆς)."[18] This is one of several passages in the corpus in which Aristotle refers to the *Dissections*. Whereas many of these references can be related to actual dissections, a few others such as the above-mentioned passage refer to some sort of collection of images that was probably based on dissections per-

---

[13] See Stückelberger (1993), 136.

[14] See Netz (1999), 61.

[15] Transl. Webster in Barnes (1984), 566.

[16] Transl. Webster in Barnes (1984), 588.

[17] See Taub (2003), 104.

[18] Transl. Thompson in Barnes (1984), 833. For a similar reference see *HA* 497a31-32: "The appearance of this organ can be investigated from the diagrams (διαγραφῆς) in the *Dissections*." Transl. Thompson in Barnes (1984), 791.

formed earlier.[19] One should note that almost all references to diagrams in Aristotle are accompanied by the verb θεωρέω, instructing us that something needs to be observed from the diagram. This also applies to another term *diagramma*, which has a particularly mathematical meaning in Aristotle, signifying either a geometrical figure or a proof.[20] In the *Meteorology* (375b16-19), for example, it is used to refer to a detailed geometrical treatment of rainbows: "The rainbow can never be a circle nor a segment of a circle greater than a semicircle. The consideration of the diagram (διαγράμματος) will show this and the other properties of the rainbow."[21] The following proof with its use of letter labels is similar to the demonstrations known from ancient mathematics.

Although we cannot find any explicit references to visualizations in the *Mechanics*, diagrams were an important part of the argument of the text as is evidenced by the author's constant use of letter labels to refer to a corresponding diagram. The next section will cover the practices of the lettered diagram in the Aristotelian *Mechanics*.

4.2.2 The Lettered Diagram

The practice of applying letter labels in the text to refer to an accompanying diagram is associated particularly with Greek mathematics. The author of the *Mechanics* employs this technique in a number of the mechanical problems. In applying letter labels to mechanical objects, the author creates a relationship between the text and the diagram. In this section, we will see that text and diagram in the *Mechanics* are interdependent. The text cannot be considered a set of instructions on how to draw the diagram, but clearly presupposes the presence of a diagram and refers to specific relations in it. However, the diagram in itself is not sufficient to understand and reconstruct the author's argument. Even though most diagrams are recoverable from the text, the text does not define one diagram specifically. A comparison between the manuscript diagrams and those in the modern critical editions reveals that different diagrams can be constructed on the basis of the text. This is mainly due to the fact that many letters in the description are not clearly specified.[22] Let us take as an example the following description at lines 855b5-6: "For let there be a

[19] I am grateful to Jim Lennox for allowing me to read his manuscript *Experience, Expertise and Aristotle's Appeal to Dissections.*

[20] See Fowler (1987), 33.

[21] Transl. Webster in Barnes (1984), 604.

[22] I rely on Netz's distinction between the different natures of letter labels. See Netz (1999), 21 for the definition of four classes of letters: letters that are completely specified, underspecified, completely unspecified, and those that change their nature throughout the proposition.

larger circle ΔΖΓ, a smaller ΕΗΒ, and Α the centre of both."[23] The letter Α in this description is completely specified as the centre of both circles. The circles are, on the contrary, underspecified; we know the labels of the larger and smaller circle, but we do not know where these labels should be positioned on the circles. In the description that follows, these labels receive increased specification, and we can infer that Δ and Ε are placed at the top of both circles. Underspecified letters like ΔΖΓ and ΕΗΒ form the largest class of labels in the *Mechanics*. Often they receive complete specification in the course of an argument, but sometimes they remain underspecified.

This occurs, for example, in a description of motion in a circle in lines 849a2-3 of the text: "Let there be a circle, ΑΒΓ, and let the endpoint Β be moved to Δ; it comes at some time to Γ."[24] In this description, the circle is again underspecified, since we do not know where the labels Α, Β and Γ are positioned on the circle's circumference. From the following description, we learn that points Β and Γ are opposite one another, since line ΒΓ is further defined as a diameter of the circle. However, there are still numerous possible positions for these points, all in agreement with the text, and thus these points remain underspecified. In these lines, we find a label that is again of a different nature, namely Δ. This letter is completely unspecified by the text; its position is unknown, whether on the circumference of the circle, or at some point inside or outside the circle. We only know the position of Δ because we see it in the diagram. If we were to draw this diagram based on the description alone, we could not know how the diagram configured by the author would look. This is precisely the situation in the critical editions of the *Mechanics*, since the editors have drawn their own figures independently of the manuscript tradition. The underspecified and completely unspecified labels are, then, responsible for many differences between diagrams that are constructed based on the text. As we will see in the next chapter, in most cases this un(der)specification results in the figures having different orientations in the modern critical editions of the treatise.

Un(der)specification is not only found on the level of letter labels, but it sometimes even applies to a figure as a whole. This does not become apparent until the diagram is detached from the text and one tries to make sense of the text only. When reading a text while simultaneously referring to a diagram, it seems only natural for a diagram to have a specific form, but sometimes the shape is not defined by the text. Lines 848b13-23 of the text describe a diagram that illustrates the principle of rectilinear motion. The figure is described by the author as the ratios that certain lines bear to one another, without referring to the specific form of the figure. In order to discover what the diagram should look like, we have to read through to the

[23] ἔστω γὰρ κύκλος ὁ μείζων μὲν ἐφ' οὗ τὰ ΔΖΓ, ὁ δὲ ἐλάττων ἐφ' οὗ τὰ ΕΗΒ, κέντρον δὲ ἀμφοῖν τὸ Α.

[24] ἔστω κύκλος ὁ ΑΒΓ, τὸ δ' ἄκρον τὸ ἐφ' οὗ Β φερέσθω ἐπὶ τὸ Δ· ἀφικνεῖται δέ ποτε ἐπὶ τὸ Γ.

end of the description: "For the small quadrilateral is in the same ratio [to the larger] (848b19-20)."[25] From this statement, we notice that the author simply presupposed the presence of a figure with four sides and did not conceive of the diagram as being established step by step from the description.

The author of the *Mechanics* employs two different techniques in applying letter labels to mechanical objects. The first is direct identification of a letter with an object, for example, in problem 5 of the text on the working of a rudder (851a17-18): ἔστω γὰρ ἡ ΑΒ κώπη; "For let AB be the oar." The second technique, which is commonly used by Aristotle, involves a periphrastic use of the preposition *epi* with the genitive or dative. In the *Mechanics* we find this use of *epi* with the genitive or dative 50 times,[26] alternating with the occasional use of the preposition *en* with the dative, or an adverbial use of οὗ. Sometimes every element in a description is defined by means of *epi*, as for instance in problem 3 on the lever (850b6-9): ἔστω μοχλὸς ἐφ' οὗ ΑΒ, βάρος δὲ ἐφ' ᾧ τὸ Γ, τὸ δὲ κινοῦν ἐφ' ᾧ τὸ Δ, ὑπομόχλιον ἐφ' ᾧ τὸ Ε, τὸ δὲ ἐφ' ᾧ τὸ Δ κινῆσαν ἐφ' ᾧ τὸ Η, κινούμενον δὲ τὸ ἐφ' οὗ Γ, βάρος ἐφ' οὗ Κ. I was unable to discover a pattern in the author's use of direct identification of diagrammatic labels with a mechanical object on the one hand, and the periphrasis using *epi* on the other. Both approaches appear to be interchangeable and can even be combined in one and the same line (849a23-25): καὶ ἐκβεβλήσθωσαν αἱ διάμετροι, ἐν μὲν τῷ μεγάλῳ, ἐφ' ὧν ΓΔ καὶ ΒΕ, ἐν δὲ τῷ ἐλάττονι αἱ ΜΧ ΝΞ. The diameters in the larger circle are described periphrastically, whereas the diameters in the smaller circle are identified directly by their labels. From the interchangeability of the two techniques, we can infer that at the time the *Mechanics* was written a systematic manner of applying letter labels to objects had not yet been developed. Periphrastic descriptions are characteristic of the earliest Greek texts which use the lettered diagram.[27] Generally, we notice that in texts that were written after Euclid such expressions by means of *epi* with the genitive or dative are rare.[28] Direct identification appears to be the more general case here, only occasionally interspersed with a periphrastic use of *epi*. Accordingly, the practice employed by the author of the *Mechanics* to frequently use periphrastic expressions corresponds well with Aristotle and the early tradition. Although we should not exclude the possibility of archaism on the part of our author, that he resolved to use earlier techniques at a time when new ones were also available, this practice may support an early dating of the treatise.

This raises the question of what exactly is meant by the technique of periphrasis. Netz suggests that in ancient Greek mathematics the expression *epi* with the geni-

[25] ὅμοιον ἄρα ἐστὶ τῷ λόγῳ τὸ μικρὸν τετράπλευρον.

[26] In the texts of the *Corpus Aristotelicum* that refer to lettered diagrams, we find this periphrastic use of *epi* with the dative or genitive often in *APr.* and *APo.* (over 100 times), *Ph.* (about 70 times), *Mete.* (about 40 times) and *Cael.* (about 30 times); this way of applying letter labels to objects is therefore a general technique used by Aristotle.

[27] Netz (1999), 45.

[28] See Heiberg (1904), 17.

tive or dative has a markedly spatial usage.[29] The preposition describes spatial relationships between the letters in the diagram and the objects referred to. Whenever a spatial reading of the term *epi* is possible, this reading should be preferred. In the mathematical setting which is focused on the spatial object of the diagram, the spatial reading is therefore the natural reading.[30] There are, however, contexts in which we may find usages of *epi* that denote a relationship between letters and objects different than an exclusively spatial one. Aristotle offers such an exception; as the Aristotelian corpus provides the earliest evidence for the lettered diagram, it often presents a somewhat different case.[31] In the *Prior Analytics* we can find the following example (30a30): ἐφ' ᾧ δὲ τὸ Γ ἄνθρωπος. Netz gives two different translations: "And [if that] on which Γ [is] man / and [if that] which Γ stands for [is] man."[32] The second translation should be preferred, since Γ does not spatially stand on the class of human beings. The use of *epi* with the genitive or dative should be understood here in a non-spatial sense. Netz argues that this expression in the *Prior Analytics* "is a bold *metaphor*, departing from the spatial mathematical usage."[33]

When we consider the usage of *epi* with the genitive or dative in the *Mechanics*, both a spatial and a non-spatial reading seem plausible. In the context of mechanics, the spatial reading in a sense suggests itself. Let us take a look at problem 17 of the text describing the working of a wedge (853a27): ἔστω σφὴν ἐφ' ᾧ ΑΒΓ; "Let there be a wedge next to which [is] ΑΒΓ." In the spatiality of the diagram, the letters ΑΒΓ literally stand next to the wedge rather than representing this object. Passages like these nonetheless reveal a complication that is not present in ancient mathematics. The field of mechanics deals with physical objects such as wedges, oars and balances. However, in the diagrams these objects are represented by mathematical lines and points. Thus, a situation is created in which the diagram does not resemble the mechanical object. Authors may have been unwilling to use the first technique of direct identification in applying letter labels to objects: AB for instance is not a balance, but a line that represents a balance. The second technique then offers an alternative, by which *epi* is used metaphorically. In order to avoid the difficulty that the diagrams do not resemble the mechanical objects, authors may have deliberately chosen to use the periphrastic use of *epi*.

In Greek mathematics there is no regular relation between signifier and signified.[34] Specific letters are not used for fixed objects, in the sense that the letter *r*, for example, always signifies radius or *d* is used as a symbol for diameter. The ordering of the letters is generally alphabetical, each description beginning with the letters A, B, Γ, etc. These letters may occur in permutations of the alphabetical, e.g. B, Γ, A, but in every case the first *n* letters of the alphabet are employed.[35] This means that

[29] See Netz (1999), 47.

[30] See Netz (1999), 47.

[31] See Netz (1999), 61.

[32] Netz (1999), 48.

[33] Netz (1999), 48–49.

[34] See Netz (1999), 79.

[35] See Netz (1999), 71.

letters are arbitrarily assigned to objects, since the first object in a description is always signified by A, regardless of the nature of the object. There is an exception to this practice, as Netz points out, namely the relation between the letter K and the centre of a circle.[36] This is found, for example, in Aristotle's *Meteorology* in a geometrical treatment of the rainbow (375b20). In this passage, Aristotle demonstrates that the rainbow cannot be greater than a semicircle. After beginning the proof in the usual manner with the letter A, the second label used is K to denote the κέντρον (centre) of a circle. This is not the only departure from the alphabetical sequence of the description; it is probably not a coincidence that the next letter, H, signifies the sun (ἥλιος). These examples show that, at least sometimes, there is a relation between signifier and signified, in which letters are used as actual symbols. In the *Mechanics*, we find one such occurrence of the letter *kappa*. The manuscripts contain a diagram to the first problem that is not referred to in the text. It depicts two concentric circles with a K at the centre. Again, rather than being arbitrarily assigned, the letter K functions as a symbol for κέντρον. Since it is such a specific point, the centre of a circle is especially suited for a symbolic use.

Generally in the text of the *Mechanics* the letters are ordered alphabetically. For example, the circles in the diagram to the introduction of the text are named by their diameters AB, ΓΔ and EZ, respectively. The diagram to the first problem depicting circular motion contains the alphabetical series ABΓΔE. Finally, in problem 17 on the working of a wedge, the following letters appear in the text: ABΓΔEHZ, containing one switch of H and Z, but otherwise being strictly alphabetical. Analogous to ancient mathematical practice, we recognize a certain pattern in the ordering of the letters in the *Mechanics*. Most labels in the text are arranged alphabetically, or are at least following permutations of the first *n* letters of the alphabet that are used. However, in some problems the order of the letters appears to be completely mixed up. For instance, problem 24 on the motions of two different concentric circles contains the sequence ΔZΓEHBAIKΛΘ. Here we find the first eleven letters of the alphabet in an apparently arbitrary order. If we look at the corresponding diagram, however, a systematic order arises: the author has introduced the labels according to their position in the diagram instead of assigning the first label mentioned in the text an A, the second a B and so forth. This practice shows that the diagram likely preceded the writing down of the text and has a significant meaning for the status of the diagrams in the *Mechanics*. The fact that the labels as they appear in the text are not always ordered alphabetically, but are sometimes structured according to the actual process of the drawing of the diagram, suggests that these diagrams were in some form already part of the text of the *Mechanics* at the author's time. This finding fits well what we discovered earlier about the diagrams. Diagrams are clearly presupposed by the text, for instance in the case of the parallelogram in problem 1, where the author does not concern himself to provide information about the actual form of the diagram. Moreover, some of the descriptions are rather complicated and employ nearly all letters of the alphabet. Here, it would have been impossible to formulate the text free from errors, unless being able to check the variants in a corresponding

[36] See Netz (1999), 80.

diagram. Even though we cannot say with certainty which form the diagrams took, if they were contained in a manuscript or drawn by the author on a whiteboard in front of his students, it is, the very least, likely that they were part of the text in antiquity.

4.2.3 *Characteristics of Diagrams*

The diagrams that are present in the extant manuscripts of the *Mechanics* look somewhat different from what we would expect today. Most importantly, they share two characteristic features with the diagrams from the ancient mathematical tradition that distinguish them from their modern counterparts. These characteristics are known as "overspecification" and "indifference to metrical accuracy".[37] Overspecification means that the geometrical objects depicted by the manuscript diagrams are often less general than is required by the text. We may find for example isosceles triangles depicting arbitrary triangles, or squares representing rectangles. The first problem of the *Mechanics* presents such a case of overspecification. The text in line 848b20 defines the figure of a quadrilateral that is depicted in all manuscript diagrams by a rectangle. Unlike ancient practice, modern representations would have depicted here an unspecialized diagram to underscore the generality of the proof.

The second feature of ancient diagrams is the indifference to metrical accuracy. Greek mathematical diagrams do not give a metrically correct depiction of the geometrical objects, which means, for example, that straight lines may be represented by curves or equal lines by unequal lines. Problem 3 of the *Mechanics* offers a good illustration of this principle. In this question on the working of the lever, it is stated that the further the force is from the centre, the more easily it will move the weight. The greater part of the manuscript diagrams, however, overlook this detail and depict the fulcrum at the exact centre of the lever. Another example of metrical inaccuracy can be found in problem 23 on the movements of the extreme points in a rhombus, of which one angle is described as acute and the other as obtuse. In a modern representation we would be able to distinguish the acute from the obtuse angle. In the manuscript diagrams, however, it is irrelevant whether one angle appears greater than the other, or whether they appear equal. Thus, we cannot infer the correct metrical properties from the manuscript diagrams, which makes these diagrams more "schematic" than their modern counterparts.[38] Instead, ancient diagrams convey topological properties only, they provide, for example, information on the position of points, whether they are contained inside or outside a certain figure.

The schematic nature of diagrams should not be confused with the errors in truly flawed diagrams that are often found in the manuscripts of the *Mechanics*. Freehand drawing is an important factor in many poorly drawn diagrams, and partly accounts for the large number of errors they contain. The diagram to the motions in circles with longer and shorter radii from the first problem of the text as it is contained in

[37] See Saito and Sidoli (2012), 140–148.

[38] See Netz (2004), 46.

codex L, effectively illustrates that many errors can be related to the freehand sketching of the figures (Fig. 4.3). The copyist of L did make use of compasses to draw the circles, but the lines in this diagram are characterized by very inaccurate freehand drawing.

In addition to the problems inherent in freehand drawing, there is also the fact that some of the scribes were very poor draftsmen. The copyist of P, for example, sometimes made several attempts before being able to draw his final diagram. This may be illustrated by the bottom two diagrams on f.289r that are included to problem 23 of the text (see Fig. 3.3). While the first diagram depicts an attempt at the shape of a rhombus, the second presents the actual diagram of a rhombus complete with letter labels. Furthermore, the diagrams contained in codex M^1 did not turn out well, as is evident from the diagram of a series of congruent circles (Fig. 4.2). The badly executed diagrams contained in P and M^1 might be related to the scribes' lack of artistic skill, and reflect the fact that they were trained in writing, rather than in drawing geometrical figures. Another reason may be the lower status accorded to the diagrams due to their marginal position in the text. Whereas the scribes' primary goal was to copy the text as faithfully as possible, the marginal diagrams may have seemed less authoritative to them in comparison to the text. This is demonstrated by the number of diagrams in M^1, since the scribe of this manuscript only included the diagrams to the introduction and the first problem of the text, then apparently gave up because he did not quite succeed in his intentions.

4.2.4 *Byzantine Origins*

Diagrams in Greek manuscripts were generally drawn after the copying of the text and could occupy several different spaces. Often they were integrated into the text when the author prescribed the presence of a diagram. Diagrams were also included at the end of the complete text as an appendix or even in a separate book which contained only diagrams. Another physical location for diagrams was in the margins of the text, which is where they are mostly found for the *Mechanics*. Because they are rarely placed differently in the extant manuscripts, the few exceptions require closer examination. In codex L we occasionally find an indentation of a few lines to create sufficient space for a diagram. This shortening of lines shows that the scribe of L already took account of the presence of figures at the stage of copying. Both codex U^m and F^2 contain only a single text-integrated diagram. Whereas all of the other diagrams are placed in the margins, this diagram has a marked position, and confirms the father-child relationship between these two manuscripts. For a greater part of the text, codex P^t contains windows in the text for diagrams that were never executed. This example illustrates that, at least in some of the manuscripts, the diagrams were not drawn immediately after the scribe had finished the copy of the text. Most likely, the scribe and illustrator were not one and the same person, and the diagrams were due to be carried out by a distinct illustrator. When diagrams were added at a later

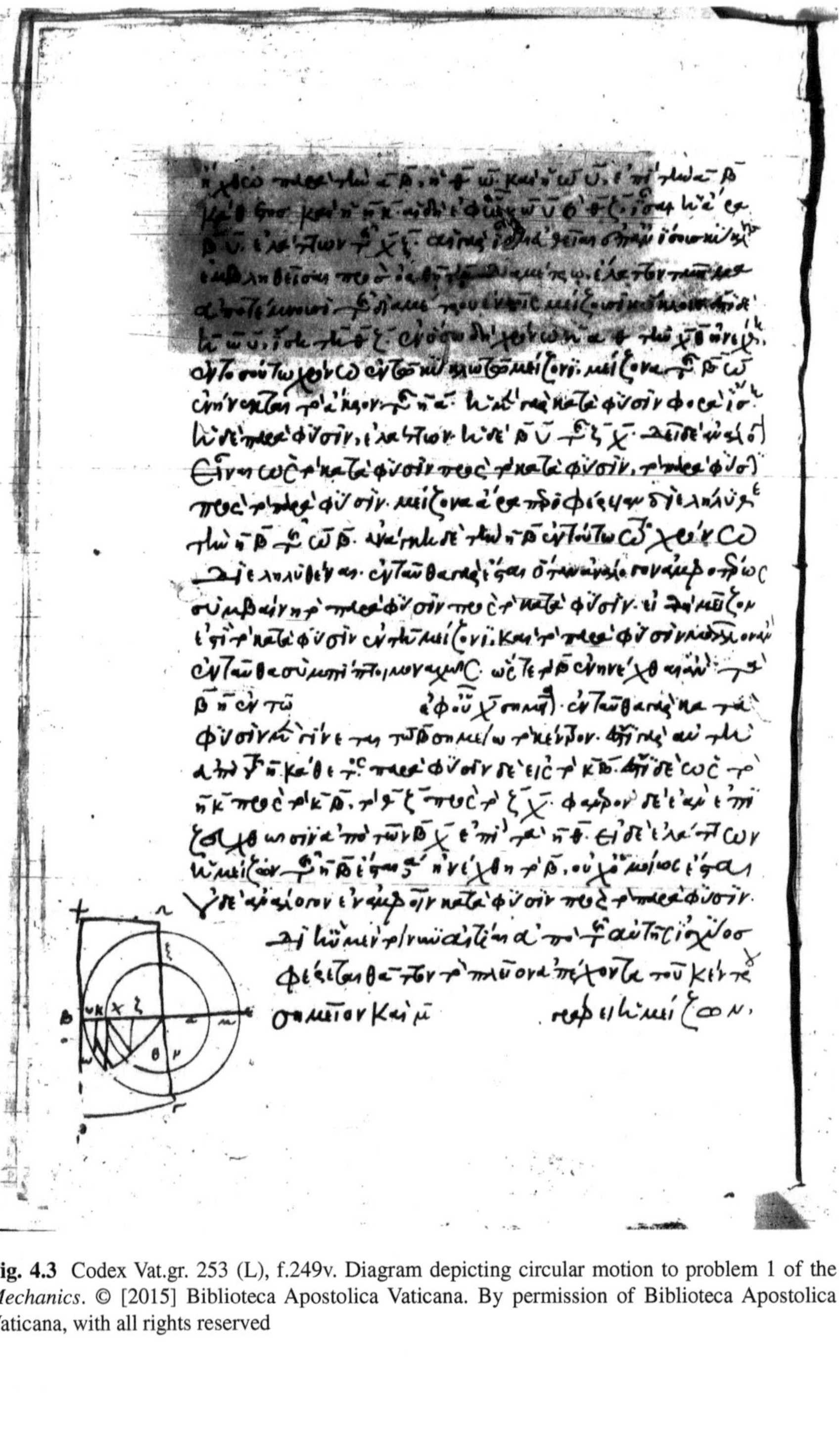

Fig. 4.3 Codex Vat.gr. 253 (L), f.249v. Diagram depicting circular motion to problem 1 of the *Mechanics*. © [2015] Biblioteca Apostolica Vaticana. By permission of Biblioteca Apostolica Vaticana, with all rights reserved

stage, it occurred that the illustrator possessed a different exemplar than the one from which the text was initially copied. This option should always be considered when editing diagrams as it would complicate the construction of a *stemma*.

Diagrams could change depending on their position on the page. In codex L, for example, the figure of a parallelogram is turned 90 degrees, because of the limited space in the right margin of the folium (see Fig. 3.2). In addition to their marginal position, many diagrams are contained at the wrong place in the text, sometimes even to a completely different chapter, as exemplified by the table on the correspondence of diagrams (Table 4.2). We have now seen a set of diagrammatic features that applies to the majority of the Greek manuscripts of the *Mechanics*: nearly all diagrams are characterized by imprecise freehand drawing, they are mostly located in the margins of the page, often in combination with an incorrect positioning in relation to the text. These aspects point to an archetype in which the diagrams were likely marginal as well. When compared with diagrams in other scientific texts, a rather different picture emerges. In the mathematical tradition diagrams generally appear in the textual space at the end of the relevant proposition. Moreover, they are mostly well executed with the aid of instruments such as ruler and compass. In those cases in which freehand drawings appear in the margins, these are often additional to the diagrams already present in the text. Such diagrams can be considered alternative diagrams that may have been copied later from other manuscripts, as Saito has shown for diagrams in Euclid's *Elements*.[39] Alternatively, marginal diagrams also appear in texts in which no diagrams were prescribed.[40] Whereas the diagrams that are found in the textual space likely derive from a late ancient archetype, possibly even resembling the author's own diagrams, the marginal sketches appear to be mostly later additions that are less authoritative than those text-integrated diagrams. Thus, the evidence from other scientific texts seems to point against the authenticity of the diagrams contained in the manuscripts of the *Mechanics*. The fact that we find almost exclusively marginal freehand sketches makes it unlikely that the diagrams derive ultimately from the author himself. Rather, it appears that they were added for pedagogical reasons somewhere in the Byzantine period by the scribe of the archetype. The diagrams may have been included to render a better understanding of the mechanical principles, and in this sense they comprise a commentary on the text. The Byzantine origins of the diagrams are further enforced by the fact that the labelling is in minuscule, which is a Byzantine introduction.[41]

Apart from the physical location and appearance of the diagrams, certain incongruities in the text-image relationship also indicate that the diagrams in their present form cannot be traced back to antiquity. There are a number of diagrams contained in the manuscripts that are not prescribed by the author. For example, in the introduction to the text we find a diagram of a circle AB with radius ΔΓΕ. Since there is no textual reference to such a diagram, it was most likely not included in the origi-

[39] See Saito (2006), 94.

[40] See Eastwood (2007), 375 on Carolingian copies of Roman astronomy.

[41] See Hunger (1989), 64–65.

nal text but added later together with some glosses and comments on the text. Something similar applies to the second problem that is concerned with a question on balances. Two diagrams are specifically defined by the text, one of a balance that is attached from above and the other from below. Part of the manuscripts, however, contain an additional diagram illustrating the first case that was probably inserted for a better understanding of the principle. Whereas this diagram of a balance might have been added by a later scribe and thus was not contained in the archetype, the case is different for the example of the circle diagram. As the circle appears in the complete manuscript tradition, it was likely to have been present in the archetype, thus showing that the diagrams from the archetype are not identical to the ones conceived by the author of the text.

We also find diagrams that do not fully coincide with the description given in the text. The diagram of circular motion included in the manuscripts to the first problem of the text is not a perfect representation of the author's intentions. In lines 849a14-16, it is argued that one of the components of which circular motion consists is directed toward the centre of the circle and the other along the tangent. This aspect is completely ignored in the manuscript tradition as these orientations are absent from all diagrams. Thus, the diagram in the manuscripts is not likely to represent the ancient authorial diagram and was probably inserted by the scribe of the archetype in an attempt to make sense of this difficult passage. As will be explained in detail in the next chapter, the text of problem 1 cannot be considered a set of instructions on how to draw the diagram to circular motion. It is by no means an easy task to recover the original diagram from the text, which explains the difficulties scribes must have faced.

Now, on the one hand, we have seen that the diagrams in the manuscripts of the *Mechanics* are probably of Byzantine origins. On the other hand, I have pointed out that the author, by his constant use of letter labels in the text, refers to corresponding diagrams that were part of the text in antiquity. From this, we can infer that the text circulated without diagrams during at least part of antiquity. Thus, the original diagrams were lost at some point in the course of the transmission, leaving open the question of what could have accounted for their absence. One possible explanation may be deduced from the character of the transmission. Scientific and philosophical texts were not collected in antiquity for their own intrinsic value as works of science, but rather for other reasons, one being their literary value, as Netz points out.[42] If the *Mechanics* was collected and transmitted in antiquity as a literary work, it would not be surprising that it circulated without diagrams. The presence of diagrams is a conspicuous feature of works of science, whereas ancient literary texts would usually not contain any diagrams. Another explanation could be that the diagrams in antiquity were transmitted separately from the text. It is not uncommon to find the diagrams at the end of a manuscript, or sometimes even in separate diagram books. Alexander Jones, for example, mentions the existence of two such diagram collections in the case of the mathematical works of Pappus of Alexandria.[43] At

[42] See Netz (2011), 248.

[43] See Jones (1986), 61.

several occasions, Aristotle himself also refers to a collection of anatomical diagrams contained in his *Dissections*, a work that unfortunately is not preserved.[44] If the diagrams to the *Mechanics* circulated in separate diagram collections, this could explain how they came to be lost at some point during antiquity.

The conclusion that we probably do not possess the original diagrams to the *Mechanics* does not automatically mean, as previous scholars have assumed, that these figures have lost their relevance. Despite the fact that the diagrams are of later Byzantine origins, they resemble truly ancient visualizations as seen in other scientific texts, especially those from Greek mathematics. We have seen that the diagrammatic practices of overspecification and metrical inaccuracy present in the manuscript diagrams reflect ancient practice. In this sense, the manuscript diagrams give us a much closer sense of what the original diagrams might have been than the modern interpretations found in the critical editions of the *Mechanics*. Moreover, the diagrams are valuable for our knowledge of the Byzantine reception of the treatise. The figures were probably inserted by the scribe of the archetype for didactical reasons and offer his understanding of the mechanical principles described in the text. Finally, as will be expounded in the next chapter, a comparison of the relationship between manuscript text and diagram enables us to discover many authentic manuscript readings.

4.3 Critical Edition

4.3.1 Introductory Remarks

The purpose of the critical edition is to make the evidence from the Greek manuscripts of the *Mechanics* available and to reconstruct the archetypal diagrams. The edition offers reconstructions of all diagrams in the Byzantine archetype; flawless diagrams are printed, in keeping with the intentions of the scribe of the archetype, regardless of potential errors. Occasionally, the edition depicts a reconstruction of a diagram from one of the hyparchetypes, in cases where a certain family contains a figure that greatly deviates from the one in the archetype.

In studying the Byzantine reception of the *Mechanics*, it is important to study all of the diagrams. Accordingly, the appendix contains reproductions of all diagrams from 13 manuscripts, together with my reconstructions of the diagrams in non-extant manuscripts, including the archetype.[45] These diagrams are arranged according to their position in the *stemma codicum*. This presentation, in which each family of related manuscripts is grouped separately, allows the reader to readily verify the relations between different diagrams and to obtain a clear picture of the material.

[44] For these references see e.g. *HA* 497a32, 525a9 and 566a15.

[45] The appendix does not offer reproductions of the diagrams from the manuscripts that are isolated from the tradition, Par, P^t, P^s, B^e and those contained in the paraphrase by Pachymeres. These diagrams do not contribute to the recovery of the earliest diagrams and therefore are not considered in the critical edition. However, they are of particular relevance in the broader context of the transmission of mechanical knowledge, and will be treated separately in Sect. 4.5 and Chap. 6.

The program DRaFT, developed by Ken Saito, was used in reproducing the diagrams. For the sake of convenience, the figures and labels were reproduced in approximately the same size. Consequently, the edition leaves out some of the information related to the different scales of diagrams in a single manuscript or between manuscripts. Following the practice of the scribes of the Greek manuscripts, I have used minuscules for all letter labels.

The appendix prints exact reproductions of the diagrams from the manuscripts; no reconstructions are made of lost or illegible labels and parts of a diagram. Sometimes a folium has been trimmed, and parts of a diagram are therefore depicted as missing. In such cases, dashed lines in the reproductions show where the page was cut off. Black lines in a figure indicate that a diagram was not drawn on one folium, but that different parts of the figure were divided among several pages. The labels in the figures are generally coloured in black. In those cases where a letter is illegible, due to the poor condition of the manuscript or microfilm, a question mark is added. Sometimes it is difficult to decide if a certain label was contained in a manuscript; those labels are printed in grey. This is especially liable to occur when a family is represented by a diagram in a single manuscript. In the case of an erroneous label or a missing letter in the manuscript, it remains unclear whether this error was already contained in the exemplar, or constituted a new mistake made by the copyist.

4.3.2 Case Study: Problems 3 and 23

In this section, I will take the manuscript diagrams for problems 3 and 23 of the *Mechanics* as a case study to highlight important issues that arise in making a critical edition of the diagrams.[46] 10 manuscripts contain a diagram to the third problem: codex L in family *a*; P and M^{v} in family *b*; codices W^{a}, N^{h}, F^{1}, U^{m}, F^{2}, N^{b} and q in family *c*. In order to recover the authentic diagram from the archetype, one first needs to make a reconstruction of the figures in each of the hyparchetypes, namely codices α, β and γ. The reconstruction of the figure in hyparchetype α is based solely on the diagram contained in codex L. From the diagram in L, the labels A and Δ are missing; given that this is due to the fact that the right side of the folium in L has been trimmed, these letters were most likely present in the hyparchetype. We also see that the figure of the lever in L has long legs and that the force Δ is depicted somewhat further from the fulcrum E than the weight Γ. The scribe of L might not have drawn it this way on purpose, since such divergences in diagrams are often caused by inaccurate freehand drawing. However, if we compare the same diagram contained in the manuscripts of Pachymeres, we see that the label E is not placed at the exact centre of the bar here as well. For that reason, it is likely to be a characteristic of all manuscripts in family *a*, and in my reconstruction of codex α the force Δ is at a further distance from the fulcrum than the weight Γ.

[46] Reproductions and reconstructions of the diagrams to these problems are included at the end of this section.

Both diagrams in P and M^v are similar and enable a straightforward reconstruction of the diagram in the hyparchetype β: the figure of the lever has much shorter legs than in family *a* and the fulcrum E is depicted at the centre of the lever.

Before one can make a reconstruction of the diagram in the hyparchetype γ, one must first reconstruct the figures in the missing manuscripts δ, ε and ζ. We notice that the figures in the manuscripts deriving from ε and ζ are very similar, apart from some new errors introduced in each copy, such as for example the wrong label M instead of K in codex F^2. A reconstruction of the diagram in manuscripts ε and ζ, and hence of codex δ, results in a diagram with the following characteristics: a figure of a lever with long legs, from which the weight Γ is missing, and in which a second label A has been inserted at the centre of the lever. In the next step, we compare it with the diagram contained in W^a, in order to reconstruct the diagram from the hyparchetype γ. The weight Γ is absent in both, and therefore it was likely to be missing from the hyparchetype as well. The second *alpha* is also contained in W^a, although here it is not placed at the centre of the lever, but floating next to the figure. It is possible that in the hyparchetype this letter label referred to a remark or scholium and was mistakenly interpreted by the copyists of W^a and δ as belonging to the figure. We cannot be certain about its exact position in the hyparchetype, but it was probably not part of the figure and was contained at a position next to the diagram, where it is also placed in W^a. This is supported by the fact that it would be highly unlikely for the same letter label to occur twice in a single diagram.

Having compared the diagrams in the three hyparchetypes, we can now make a reconstruction of the figure from the archetype. Two general principles are important to the recovery of the authentic figure. The first principle is known from textual criticism as *lectio difficilior potior*. When applied to diagram traditions, the principle suggests that in those cases where different manuscripts disagree on a particular diagram, the variant diagram that is more difficult, is more likely to represent the authentic figure. The second principle is related to the evidence from the manuscripts. Since we possess diagrams from three different manuscript families, in general, when two families agree against the third, those two families are more likely to contain the authentic diagram. If we apply the latter principle to the reconstructed hyparchetypes of families *a*, *b*, and *c*, we notice that the diagram in family *c* contains two errors which are not present in the other manuscript families. These errors, the missing label Γ and the second *alpha* inserted in the figure, do not occur in families *a* and *b*. Therefore, families *a* and *b* probably contain authentic variants of the archetype. In regard to the form of the lever, the figure in the archetype most likely had long legs and the fulcrum was probably depicted at the centre of the lever, since this is where it was placed in hyparchetypes β and γ. This would be in perfect agreement with a characteristic feature of ancient diagrams that we have seen before. Ancient diagrams do not provide a metrically correct picture, and therefore it would not be surprising if most copyists were ignorant of the fact that the force should be further away from the centre of the lever than the weight.

As another example, I will describe my reconstruction of the archetypal diagram to problem 23. This diagram is contained in twelve manuscripts: codex L in family *a*; codices P, M^v, D^1 in family *b*; and, finally, in the following manuscripts from family *c*: W^a, V^1, N^h, F^1, U^m, F^2, N^b and q. Codex L is important for the reconstruction of the hyparchetype α. Again, the page in L has been trimmed which results in the labels Γ, Z and I missing from the endpoints of the rhombus. We may assume that the labels were present in the source of L, although we see that a letter Γ is wrongly placed at the centre of the figure. Another interesting feature, which is not defined by the text, can be found in the circle drawn at the centre of the rhombus. When we compare the diagram in L with the diagram contained in the paraphrase by Pachymeres preserved for this problem, we notice that there is no such circle in Pachymeres; rather, the label *theta* drawn at the centre of the figure is somewhat larger than the other labels. From this, it may follow that there was no such circle in the hyparchetype, but rather an enlarged label at the centre of the rhombus.

For the reconstruction of the hyparchetype β we possess diagrams from all three of its direct copies. Codex P contains three different diagrams of rhombi instead of the one that is prescribed by the author. Two figures were drawn together on the same folium, the first of which is completely without any letter labels, and appears to be a failed attempt by the scribe to draw the shape of a rhombus. The third diagram was inserted at a different page, which may be related to the fact that there is a further reference to the same diagram later in the text of problem 23. The labelling of the diagrams in P is erroneous, for instance the letter Θ is missing from all of them. In comparison to codex P, both M^v and D^1 contain only one rather precise diagram of a rhombus in which all labels are present. In the analysis of the manuscript text, we have seen that the exemplar β was a highly corrupted codex. Thus, it seems likely that the diagrams in the hyparchetype β were also defective, and the copyists of M^v and D^1 might have been themselves responsible for making the corrections. However, in that case we cannot be certain about the diagram in the hyparchetype, since we do not know which corrections were made by the scribes of M^v and D^1 or which errors in P were already present in its source. It may be questioned whether the hyparchetype contained the same diagram twice; also the presence of the label Θ remains uncertain.

All diagrams in family *c* are to a high degree similar, which makes a reconstruction of the non-extant codices δ, ε and ζ, and, subsequently, of the hyparchetype relatively straightforward. Three codices contain two identical diagrams of a rhombus, N^h, F^1 and q, which may again be related to the fact that there are two references in the text to the same diagram. The majority of the manuscripts, however, contains only a single diagram to this problem, which likely represents the situation in the hyparchetype. We further notice some small errors, for example, the letter label E instead of I in F^1, as well as different shapes of a rhombus. The figures in V^1 and N^b, for instance, are somewhat diverse, as a result of the freehand drawing of the scribes. Characteristic of family *c* is further the absence of the letter *theta* from the centre of the rhombus in all diagrams.

On the basis of this analysis, we are now able to reconstruct the archetypal diagram of a rhombus, in which all endpoints and the centre of the figure are labelled. Interestingly, one of the endpoints is labelled with an *iota*, a letter that is not mentioned in the text. The fact that all diagrams, including those in the paraphrase by Pachymeres, contain this label, demonstrates that it was already present in the archetype from which all diagrams were faithfully copied. The manuscript diagrams further neglect the general statement in the text of what happens with the two components of motion, when one of the angles is more acute and the other more obtuse. The angles in the diagrams appear almost equally and it is impossible to distinguish between acute and obtuse angles. This feature of metrical inaccuracy is inherent in ancient diagrams and shows that we should not draw any conclusions about metrical properties from the diagrams.

Problem 3

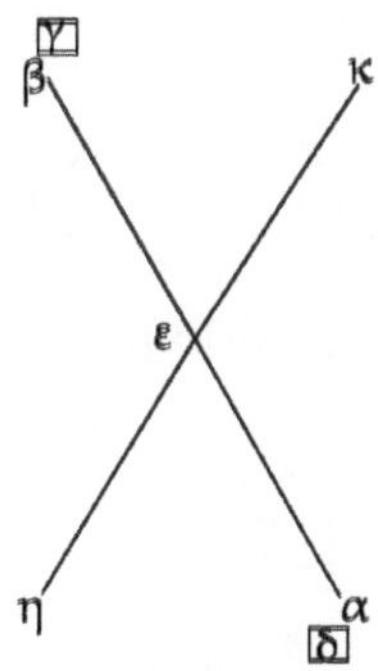

codex ω

Family *a*

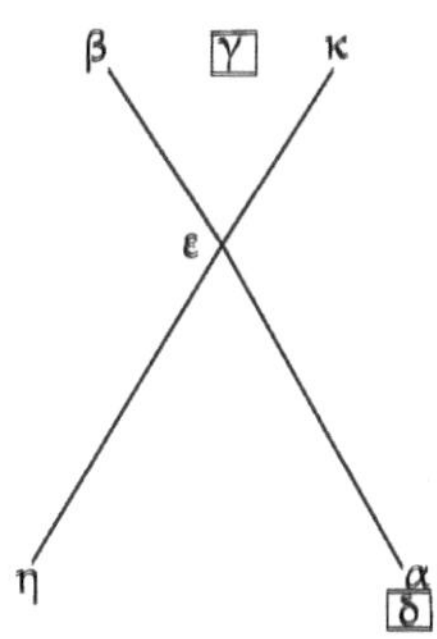

codex α

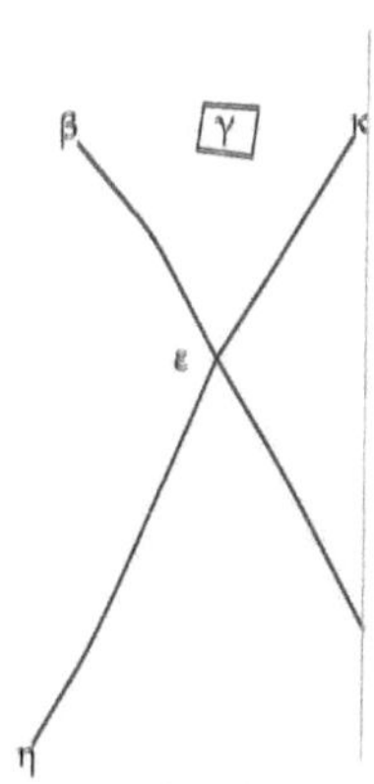

codex L

Family *b*

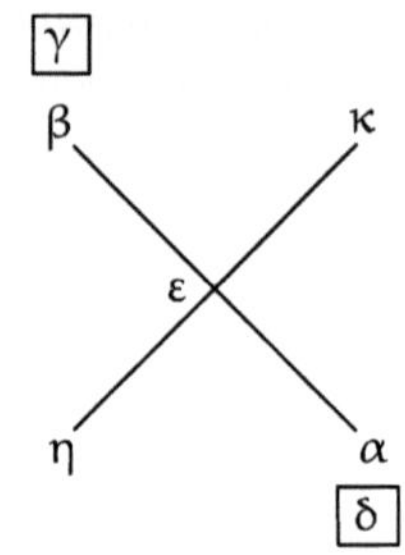

codex β

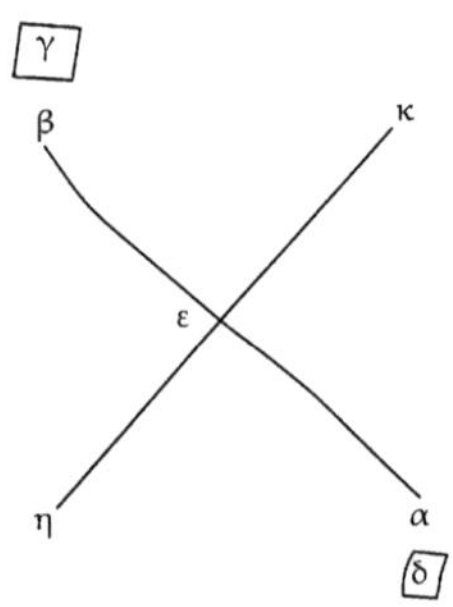

codex P

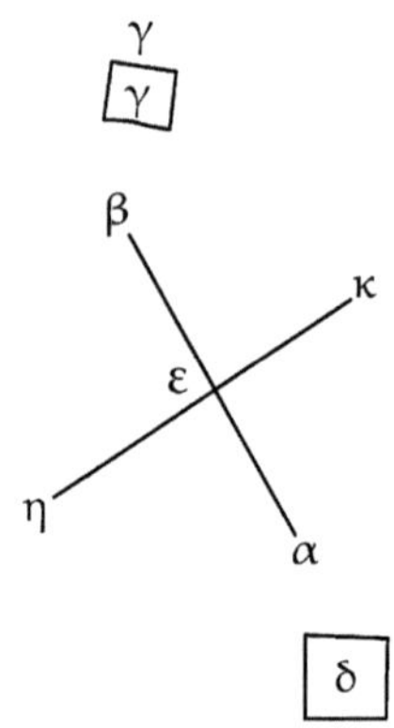

codex M^v

Family *c*

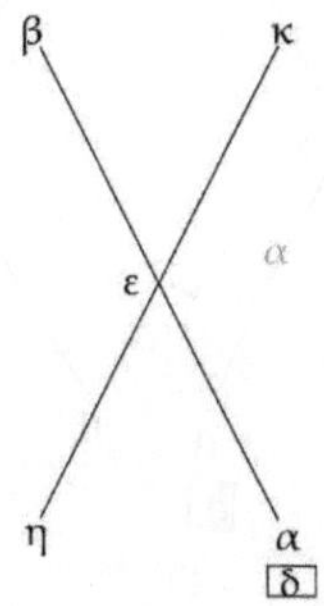

codex γ

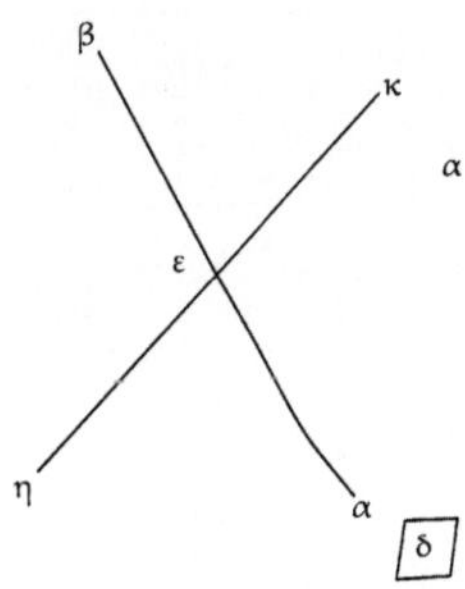

codex Wᵃ

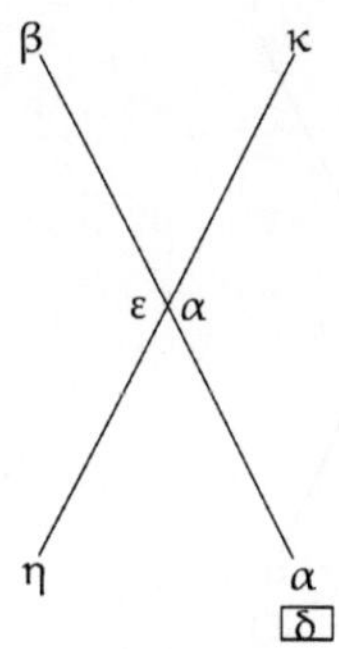

codex δ

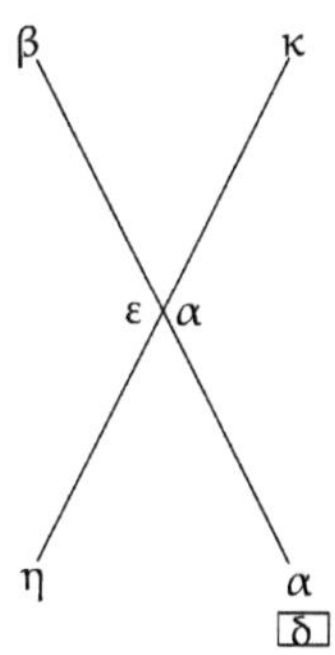

codex ε

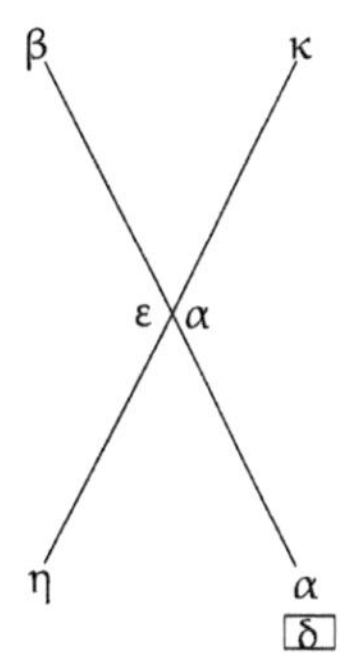

codex ζ

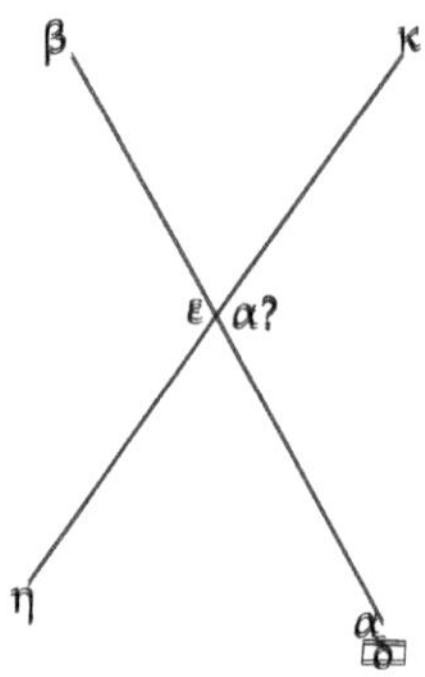

codex N^{h}

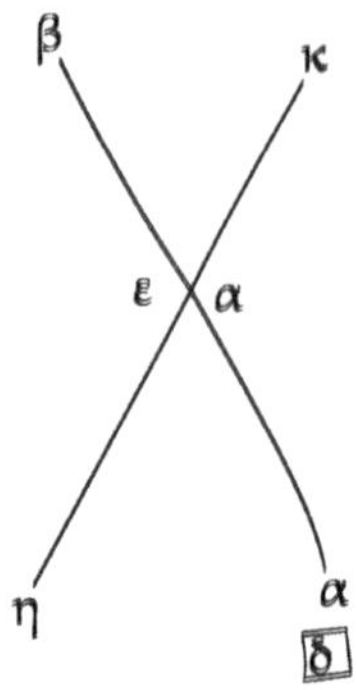

codex U^{m}

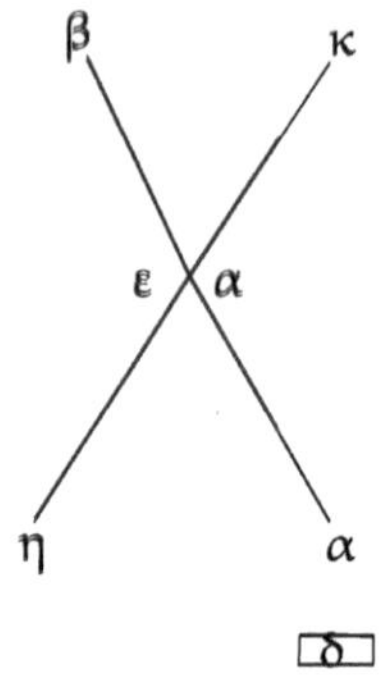

codex N^{b}

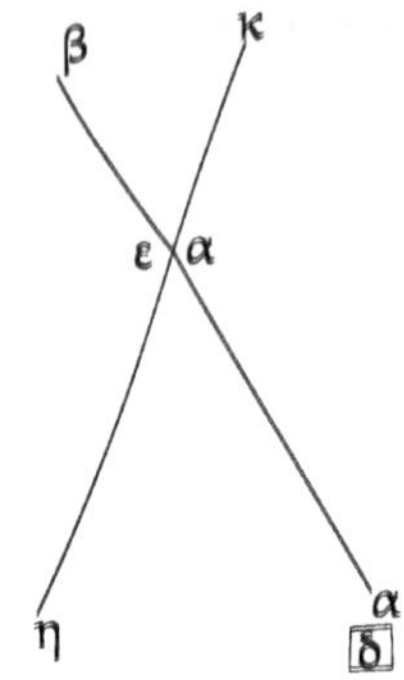

codex q

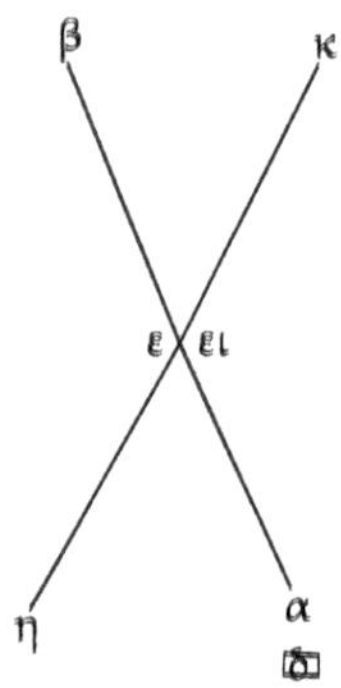

codex F^{1}

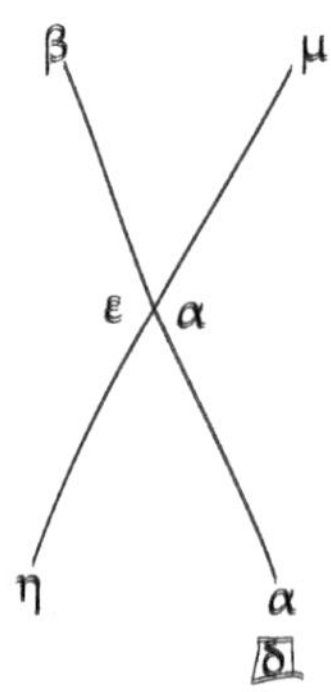

codex F^{2}

Problem 23

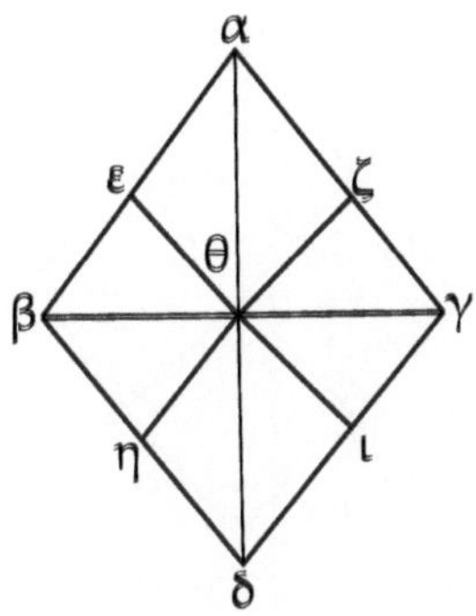

codex ω

Family *a*

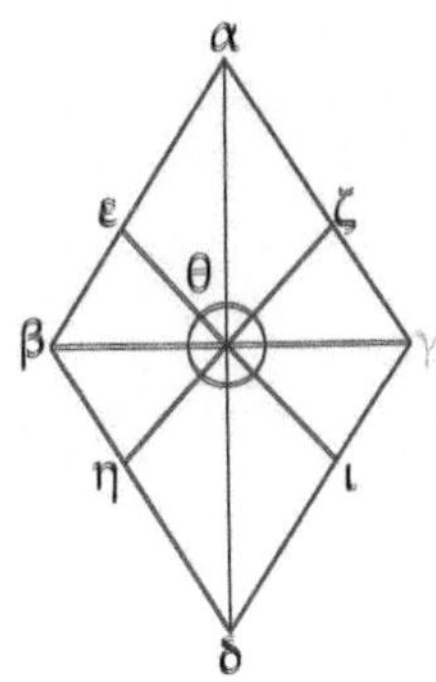

codex α

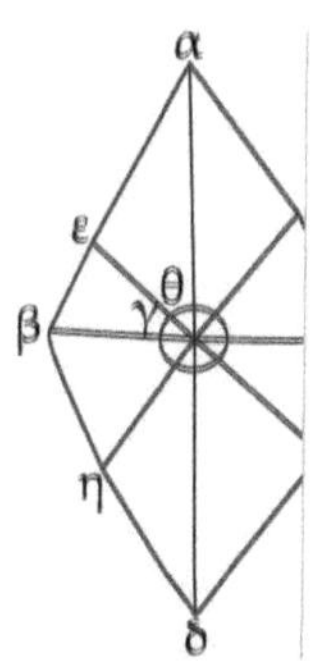

codex L

Family *b*

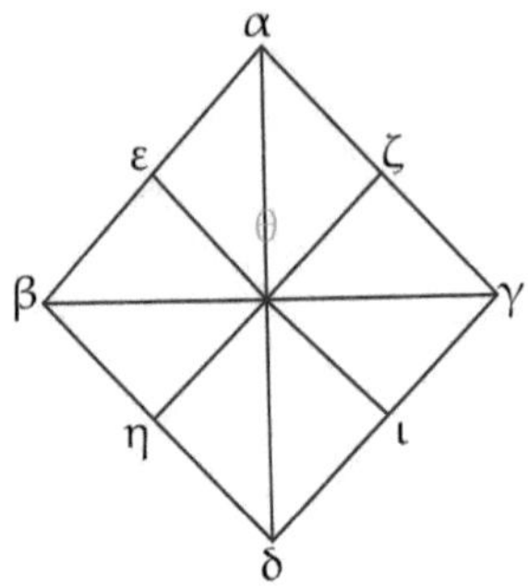

codex β

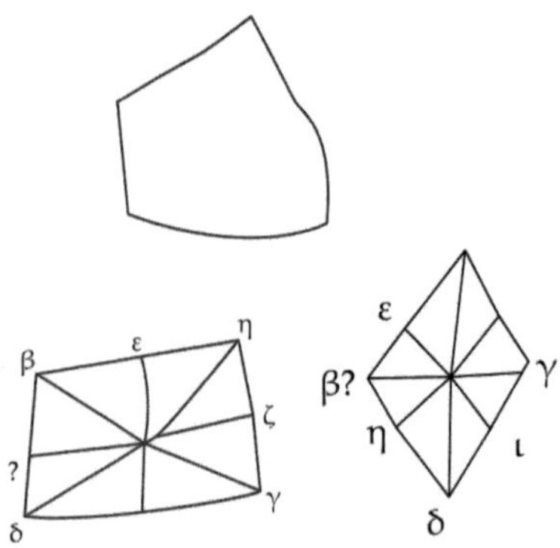

codex P

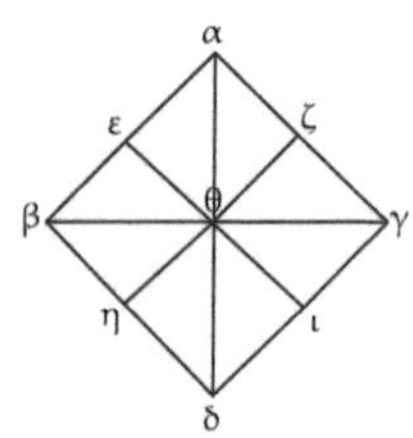

codex M[v]

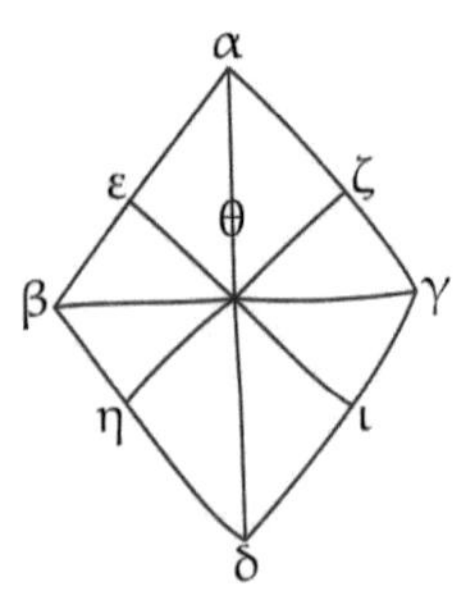

codex D[1]

Family *c*

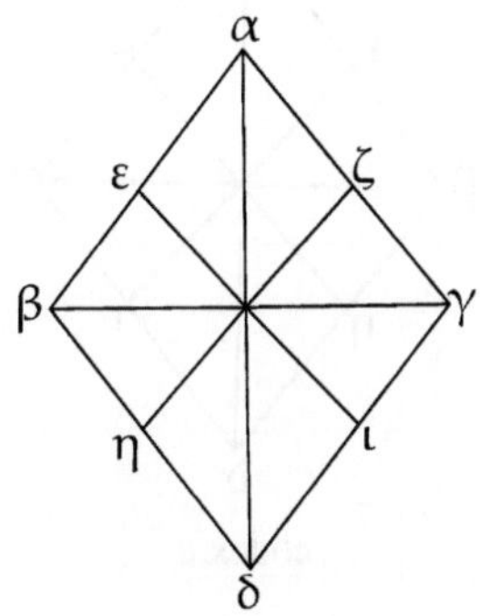

codex γ

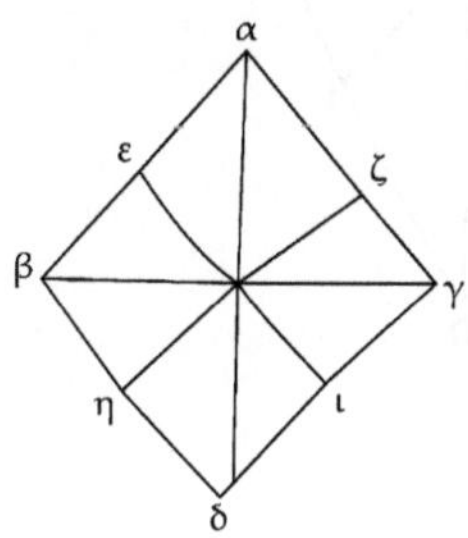

codex W[a]

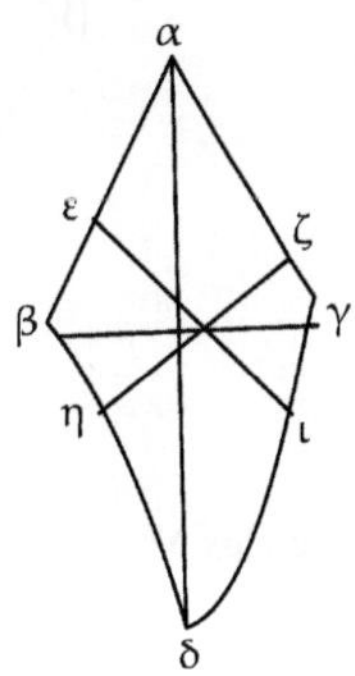

codex V[1]

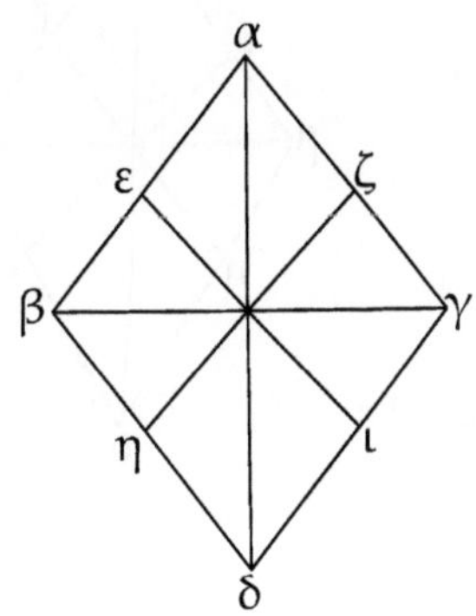

codex δ

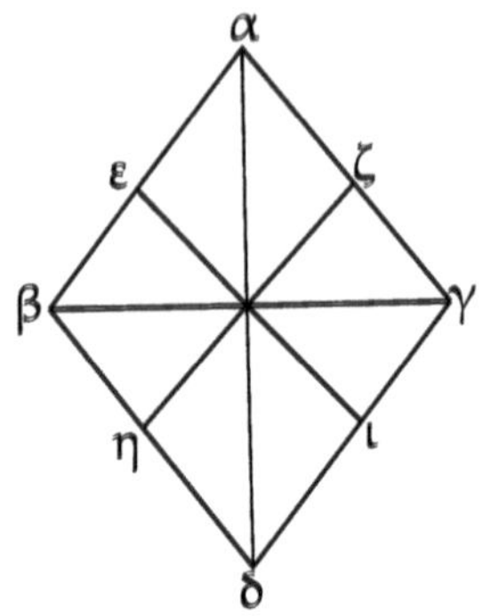

codex ε

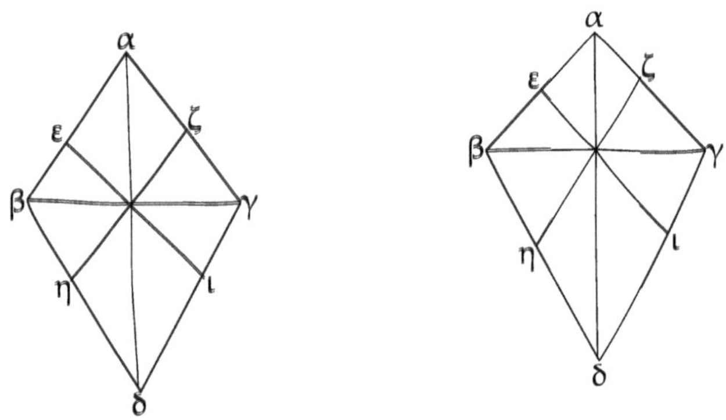

codex N[h]

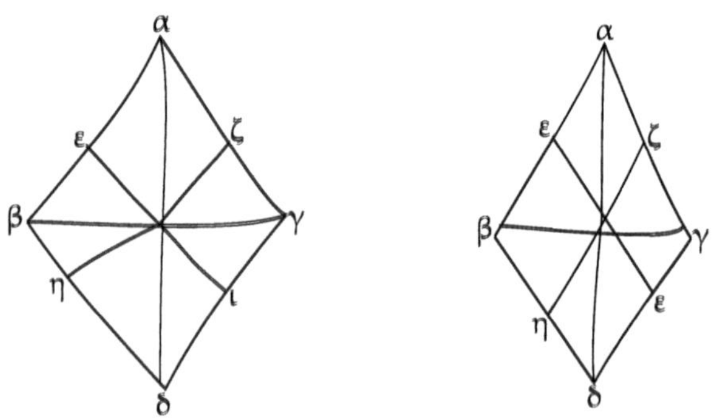

codex F[1]

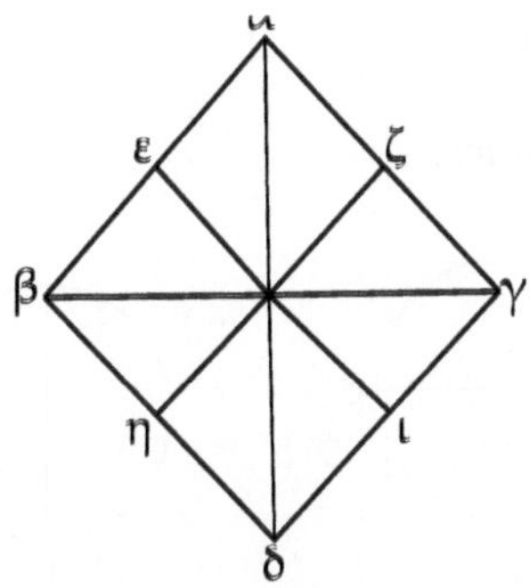

codex ζ

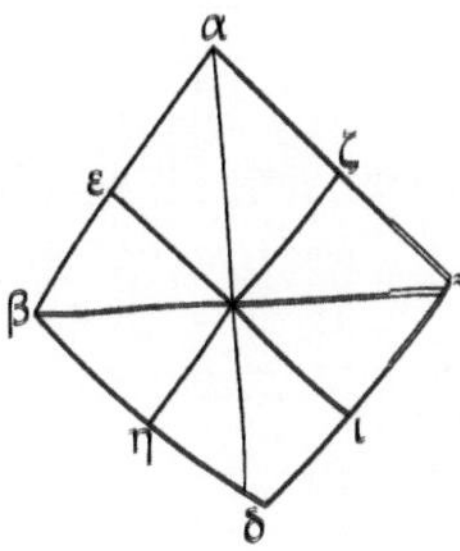

codex U^{m}

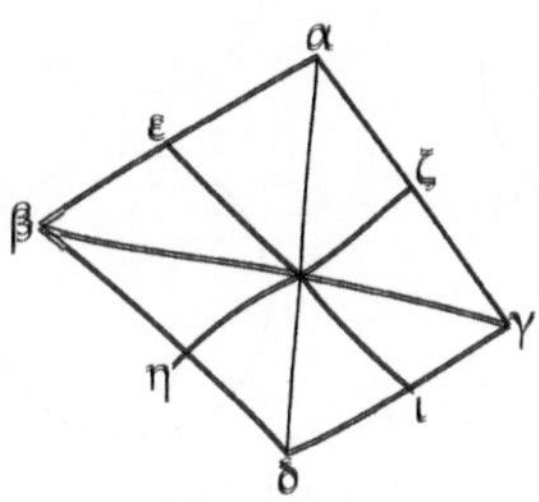

codex N^{b}

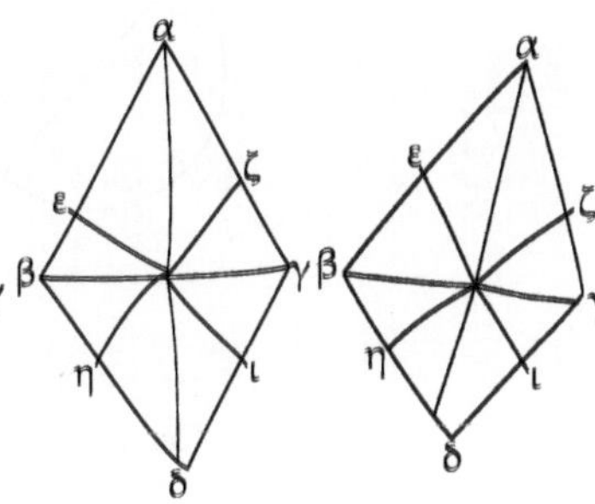

codex q

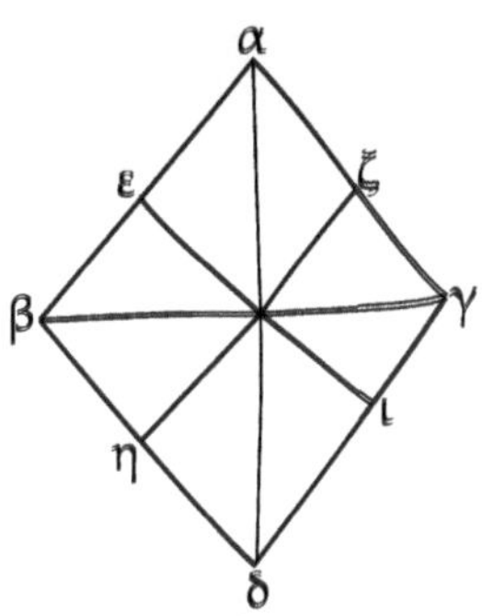

codex F^{2}

4.3.3 *Critical Edition*

Introduction to the *Mechanics*

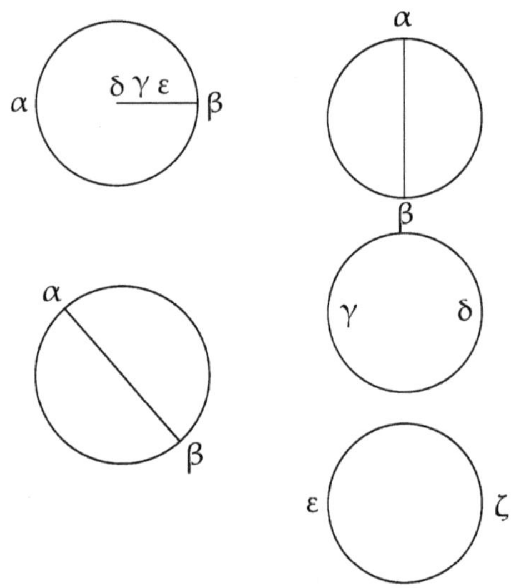

Reconstructions of the archetype

Problem 1

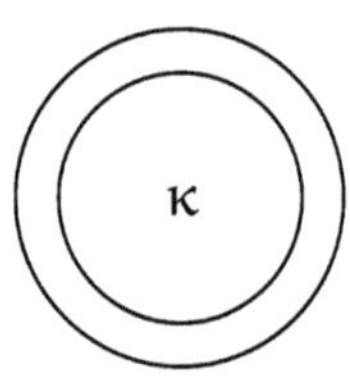

Reconstruction of the archetype

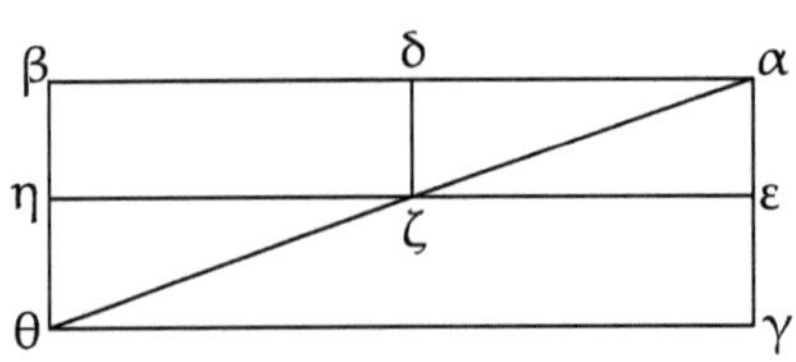

Reconstruction of the archetype

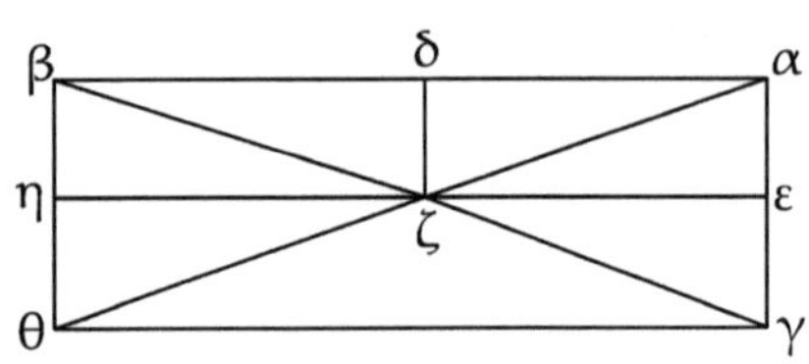

Reconstruction of hyparchetype γ

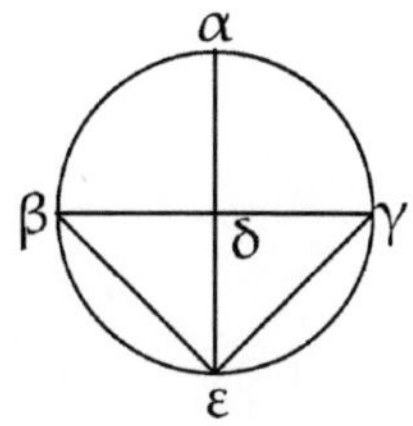

Reconstruction of the archetype

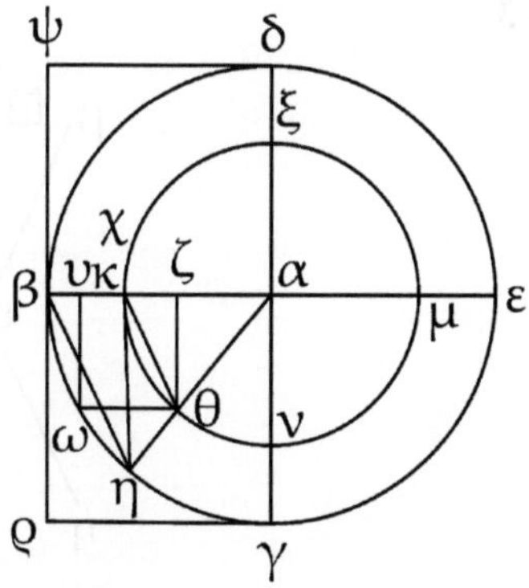

Reconstruction of the archetype

Problem 2

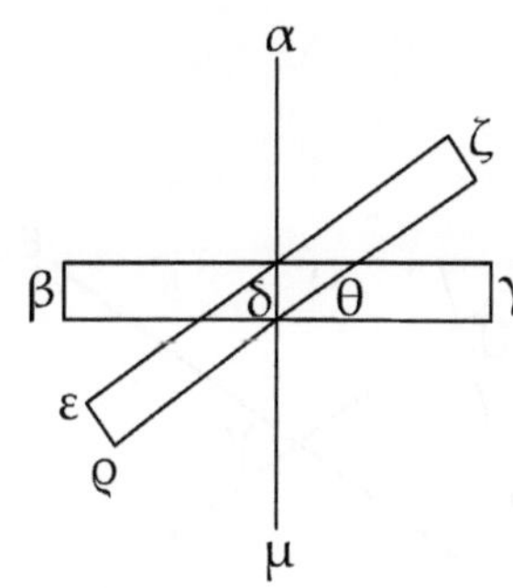

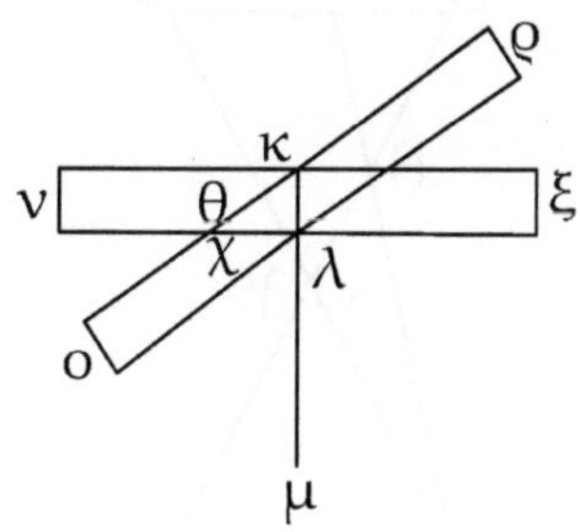

Reconstructions of the archetype

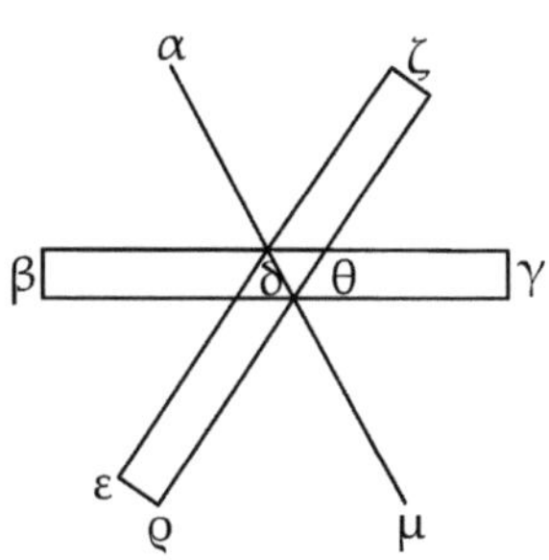

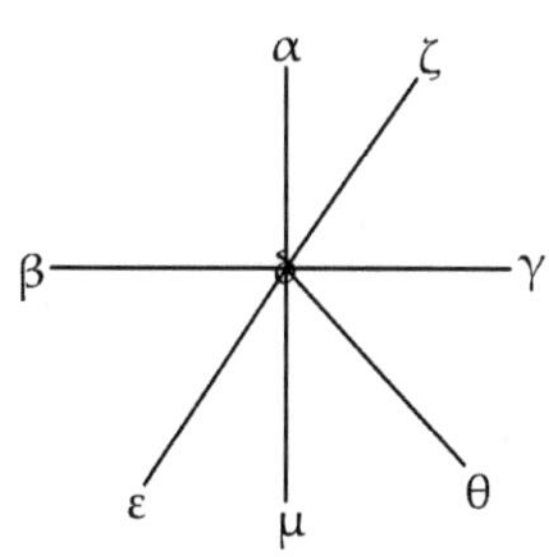

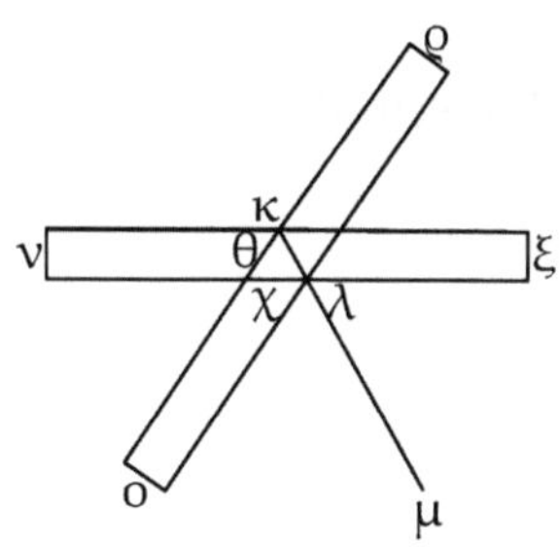

Reconstructions of hyparchetype γ

Problem 3

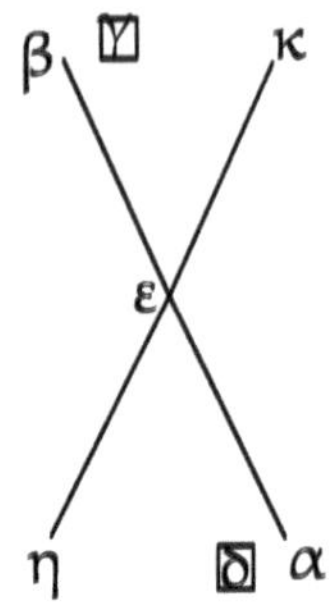

Reconstruction of the archetype

Problem 5

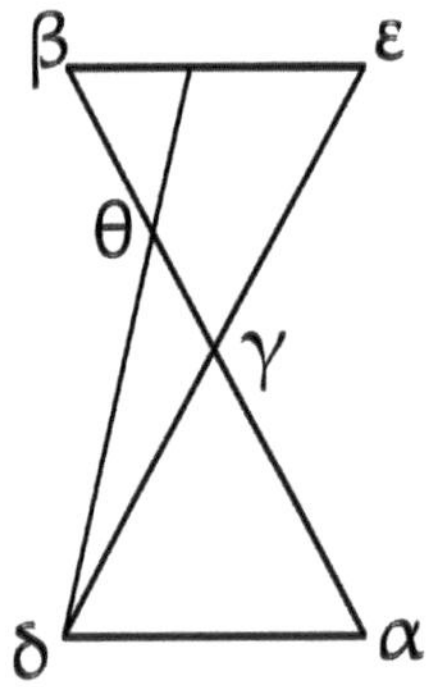

Reconstruction of the archetype

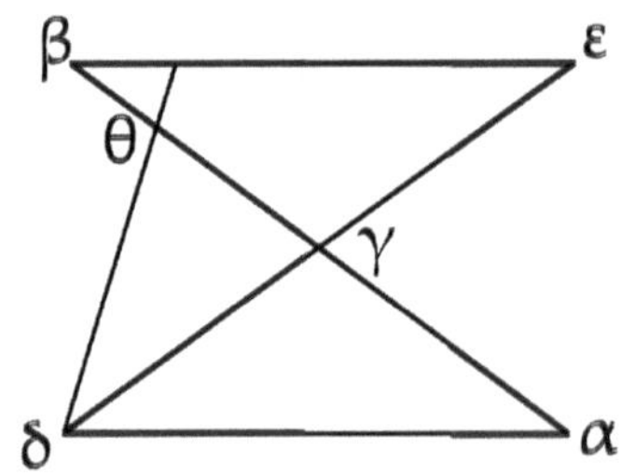

Reconstruction of hyparchetype γ

Problem 17

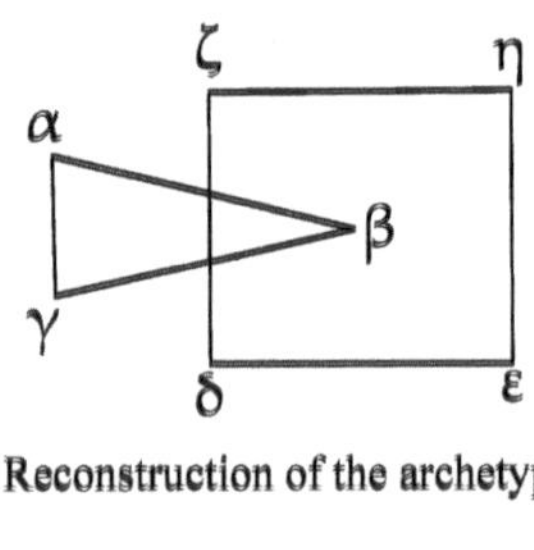

Reconstruction of the archetype

Problem 21

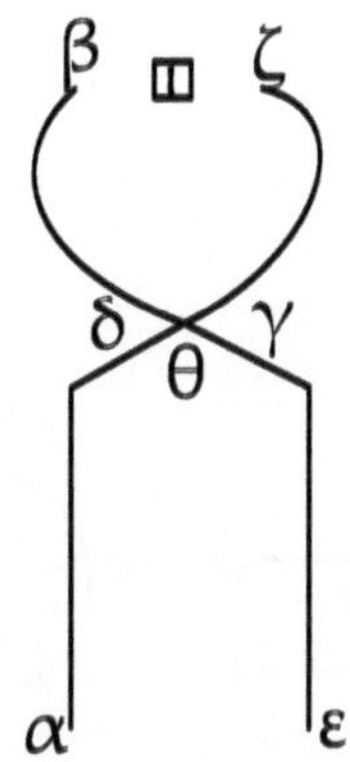

Reconstruction of the archetype

Problem 22

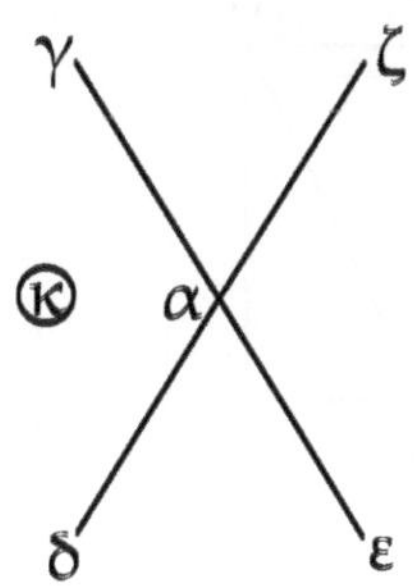

Reconstruction of the archetype

Problem 23

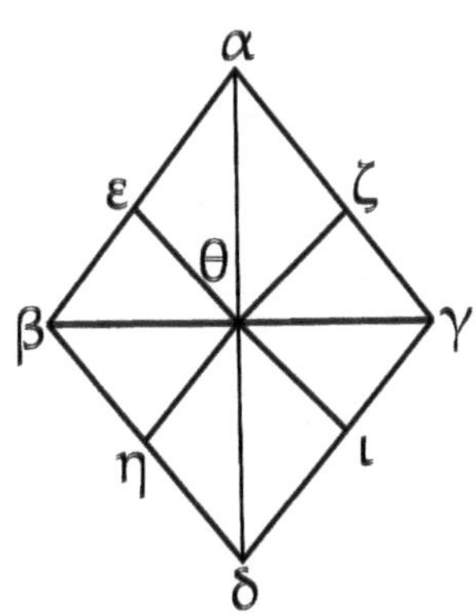

Reconstruction of the archetype

Problem 24

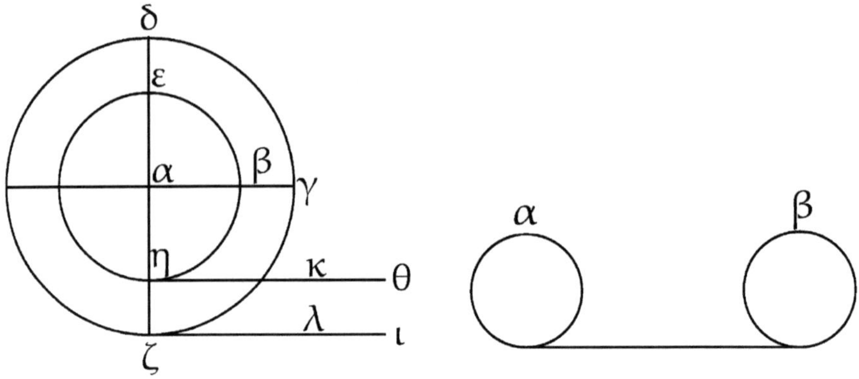

Reconstructions of the archetype

Problem 25

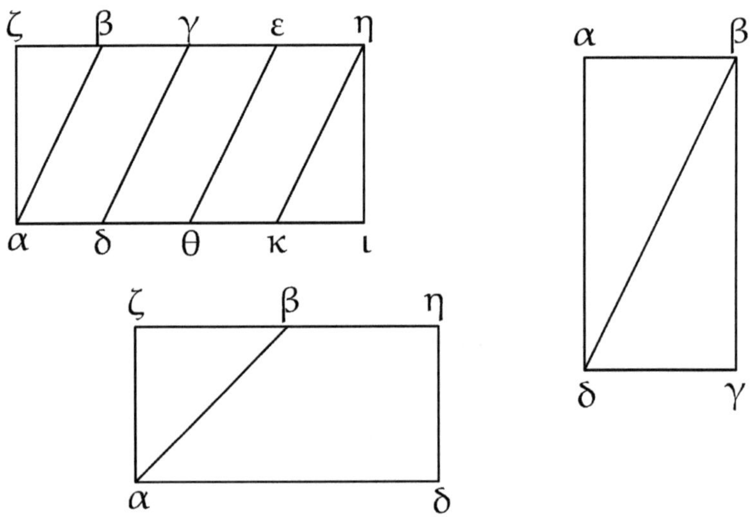

Reconstructions of the archetype

Problems 26–27

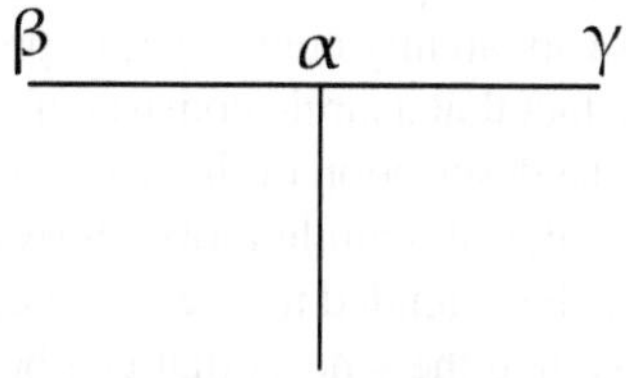

Reconstruction of the archetype

Problem 30

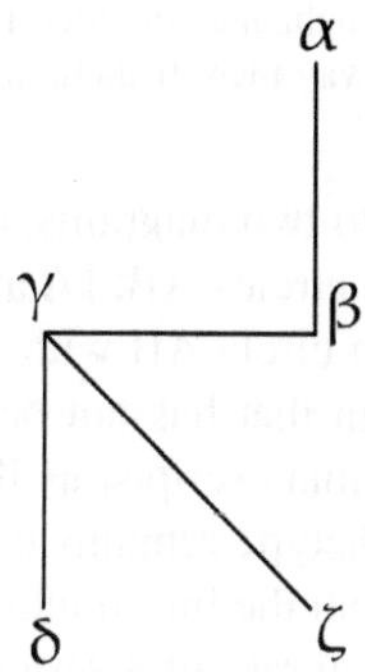

Reconstruction of the archetype

4.4 Notes on the Critical Edition

This section explains my most important decisions in reconstructing the diagrams from the Byzantine archetype of the *Mechanics*. Reproductions of all manuscript diagrams as well as reconstructions of the diagrams from the lost codices are presented in the appendix, in order to enable the reader to acquire an understanding of each individual step in the making of the critical edition.

Introduction to the *Mechanics*

In the introduction to the *Mechanics*, the circle is described as the underlying principle of all mechanical phenomena. The circle unites many properties that seemingly contradict one another. It is, for instance, made up of both a moving and a stationary component. What is more, it moves simultaneously in opposite directions; for it moves at the same time forwards and backwards. But the most remarkable property of the circle is the fact that a point on its radius that is further from the centre moves faster than a point that is closer by.

People make use of these apparently contradictory properties in the construction of mechanical devices. The fact that a circle consists of two opposite motions at the same time is illustrated by the description of the first diagram in lines 848a19-25: if one extremity A of the diameter of a circle moves forward, the other end B moves backward. This property can be extended to a series of circles that move simultaneously in opposite directions, like the wheels that can be found at the entrance to a temple.[47] The next lines of the text (848a26-34) describe the diagram corresponding to this mechanical device:

> For if there is one circle ΓΔ touching the circle AB, and the diameter of the circle AB moves forward, ΓΔ will move backward compared to the circle on which is A, if the diameter moves around the same point. Then, the circle ΓΔ will move in the opposite direction to AB; and again, the circle will itself move the next in order, EZ, in the opposite direction to itself for the same reason. In the same way then, if there are more circles, they will make this happen, when only one is moved.[48]

Whereas the text only refers to two diagrams, one of a circle with diameter AB, and the other of a series of three circles AB, ΓΔ and EZ, families *b* and *c* contain a third diagram. This diagram of a circle AB with a radius ΔΓE might have entered the text together with a scholium that has not been preserved. From the different locations of this diagram in the manuscripts, as illustrated by Table 4.2, it follows that its exact position in the archetype remains uncertain, but we know that it preceded the other two diagrams from the introduction.

Family *a* contains only the diagram of a series of circles. Whereas in the other manuscript families the circles are depicted vertically above each other, in both codex L and Pachymeres the circles are drawn next to each other. This horizontal orientation may go back to the hyparchetype α, in which the diagram was likely already included at the bottom of the folium. In contrast to the diagram in L and in the other manuscripts, the circles in Pachymeres' paraphrase are actually touching one another, as prescribed by the author.

The diagrams in family *c* are all very similar. We see some irregularities in the roundness of the circles in M^1, since these diagrams were drawn by an unskilled hand. Missing labels in N^h and its copy F^1, and in U^m and its copy F^2, show the dependences between these manuscripts. In some codices, V^1 and F^2 for example, the circles in the diagram of three consecutive circles have been divided over different folia. The scribe of F^2 was probably not aware that all three circles belonged to the same diagram, since there was sufficient space in the margin of his text to fit all circles above each other on one page. Codex N^b is exceptionally different from the other manuscripts in family *c*; it contains only the diagram of the series of circles.

[47] See Berryman (2009), 50–51 on this device, to which Hero refers as a *hagnistērion*.

[48] εἰ γὰρ εἴη τοῦ AB κύκλου ἁπτόμενος ἕτερος κύκλος ἐφ' οὗ ΓΔ, τοῦ κύκλου τοῦ ἐφ'οὗ AB κινουμένης τῆς διαμέτρου εἰς τοὔμπροσθεν, κινηθήσεται ἡ ΓΔ εἰς τοὔπισθεν τοῦ κύκλου τοῦ ἐφ'οὗ A, κινουμένης τῆς διαμέτρου περὶ τὸ αὐτό. εἰς τοὐναντίον ἄρα κινηθήσεται ὁ ἐφ' οὗ ΓΔ κύκλος τῷ ἐφ' οὗ τὸ AB· καὶ πάλιν αὐτὸς τὸν ἐφεξῆς, ἐφ' οὗ EZ, εἰς τοὐναντίον αὑτῷ κινήσει διὰ τὴν αὐτὴν αἰτίαν. τὸν αὐτὸν δὲ τρόπον κἂν πλείους ὦσι, τοῦτο ποιήσουσιν ἑνὸς μόνου κινηθέντος.

Here, the diagram is orientated horizontally, and the circles are actually attached to one another.

Problem 1

The first problem concerning the question of why larger balances are more accurate than smaller ones will be discussed at length in the next chapter. Here, I will confine myself to the reconstruction of the diagrams. Families *b* and *c* both contain a diagram that is not required by the text. This figure illustrates the following assertion in line 848b9: "For the outer circle is greater than the inner."[49] The figure shows two concentric circles of different sizes with the letter *kappa* depicted at their common centre. The *kappa* stands here as a symbol for κέντρον, the centre of a circle. This diagram was probably inserted in Byzantine times for didactical reasons. The scribe of the archetype might have added it together with a gloss or comment on this passage that has not been preserved.

The second diagram for this problem depicts a parallelogram of motions. The diagram illustrates the case of rectilinear motion, where the two components of motion move in a fixed ratio, and the resulting motion accordingly occurs in a straight line. The complete description of the diagram is given in lines 848b13-23 of the text, but it takes until line 848b20 until we discover that the described figure is in fact a τετράπλευρον or quadrilateral. Instead of depicting the more regular figure of a parallelogram, all manuscripts contain a rectangle, thus representing a case of overspecification. Furthermore, the manuscript diagrams all contain the label H which is not mentioned in the text. This label was probably inserted by the scribe of the archetype in order to avoid any gaps in the alphabetical order of the lettering, and it was subsequently copied by all later scribes.

The diagram in L is turned 90° counterclockwise, presumably because the scribe did not have sufficient space in the margin to draw the parallelogram horizontally. The same folium in codex P contains two similar diagrams of a parallelogram, differing only in size. These diagrams largely coincide with the figure in the other manuscript in family *b*, M^v, with the difference that the scribe of M^v inadvertently drew line ΔZ through to the bottom of the parallelogram. The diagrams in family *c* all contain a second diagonal BΓ that is neither mentioned in the text, nor has any special function in the diagram. Since it is not contained in the other manuscript families, it was likely an error by the scribe of codex γ that was copied into all dependent manuscripts. Furthermore, we see that all figures dependent on codex δ depict line ΔZ by a double line; other errors that confirm relations between manuscripts include, for example, the second letter B at the bottom of the parallelogram in both V^1 and its copy M^1.

The next diagram for this problem illustrates the case of circular motion, in which the two components of motion move in no fixed ratio. This diagram is present in families *b* and *c* and can be reconstructed without complications. The figure in P

[49] ὁ γὰρ ἐκτὸς μείζων τοῦ ἐντός.

erroneously contains a second B at the bottom of the circle, but from the figure in M^v we see that the correct label should be an E. The diagrams in family *c* are similar to those in family *b*, except for the letter Δ at the centre of the circle that is missing from all manuscripts in this family. The bottom of the page in U^m is trimmed, therefore the label E is absent, as it is in its copy F^2. Lines BE and EΓ, finally, are depicted as curves rather than straight lines in V^1 and its copy M^1.

The final diagram for the first problem is the most complicated one in the *Mechanics*. It is described by a long argument in lines 849a21-b19, comparing the motions of longer and shorter radii in concentric circles. The diagrams in the manuscripts are all highly flawed, therefore I will only sum up some frequent errors: letters are incorrect or missing; labels are unreadable; lines are missing from the diagram or are incorrect due to inaccurate freehand drawing; the rectangle ΔΨPΓ is not tangential to the outer circle; the size of the smaller circle in relation to the larger circle varies greatly. The diagram contained in M^v highlights many of the difficulties the scribes faced in copying. This diagram was probably corrupted in the exemplar, codex β, and the copyist of M^v, rather than attempting to complete the different lines and labels, simply left the diagram unfinished. W^a is the only codex that contains a correct and neatly drawn diagram. As I have previously argued, the person who later added the diagrams in W^a was probably responsible for correcting the diagrams he found in his exemplar. All other diagrams deriving from codex γ contain similar corruptions, for example, the missing line HK.

The fact that all manuscripts contain numerous errors in this diagram complicates the reconstruction of the archetypal diagram. It is impossible to determine which errors were already contained in the exemplar, and which were new ones introduced by the copying scribe. For that reason, the diagram in my critical edition includes all letter labels mentioned in the text and thus presents a perfect reconstruction of the author's intentions.

Problem 2

Problem 2 of the *Mechanics* asks why it is that a balance that is suspended from above returns to its original position after the weight is removed, while a balance that is supported from below does not.[50] The author begins with an explanation of the first case. In a balance that is suspended from above, the longer part of the beam is above the perpendicular; for this reason, the balance returns to its original position after the weight is removed. Lines 850a11-19 of the text describe the diagram for this first case. The beam of the balance is BΓ, and the cord AΔ. When the cord is extended in a downward direction the perpendicular will be represented by AΔM. The weight is placed on the balance at B, such that B is displaced to E, and Γ moves towards Z. Thus, the line bisecting the beam was first in the position of the perpendicular ΔM, but has now taken the position of ΔΘ. The part of beam EZ that is

[50] This question will be treated in more detail in Chap. 6 in the context of the early modern transformations of the *Mechanics*.

outside the perpendicular will exceed half of the beam by ΘΠ.[51] If the weight is taken away, the balance will then return to its original position, since the arm of balance E is shorter than Z.

But in the second case, in which the balance is supported from below, the contrary takes place, as is explained by the description of the second diagram in lines 850a24-29. Here the beam of the balance is NΞ, and the perpendicular bisecting the beam KΛM. When the weight is placed at N, N is displaced to O, and Ξ to P. The perpendicular KΛ moves towards ΘΛ, such that the part of the beam KO is greater than ΛP by ΘKΛ. When the weight is removed, the balance does not return to its original position, because the longer half of the beam acts as a weight and holds down the balance.

While the text refers to two figures, one for the balance suspended from above, and another for the balance supported from below, part of the manuscripts in family *c* contain an additional diagram illustrating the first case. This second figure depicts the balance with single rather than double lines, and, unlike the first figure, the perpendicular of the moved balance is now represented (line ΔΘ). This diagram was probably a later addition that might have entered the hyparchetype γ in combination with a scholium that has not been preserved.[52] This doubling of the first case was not contained in the archetype, because this diagram is not required by the text and, most importantly of all, neither of the remaining two families contain it.

Family *c* is the only family where lines AΔM and KΛM in the figures to the different cases are not represented as being perpendicular to the straight balance. The text describes these lines as the extensions of the chords of the balances; hence, lines AΔM and KΛM should be perpendicular to the straight balance. This error was probably contained in the hyparchetype γ. W^a is the only codex in family *c* that depicts the lines as perpendiculars to the straight balance. The second hand that added the diagrams in this manuscript probably corrected the figures he found in his exemplar. One should therefore be careful in consulting the figures from W^a in reconstructing the hyparchetype. Furthermore, the perpendiculars in W^a do not appear to represent the authentic diagrams in codex γ.

The figures in family *c* show that copyists tended to remain faithful to their exemplar. The pages of the manuscript U^m have been trimmed, and consequently parts of a figure, along with certain labels, are missing. In the second figure for the balance suspended from above, the labels E and M at the bottom are missing. The copyist of F^2 did not add these letters to his figure, something he could easily have done with the help of the text, but rather stayed close to his source and left out E and M as well. Something similar applies to a group of manuscripts in this family, N^h, F^1, U^m and q. The letter Z at the top of the first diagram is missing from these manuscripts, and none of the copyists included this label. These kinds of diagrammatical errors are very helpful to editors and provide them with a good tool for verifying the

[51] The reading ΘΠ is an emendation by van Cappelle; all manuscripts read ΦΠ, but none of these letters is contained in the manuscript diagrams.

[52] Family *c* is the only manuscript family of the *Mechanics* that contains scholia. The star symbol in the figures to the second case in U^m, F^2 and q might also refer to a lost scholium.

stemma codicum of a text. New errors that were made in almost every copy of a manuscript are equally helpful in ascertaining dependences between manuscripts, as a comparison between the figures in N^h and F^1 shows. In the second diagram to the first case, we see, for example, that the scribe of F^1 forgot to copy line ΔΘ from his source N^h, and in the diagram to the second case he did not extend line ΚΛΜ to point K.

The manuscripts in family *b* present another difficulty in the reconstruction of the ancestor. The scribe of codex P depicted the beams of the balance with double lines, whereas in M^v they are represented by single lines. Which of these two cases is more likely to coincide with the hyparchetype? When we apply the principle of *lectio difficilior potior* here, the diagram in P probably represents the original diagram. In modifying the diagram, the most likely explanation is that the scribe of M^v simplified the figure by drawing single instead of double lines. The assumption that the diagram in P represents the figure from the hyparchetype is also confirmed by the other manuscript families, where double lines are the standard for representing balance beams.

Problem 3

Why is it that small forces move great weights with the help of a lever, although the weight of the lever is added? In answering this question, the author first compares a lever with a balance beam and refers to the principle that given the same weight the longer radius moves quicker than the shorter one. Next he formulates a pre-Archimedean law of the lever (850a39-850b2): "Therefore, as the weight moved is to the weight moving it, so, inversely, is the length of the one arm to the length of the other."[53] The further the force is from the centre, the more easily the weight will move. The reason for this was given in the introduction to the treatise, namely that the point on the radius that is further from the centre of the circle moves faster than the point that is closer. The working of a lever is explained by the following description of the diagram (850b6-9): "Let AB be the lever, Γ the weight, Δ the mover, and E the fulcrum; and Δ will be moved to H, and Γ, the weight moved, to K."[54]

The diagram as it is contained in all of the manuscripts can only be understood properly if it is rotated 90° counterclockwise, since otherwise, when the weight Γ moves towards K, it passes beyond the turning point, and from there it would no longer require a force to move any further. Most diagrams do not consider the fact that the force is further from the centre than the weight and place point E at the centre of the bar. Only the figures in family *a* (codex L and Pachymeres) appear metrically correct, depicting the force somewhat further from the fulcrum than the weight.

[53] ὃ οὖν τὸ κινούμενον βάρος πρὸς τὸ κινοῦν, τὸ μῆκος πρὸς τὸ μῆκος ἀντιπέπονθεν.

[54] ἔστω μοχλὸς ἐφ' οὗ AB, βάρος δὲ ἐφ' ᾧ τὸ Γ, τὸ δὲ κινοῦν ἐφ' ᾧ τὸ Δ, ὑπομόχλιον ἐφ' ᾧ τὸ E, τὸ δὲ ἐφ' ᾧ τὸ Δ κινῆσαν ἐφ' ᾧ τὸ H, κινούμενον δὲ τὸ ἐφ' οὗ Γ βάρος ἐφ' οὗ K.

The figure of the lever in family *a* has long legs; the right side of the page in L is trimmed, and for that reason the labels A and Δ are missing from the figure. Omitted from the text in all manuscripts in family *a* are lines 850b7-8: ὑπομόχλιον ἐφ' ᾧ τὸ E, τὸ δὲ ἐφ' ᾧ τὸ Δ. Due to this omission, it is impossible, based on the text, to reconstruct the correct diagram with label E at the fulcrum of the lever. However, since both diagrams in L and Pachymeres depict the label E in the diagram, such textual omissions serve as a proof that the diagrams were not inserted independently at a later point in the tradition, but can be traced back to the archetype.

The figures in family *b* do not contain any distortions; all labels are present in P and M^v, and both figures depict a lever with much shorter legs than those in other families. The diagrams of the lever in family *c* contain some characteristic errors that confirm the connection between these manuscripts. The label of the weight Γ is absent from all manuscripts. Furthermore, we notice a second *alpha* in every diagram that is not required by the text. Whereas this label in W^a is floating next to the figure, in all other manuscripts it is inserted at the fulcrum of the lever. All manuscripts that were copied from codex δ (U^m, F^2, N^b, q, N^h, F^1 and Par) are missing lines 850b7-8: τὸ δὲ κινοῦν ἐφ' ᾧ τὸ Δ, ὑπομόχλιον ἐφ' ᾧ τὸ E, τὸ δὲ ἐφ' ᾧ τὸ Δ κινῆσαν ἐφ' ᾧ τὸ H. Again, much of the information needed for the construction of this diagram is omitted from the text. Since all labels are present in the diagram, we can trace it back to an earlier common form. It further shows that the diagrams are generally faithful copies of their exemplars, rather than later additions or idiosyncratic readings by a single scribe.

Problem 5

This problem asks why it is that a small rudder has the power to move a whole ship.[55] In the following explanation, the author reduces the working of a rudder to that of the lever. The rudder itself can be compared to a lever: the point at which the rudder is attached to the ship is the fulcrum, the sea acts upon it as a weight and the steersman is the moving force. This principle is explained by the following diagram (lines 851a17-28):

> For let AB be the oar, Γ the pin, and A the end of the oar inside the ship, while B is the end in the sea. Then if A be moved to Δ, B will not be at E; for BE is equal to AΔ, and its displacement would be the same. But it is smaller, and it will be at Z. Θ then cuts AB not at Γ but below it. For BZ is less than AΔ, so that ΘZ is less than ΔΘ; for the triangles are similar. The centre Γ will also be displaced; for it moves in a contrary direction to B, the end of the oar in the sea, and it moves in the same direction as A, the end in the ship, and A will be at Δ. So the ship will also be moved there where the end of the oar is displaced.[56]

[55] The same question is considered by Hero in his *Mechanics* II.34, question *p*: "Why is it that a small rudder displaces a large ship?" See Nix and Schmidt (1900), 186–187.

[56] ἔστω γὰρ ἡ AB κώπη, τὸ δὲ Γ ὁ σκαλμός, τὸ δὲ A τὸ ἐν τῷ πλοίῳ, ἡ ἀρχὴ τῆς κώπης, τὸ δὲ B τὸ ἐν τῇ θαλάττῃ. εἰ δὴ τὸ A οὗ τὸ Δ μετακεκίνηται, τὸ B οὐκ ἔσται οὗ τὸ E· ἴση γὰρ ἡ BE τῇ AΔ.

This diagram can only be reconstructed based on families *b* and *c*, since it is missing from family *a*. The rudder in family *b* has a rather oblong shape when compared with family *c*, but otherwise the diagrams in both families are nearly identical. I believe that the diagram in family *b* should be favoured in the reconstruction of the archetypal diagram, since this family stems from an older exemplar. The label Z is absent from all figures; it is remarkable that none of the scribes added this letter to the diagram, which again shows how faithfully the diagrams were generally copied. Furthermore, some diagrams in family *c* depict line ΔZ incorrectly, for it does not extend to point Δ, but to some other point on AΔ.

Problem 17

Problem 17 explains how a wedge works by referring this device to two levers that work in opposite directions. Lines 853a27-31 of this short problem describe the corresponding figure: "Let there be a wedge ABΓ, and that which is to be cleaved ΔEHZ. AB then is the lever, and the weight is below at B, and ZΔ is the fulcrum. Opposite of this is the lever BΓ. When AΓ is struck it uses each of these as lever; for B pulls up."[57] A very similar diagram is contained in all three manuscript families. We notice that in codex L the letters H and E are missing from the right side of the diagram due to page trimming. The figure to this problem is absent from U^{m}, F^{2}, N^{b} and q, and thus the diagram must have been missing from codex ζ as well. A comparison of the figures in N^{h} and F^{1} reveals that these manuscripts are related. On the microfilm it appears as if the scribe of N^{h} at first drew the piece of wood ΔEHZ much larger, only to rub out these lines afterwards. From the diagram in F^{1} we learn that this was in fact the interpretation of the scribe copying from N^{h}, since both lines ZH and ΔE are extended in his copy.

ἴσον γοῦν μετακεχωρηκὸς ἔσται. αλλ᾽ ἦν ἔλαττον. ἔσται δὴ οὗ τὸ Z. τὸ Θ ἄρα τέμνει τὴν AB, καὶ οὐχ ἡ τὸ Γ, καὶ κάτωθεν. ἐλάττων γὰρ ἡ BZ τῆς AΔ, ὥστε καὶ ἡ ΘZ τῆς ΔΘ · ὅμοια γὰρ τὰ τρίγωνα. μεθεστηκὸς δὲ ἔσται καὶ τὸ μέσον, τὸ ἐφ᾽ οὗ Γ · εἰς τοὐναντίον γὰρ τῷ ἐν τῇ θαλάττῃ ἄκρῳ τῷ B μεταχωρεῖ, ᾗπερ τὸ ἐν τῷ πλοίῳ ἄκρον τὸ A μετεχώρει δὲ τὸ A οὗ τὸ Δ. ὥστε μετακινηθήσεται τὸ πλοῖον, καὶ ἐκεῖ οὗ ἡ ἀρχὴ τῆς κώπης μεταφέρεται.

οὖν B^{e} and editions.

Z. τὸ Θ ἄρα τέμνει Par; Z ἢ τὸ Θ. ἄρα τοίνυν *cett.*

καθεστηκὸς Pachymeres and editions.

μετεχώρει δὲ τὸ A Par; μὴ ἐχώρει *cett.*

[57] ἔστω σφὴν ἐφ᾽ ᾧ ABΓ, τὸ δὲ σφηνούμενον ἐφ᾽ ᾧ ΔEHZ. μοχλὸς δὴ γίνεται ἡ AB, βάρος δὲ τὸ τοῦ B κάτωθεν, ὑπομόχλιον δὲ τὸ ZΔ. ἐναντίος δὲ τούτῳ μοχλὸς τὸ BΓ. ἡ δὲ AΓ κοπτομένη ἑκατέρᾳ τούτων χρῆται μοχλῷ· ἀνασπᾷ γὰρ τὸ B.

om. families *b* and *c*.

Problem 21

Why is it that dentists extract teeth more easily with a forceps than with their bare hands? In answer to this question the author explains that a forceps consists of two levers that work in opposite directions, with the fulcrum at the point where the pincers are joined. Lines 854a25-31 of the text describe this by means of a diagram: Now one of the levers is ΑΔΖ, the other ΒΓΕ, and the fulcrum ΓΔΘ. The tooth is at point I, where the extremities of the forceps come together. The dentist holds the tooth with BZ and moves it at the same time.

This diagram is contained in all manuscript families. The diagrams reproduced in the appendix illustrate that the shape of the forceps very much differs from one codex to the next, because most of the figures were drawn freehand. The reconstruction of the letter labels in the diagram from the archetype is unproblematic; all letters are contained in families *a*, *b* and *c*, only the letter Θ is missing in those manuscripts that derive from ζ. M^v and D^1 both contain the label *iota* for the tooth, but do not depict the tooth itself. This pictorial element was probably omitted by the scribe of these manuscripts, since not only in P, but in all other manuscripts as well, the tooth is represented by a square.

Problem 22

Problem 22 discusses the working of a nutcracker, which functions in a similar way to the dentist's forceps in the previous question. Again, we recognize two levers that work in opposite directions (lines 854a38-b15):

> For the instrument is composed of two levers, with the same fulcrum, that is A, the point of connection. As, then, if E and Z have been set apart, they could easily be brought together by a small force, by which the extremities at Γ and Δ are moved. So EΓ and ZΔ being levers exert the same or greater force than which the weight exerts in a blow; for in raising they are raised in opposite directions, and, when pressed together, they break that which is at K. For the same reason the closer K is to A, the quicker it will be crushed. For the further distance the lever is from the fulcrum, the more easily and the greater does it move by the same force. A, then, is the fulcrum, and ΔAZ and ΓAE the levers. The closer, then, K is to the angle at A, the closer it is to the point of connection at A; and this is the fulcrum. Therefore, by applying the same force, ZE must be raised more. So that, when the raising is from opposite directions, it must be more compressed; and that which is more compressed, will break sooner.[58]

[58] τὸ γὰρ ὄργανον ἐκ δύο σύγκειται μοχλῶν, ὑπομόχλιον ἐχόντων τὸ αὐτό, τὴν συναφὴν ἐφ' ἧς τὸ Α. ὥσπερ οὖν εἰ ἦσαν ἐκβεβλημέναι, ὑφ' ὧν κινουμένων εἰς τὰ τῶν ΓΔ ἄκρα αἱ ΕΖ συνήγοντο ῥᾳδίως ἀπὸ μικρᾶς ἰσχύος· ἣν οὖν ἐν τῇ πληγῇ τὸ βάρος ἐποίει, ταύτην ἡ κρείττων ταύτης, ἡ τὸ ΕΓ καὶ ΖΔ, μοχλοὶ ὄντες ποιοῦσι· τῇ ἄρσει γὰρ εἰς τοὐναντίον αἴρονται, καὶ θλίβοντες καταγνύουσι τὸ ἐφ' ᾧ Κ. δι' αὐτὸ δὲ τοῦτο καὶ ὅσῳ ἂν ἐγγύτερον ᾖ τῆς Α τὸ Κ, συντρίβεται θᾶττον· ὅσῳ γὰρ ἂν πλεῖον ἀπέχῃ τοῦ ὑπομοχλίου ὁ μοχλός, ῥᾷον κινεῖ καὶ πλεῖον ἀπὸ τῆς ἰσχύος τῆς αὐτῆς. ἔστιν οὖν τὸ μὲν Α ὑπομόχλιον, ἡ δὲ ΔΑΖ μοχλός, καὶ ἡ ΓΑΕ. ὅσῳ ἂν οὖν τὸ Κ ἐγγυτέρω ᾖ τῆς γωνίας τῶν Α, τοσούτῳ ἐγγύτερον γίνεται τῆς συναφῆς τῶν Α· τοῦτο δέ ἐστι τὸ ὑπομόχλιον. ἀνάγκη τοίνυν ἀπὸ τῆς αὐτῆς ἰσχύος συναγούσης τὸ ΖΕ αἴρεσθαι πλέον. ὥστε ἐπεί ἐστιν ἐξ ἐναντίας ἡ ἄρσις, ἀνάγκη θλίβεσθαι μᾶλλον· τὸ δὲ μᾶλλον θλιβόμενον κατάγνυται θᾶττον.

The fact that the further away the fulcrum is from the centre of the lever, the more easily it is moved by the same amount of force, is not accounted for in the manuscript diagrams. All of them depict the fulcrum at the middle of the nutcracker.

The diagram is missing from family *a*. In family *b* we have a situation similar to the previous diagram: P contains the letter K inside a small circle that represents the nut, whereas both M^v and D^1 depict only the label K. The scribe of M^v and D^1 probably emended the diagram from his exemplar, and omitted the symbol of the nut. This is confirmed by the diagrams in family *c*, which, like codex P, contain the same pictorial element. The figures in W^a and V^1 illustrate that the shape of the nutcracker varies in the manuscripts from family *c*. In addition, the label K and the nut are missing from W^a; they are, however, depicted in all other manuscripts in this family, and are therefore also likely to be present in the hyparchetype γ.

Problem 23

The author raises the question of why it is that in a rhombus, when the points at the extremities travel in two motions, they do not complete equal straight lines, but one of them a much longer line than the other? In a long description given in lines 854b22-855a7, the parallelogram of motions from the first problem is now applied to the case of a rhombus, in which one diagonal is longer than the other.[59] Depending on the inclination of the angle, the resulting motion is longer or shorter. If point A, traveling from an obtuse angle, is moved in two motions, it follows the shorter diagonal AΔ. The motion of point B on the contrary, starting from an acute angle, describes the longer diagonal BΓ. The reason that the point starting from an acute angle travels further, is that the lines in an acute angle tend to approach the same direction. The two motions BA and BΔ in the acute angle B are closer together; the one motion is, so to speak, assisted by the other, and therefore the resulting motion is longer. Lines 855a19-27 of the text refer back to this diagram by explaining that in the obtuse angle A the directions AΓ and AB are more opposite to each other, and therefore the distance travelled is shorter. The more obtuse the angle, the more opposed to each other the two motions become, which results in a reduced distance.

The diagram to this problem is contained in all manuscript families; some codices (P, N^h, F^1, and q) contain two diagrams of a rhombus, which might be connected to the fact that there are two references in the text to the same diagram. This second diagram is helpful to the manuscript's readers, in that it saves the user from the need to leaf back to the previous diagram, but it is not required by the text. All diagrams, including those in the paraphrase by Pachymeres, further contain an *iota* that is not mentioned in the text. This letter does not serve any function in the diagram and was probably added by the scribe of the archetype, to ensure that all points on the rhombus were labelled. None of the copyists omitted this label from their figures, which again demonstrates that scribes generally made faithful copies of their exemplars.

[59] See Heath (1949), 245.

As one of the characteristics of ancient diagrams, I pointed to their schematic nature, which means that in general they do not give a metrically correct picture of the objects. In this diagram, we notice that none of the scribes considered the fact that one of the angles, B, is acute, whereas the angle at A should be obtuse.

The right side of the page in manuscript L is trimmed, for which reason the letters Z, I and Γ are missing, although a *gamma* is contained in the wrong position at the centre of the rhombus. In L we further see a circle drawn around the centre of the rhombus; upon comparison with the diagram in the manuscripts of Pachymeres, we notice that there probably was no such circle in codex α. In Pachymeres, we find a large Θ at the centre of the rhombus, which might very well have been contained in the hyparchetype as well, and was subsequently interpreted as a circle by the copyist of L. Codex P contains three rhombi, the first of which is clearly an attempt at drawing a rhombus. These figures show that the copyist of P was a very poor illustrator; furthermore, the labelling of the diagrams is incomplete. The other two manuscripts in family *b*, M^v and D^1, contain one correct diagram of a rhombus. It is likely that the hyparchetype β already contained some errors that were emended by the scribe of M^v and D^1. Characteristic of all manuscripts in family *c* is that the label Θ is missing from the centre of the rhombus. Apart from differences in the shapes of the rhombi (compare, for example, the figures in V^1 and N^b), which are mainly caused by the fact that the figures were drawn freehand, all diagrams in family *c* are similar.

Problem 24

This problem raises the difficulty of why it is that in two concentric circles, the larger circle unfolds the same path as the smaller circle, and vice versa.[60] If two circles are concentric, sometimes the path along which the circles revolve is equal to the circumference of the smaller circle, and sometimes they describe a length equal to the larger circumference. This observation is illustrated by the following description (855b5-23)[61]:

> For let there be a larger circle ΔZΓ, a smaller EHB, and A the centre of both; let ZI be the line which the greater unrolls on its own, and HK that which the smaller unrolls on its own, equal to ZΛ. When I move the smaller circle, I move the same centre, that is A; let the larger be attached to it. When AB becomes perpendicular to HK, at the same time AΓ becomes perpendicular to ZΛ, so that it will always have completed an equal distance, namely HK for the circumference HB, and ZΛ for ZΓ. If the quarter unrolls an equal distance, it is clear that the whole circle will unroll an equal distance to the whole circle, so that when the line BH comes to K, the circumference ZΓ will be in ZΛ, and the whole circle will be unrolled. In the same way, when I move the large circle, fitting the small one to it, their centre being the same, AB will be perpendicular and at right angles simultaneously with AΓ, the latter to ZI, the former to HΘ. So that, when the one will have completed a line equal to HΘ, and

---

[60] For a detailed treatment of this question, the so-called problem of Aristotle's wheel, see Drabkin (1950), 162–198.

[61] Note in this description the verbs in the first person (855b9 and 855b18), the occurrence of which is unusual in the Aristotelian corpus.

> the other to ZI, and ZA becomes again perpendicular to ZΛ, and HA to HK, so that they will be as in the beginning at Θ and I.[62]

In the lines that follow, this observation is explained by the author: if a body that is naturally moved, moves itself and another body that is not naturally in motion, it will move slower than if it moves itself only. Furthermore, it is impossible for a body that is not naturally moved to move faster than the mover. This is illustrated by the second diagram to this problem (856a1-16), in which there are two circles, a larger circle A and a smaller B. If the smaller pushes the larger circle, the larger will travel a path as far as it has been pushed by the smaller circle. Similarly, when the larger circle pushes the smaller one, the latter moves exactly as far as the former. In our case, namely that of two concentric circles, the circles behave in the same way. When the larger circle moves the smaller one, the smaller circle must be moved as far as the former, and vice versa. The lengths of the paths traced by the circles depend on which circle moves the other. Both circles have the same centre, but this is only accidental.[63] When the smaller circle produces the movement, the centre belongs to the smaller circle and is only accidentally the centre of the larger circle. The opposite applies when the larger circle moves the smaller one; in this case the centre belongs to the larger circle.

The first diagram for problem 24 is contained in all manuscript families. The figure in codex L depicts all of the labels, but the letter *eta* is hardly recognizable on the microfilm. Lines HΘ and ZI, which represent the paths traced out by the circles, are rather short; this may be due to the relatively small space in the left margin of the text. The figures in family *b* are all complete and allow an unproblematic reconstruction of the hyparchetype. In family *c*, we notice some common errors that confirm the dependences between these manuscripts. Both in N^h and F^1, lines HΘ and ZI are much longer than in the other manuscripts. U^m, F^2 and q all contain a

---

[62] ἔστω γὰρ κύκλος ὁ μείζων μὲν ἐφ᾽ οὗ τὰ ΔΖΓ, ὁ δὲ ἐλάττων ἐφ᾽ οὗ τὰ ΕΗΒ, κέντρον δὲ ἀμφοῖν τὸ Α· καὶ ἣν μὲν ἐξελίττεται καθ᾽ αὑτὸν ὁ μέγας, ἡ ἐφ᾽ ἧς ΖΙ ἔστω, ἣν δὲ ὁ ἐλάττων καθ᾽ αὑτόν, ἡ ἐφ᾽ ἧς ΗΚ, ἴση τῇ <u>ΖΛ</u>. ἐὰν δὴ κινῶ τὸν ἐλάττονα, τὸ αὐτὸ κέντρον κινῶ, ἐφ᾽ οὗ τὸ Α· ὁ δὲ μέγας προσηρμόσθω. ὅταν οὖν ἡ ΑΒ ὀρθὴ γένηται πρὸς τὴν ΗΚ, ἅμα καὶ ἡ ΑΓ γίνεται ὀρθὴ πρὸς τὴν ΖΛ, ὥστε ἔσται ἴσην ἀεὶ διεληλυθυῖα, τὴν μὲν ΗΚ, ἐφ᾽ ᾧ ΗΒ περιφέρεια, τὴν δὲ ΖΛ ἡ ἐφ᾽ ἧς ΖΓ. εἰ δὲ τὸ τέταρτον μέρος ἴσην ἐξελίττεται, δῆλον ὅτι καὶ ὁ ὅλος κύκλος τῷ ὅλῳ κύκλῳ ἴσην ἐξελιχθήσεται, ὥστε ὅταν ἡ ΒΗ γραμμὴ ἔλθῃ ἐπὶ τὸ Κ, καὶ ἡ ΖΓ ἔσται περιφέρεια ἐπὶ τῆς ΖΛ, καὶ ὁ κύκλος ὅλος ἐξειλιγμένος. ὁμοίως δὲ καὶ ἐὰν τὸν μέγαν κινῶ, ἐναρμόσας τὸν μικρόν, τοῦ αὐτοῦ κέντρου ὄντος, ἅμα τῇ ΑΓ ἡ ΑΒ κάθετος καὶ ὀρθὴ ἔσται, ἡ μὲν πρὸς τὴν ΖΙ, ἡ δὲ πρὸς τὴν ΗΘ. ὥστε ὅταν ἴσην ἡ μὲν τῇ ΗΘ ἔσται διεληλυθυῖα, ἡ δὲ τῇ ΖΙ, καὶ γένηται ὀρθὴ πάλιν ἡ ΖΑ πρὸς τὴν ΖΛ, <u>καὶ ἡ ΑΗ ὀρθὴ πάλιν πρὸς τὴν ΗΚ</u>, ὡς τὸ ἐξ ἀρχῆς ἔσονται ἐπὶ τῶν ΘΙ.

ΖΛ Par; ΑΖ *cett.*

καὶ ἡ ΑΗ ὀρθὴ πάλιν add. πρὸς τὴν ΗΚ Par; καὶ ἡ ΑΓ ὀρθὴ πάλιν *cett.*

[63] Lines 856a33-35 of the text read: "Both circles have the same centre, but this is only accidental, like the musical or the white;" τὸ αὐτὸ μὲν γάρ ἐστι κέντρον ἀμφοῖν, ἀλλὰ κατὰ συμβεβηκός, ὡς μουσικὸν καὶ λευκόν. It is a characteristically Aristotelian expression to mention the musical and the white as accidental properties, which can also be found in e.g. *APo.* 83a9, *Int.* 21a13, *Metaph.* 1007b2, 1017a8, 1017b27, 1031b22, and *Ph.* 188a34, 196b25.

second letter E that was probably also contained in codex ζ. We further see in these manuscripts that the smaller circle is almost the same size as the larger one. The bottom of the page in U^m is cut off, and hence the labels H, K, Θ, Z, Λ and I are missing from the figure; none of these labels were completed by the scribe of F^2. Codex q contains two identical figures to this problem on different folia.

The second diagram is present in families *b* and *c*. In none of the figures are the circles actually touching one another, despite the text, which calls for a smaller circle B pushing along a larger circle A. Moreover, most of the diagrams do not represent the fact that one of the circles is smaller than the other, but rather depict both circles as being approximately the same size.

Problem 25

Problem 25 asks two different questions relating to beds: Why are beds constructed with a length twice their width? And why are they not corded diagonally but from side to side?[64] In answer to the first question, the author argues that the dimensions of beds correspond to the size of human bodies. As to the second question, the author gives three reasons why beds are corded from side to side rather than diagonally: the timbers are less strained this way; the ropes are less strained; less rope is used. The side to side cording of beds is explained by a complicated and incomplete passage, in which many letter labels are un(der)specified (856b11-38). AZHI is the bed, and ZH is bisected at B. The holes in ZB are equal to those in ZA, since the sides are equal. And the whole length ZH is twice ZA. The bed is strung from A to B, then to Γ, then to Δ, then to Θ, then to E, and further until the other corner of the bed is reached. From the following description, we can infer that B is at the middle between Z and H, and points Γ and E are placed at equal distances between B and H. At the other side of the bed, line AI is bisected at K, and points Δ and Θ are placed at equal distances between A and K. In the bed there are four lengths of rope equal to AB. And whatever number of holes there are in ZH, there will be half the number in ZB, which is half of ZH. In the second diagram for this problem, the author turns to the diagonal cording of beds, which he illustrates by the bed ABΓΔ (856b38-857a4). The principle explained in these lines is incomplete, and unfortunately the corresponding diagram is of no help. It appears as if there are fewer holes in this bed than in the bed that is corded crosswise, but as many as in the sides ZA or ZB, which are half of ZH.

The particular obscurity of the diagonal cording of beds is evident from the manuscript diagrams, all of which are incomplete. Apparently, the text was already corrupt in the archetype and the scribe could not give a sensible interpretation of it by means of a diagram. In the next chapters, it will be shown that later Renaissance commentators and modern editors of the text also experienced great difficulties

[64] This question will be treated in more detail in Chap. 6 in the context of the early modern transformations of the *Mechanics*.

with problem 25 and constructed a variety of diagrams based on the descriptions in the text.

The manuscripts contain three diagrams to this problem, whereas only two are required by the text. The first shows the side to side cording of beds. The principle is clearly illustrated here, although all diagrams are metrically incorrect. None of the scribes depicted the labels B and K at the middle of lines ZH and AI. The second diagram shows the same bed, but with only one cord AB. In this diagram, the letter B is depicted more or less at the centre of ZH; at the bottom right of the figure we further see the label Δ instead of I.[65] The third diagram illustrates the case of the bed ABΓΔ, which is strung diagonally, with only one cord depicted.

In codex L, the label H is illegible in the first diagram, and it is missing from the second. The third figure is not labelled clockwise, but reads ABΔΓ. Manuscript P contains four diagrams, one of which is an unlabelled rectangle, which might be a failed attempt at the third diagram. The labels in all diagrams are erroneous, and some of the letters are indecipherable. Codices M^v and D^1 in the same family *b* contain two diagrams; the scribe probably omitted from his exemplar the diagram not required by the text. The labelling of these diagrams is correct and complete; the second diagram of the bed that is corded diagonally is not depicted in an upright position, as in all other manuscripts, but has the same orientation as the first diagram. The manuscripts in family *c* all contain three diagrams, except for N^b, which only depicts the first diagram. Again, in U^m the bottom of the page of the first diagram is cut off, and hence the labelling is missing from this diagram, as well as from its copy F^2. The third diagram in the manuscripts that derive from codex ζ is not labelled clockwise: ABΔΓ. The diagrams in N^h and F^1 are similar, apart from the third diagram, where the letters are now placed clockwise: ABΓΔ. We further notice that the shape of the first diagram in F^1 is smaller and much more elongated in comparison to its source N^h, because of the limited space at the bottom of the page. W^a and V^1 confirm the clockwise ordering of the letters in the third diagram, but in W^a this figure is not represented in upright position. In the first diagram, both W^a and V^1 contain a second B instead of the correct label Θ, showing that this letter might have been corrupted in the hyparchetype.

Problems 26–27

These two problems refer to the same diagram. Problem 26 asks why it is more difficult to carry long timbers on the shoulders by the end than by the middle. The reason given for this is that when a timber is raised from the middle, the one end assists in lifting the other. The middle acts as a centre, and each end by pressing downwards raises the other end upwards. This does not apply when a timber is carried by the end, since in this case all the weight presses in one direction. Lines 857a18-21 describe the corresponding diagram: A is the centre of a timber with

[65] Note that this diagram in the Pachymeres manuscripts contains the correct label I.

ends B and Γ. When it is carried from A, end B pressing down raises Γ, while Γ pressing down raises B.

Problem 27 asks why it is that if a weight is very long it is harder to carry on the shoulder, even in the middle, than when it is shorter. This is the case because of the vibration; the longer the timber is, the more the ends shift upon the same movement. Lines 857a29-32 describe the diagram: "For the shoulder is the centre, at A – for it stays put –, and AB and AΓ are the lines from the centre. The longer the line from the centre is, i.e. AB or AΓ, the more the magnitude is displaced."[66]

The reconstruction of the diagram for these problems in the archetype is unproblematic, since the figures in all manuscripts are similar. The positioning of the diagram is, however, very different in all manuscripts; as is evident from Table 4.2, many manuscripts position this diagram alongside a later problem in the text.

Problem 30

Why do men make an acute angle between the lower leg and the thigh, and between the trunk and the thigh when rising from a seated position? The reason for this is that in standing up a man becomes perpendicular to the ground. In order to attain this position, he must have his head in line with his feet, and this occurs when he is rising. As long as he is sitting, he keeps his feet and head parallel to one another and not in one straight line. Lines 857b32-858a2 of the text illustrate this by means of a diagram:

> Let A be the head, AB the trunk, BΓ the thigh, ΓΔ the lower leg. The trunk AB is at right angles to the thigh, and the thigh to the lower leg, when someone is seated in this way. So that while keeping it like this he is not able to rise. But he must bend in his lower leg and bring his feet under his head. This will be the case when ΓΔ has come to ΓZ, and at the same time he will be able to rise, and he will have his head and feet in the same straight line. And ΓZ makes an acute angle with BΓ.[67]

The reconstruction of this diagram presents no difficulties. The diagram is contained in families *b* and *e*, but is missing from codex L. Since it is present in Pach^p^, we do not know whether this diagram was present in the hyparchetype and left out by the scribe of L, or was added by Pachymeres. An additional complexity is the fact that the figure in Pachymeres is a mirror image of the other manuscript figures. We do not know why this is the case, but we have seen more examples of diagrams where Pachymeres deviated from his source. However, Pachymeres' paraphrase did

[66] ὁ μὲν γὰρ ὦμος κέντρον, ἐφ' οὗ τὸ A μένει γὰρ τοῦτο, αἱ δὲ <u>AB AΓ</u> αἱ ἐκ τοῦ κέντρου. ὅσῳ δ' ἂν ᾖ μεῖζον τὸ ἐκ τοῦ κέντρου ἢ τὸ AB ἢ καὶ τὸ AΓ, πλέον μεθίσταται μέγεθος.

AB καὶ AΓ Pachymeres and editions.

[67] ἡ κεφαλὴ A ἔστω, θῶραξ AB, μηρὸς BΓ, κνήμη ΓΔ. πρὸς ὀρθὴν δὲ γίνεται ὅ τε θῶραξ ἐφ' ὧν AB τῷ μηρῷ καὶ ὁ μηρὸς τῇ κνήμῃ οὕτως καθημένῳ. ὥστε οὕτως ἔχοντα ἀδύνατον ἀναστῆναι. ἀνάγκη δὲ ἐγκλῖναι τὴν κνήμην καὶ ποιεῖν τοὺς πόδας ὑπὸ τὴν κεφαλήν. τοῦτο δὲ ἔσται, ἐὰν ἡ ΓΔ ἐφ' ἧς τὰ ΓZ γένηται, καὶ ἅμα ἀναστῆναι συμβήσεται, καὶ ἔχειν ἐπὶ τῆς αὐτῆς <u>εὐθείας</u> τὴν κεφαλήν τε καὶ τους πόδας. ἡ δὲ ΓZ ὀξεῖαν ποιεῖ γωνίαν πρὸς τὴν BΓ.

ἴσας families *b* and *e*; ἴσης corr. B[e] and editions.

not include any diagrams that were not present in his exemplar. For this reason I believe it likely that codex α contained this diagram, which was unintentionally left out by the scribe of L, but copied by Pachymeres.

4.5 Independent Diagrams

The four codices B[e], Par, P[s] and P[t] were not considered in the critical edition, as their diagrams were largely made independently of the manuscript tradition. These diagrams do not contribute to the recovery of the archetypal diagram, but comprise individual commentaries on the *Mechanics*. They offer alternative interpretations of the text which are relevant to the broader picture of the transmission and transformation of mechanical knowledge in the Byzantine and early modern period. Codex B[e], which was copied by the humanist Niccolò Leonico Tomeo, constitutes a special case, since it can be connected with Leonico Tomeo's Latin translation of the *Mechanics*. Leonico Tomeo's diagrams give us a unique insight into the process of transition from handwritten manuscript to printed book, and will therefore be treated in Chap. 6 on the early modern reception of the treatise. In that context, I will also return to the diagrams contained in the paraphrase by Pachymeres which may have influenced Leonico Tomeo's interpretation of the text.

Par belongs to family *c* and was copied from the lost codex ε; after its completion it received extensive emendation in which phrases were added and many variants were erased and replaced by different readings. At this stage, diagrams were added to the text that are different from the ones in the other manuscripts. The scribe paid attention to the presentation of the diagrams and used straight edge and compass in all drawings. The diagrams in Par are not connected with family *c*, but for one notable exception in problem 23. To this problem, the Par codex contains two diagrams of a rhombus: while first is a carefully executed diagram added by the subsequent corrector of the text, second is a freehand drawing that displays the characteristics of family *c*. This diagram is drawn by a different hand that can very likely be identified with the first scribe of Par. The missing label Θ from the centre of the rhombus is a feature that is only present in family *c* and underscores the attribution of the diagram to this manuscript family. We further notice the presence of the letter *iota* at one of the endpoints of the rhombus. This label, that is not mentioned in the text but represented in all manuscript diagrams including the archetype, again connects this diagram to the manuscript tradition. The other diagram that was included at a later stage, however, is purely based on the text. The letter Θ is now positioned at the centre of the figure and we do not find the letter I back in the diagram.

The fact that the subsequent scribe remained close to the text in his drawings is also evident from another aspect which strongly distinguishes these later diagrams from the manuscript tradition. While the Byzantine diagrams deriving from the archetype were following ancient practice in that they did not give a metrically correct picture of the objects, the diagrams in Par are drawn according to their metrical properties. The circles mentioned in the introduction to the text, for instance, are all touching one another, and in the diagram of the lever to the third problem the fulcrum is drawn at a position away from the centre of the device. The second scribe tried to take account of all aspects mentioned in the text, which is also attested by his attentive positioning of the letter labels. In those cases where the diagram in Par seems to contradict the Greek, it turns out that it is in agreement with the corrected version of the text. In the diagram of circular motion to the first problem, we notice, for example, that the label E is absent from the diagram in Par. A closer examination of the Greek text in Par shows that the primary reading in line 849a4 had been erased and a new text put in its place.[68] From this emendation, combined with another variant in line 849a6, it follows that the letter E was not mentioned in Par and that the diagram thus corresponds to the text. This example is a good illustration of the fact that the diagrams in Par were added at a later stage, since they are based on the emended version of the text.

The next codex P^s is a contaminated manuscript that shows a close connection with manuscript family *c*. This relationship established for the text can be confirmed by certain characteristics of the diagrams. The diagram of a parallelogram to problem 1 of the text, for example, contains the second diagonal and the double line ΔZ that are only present in family *c*. In the next diagram depicting circular motion the label Δ is omitted from the centre of the circle. Further reliance on family *c* is apparent from the diagrams to the second problem in which the balance suspended from above is depicted by single lines, whereas the balance supported from below is drawn with double lines. The diagrams contained in P^s, however, do not fit unproblematically in this manuscript family. We find many variants and idiosyncratic interpretations of the text, for which reason these diagrams were not considered in the critical edition. The letter labels often have a different orientation from the manuscript tradition and the diagrams contain peculiar errors that cannot be related to any of the extant manuscripts. The most conspicuous example can be found in the introduction to the text, where the author prescribes a diagram of three successive circles. The copyist of P^s did not depict three circles next to or above each other as in the other manuscripts, but drew four overlapping circles in one of which a diameter AB is inscribed. Because of the fact that this codex cannot be related to any of the extant manuscripts, we do not know whether the scribe of P^s copied these diagrams from his source or was himself responsible for the innovative interpretations of the text.

P^t is another contaminated manuscript which, due to its paraphrastic character, its many lacunae and peculiar readings, cannot be attributed to a specific family. Only a few diagrams are contained in P^t, most notably those to the first problem. On later pages, we find several windows in the text for diagrams that were never made. The most remarkable diagram in this manuscript illustrates the case of rectilinear motion in problem 1. Whereas all other manuscripts depict a rectangle, codex P^t shows two different-sized squares that are aligned in such a way that they have the

[68] The second reading in line 849a4 is consistent with an emendation proposed by Pachymeres that was adopted in all critical editions. In the next chapter, I will discuss in more detail this emendation and its implications for the constructed diagrams.

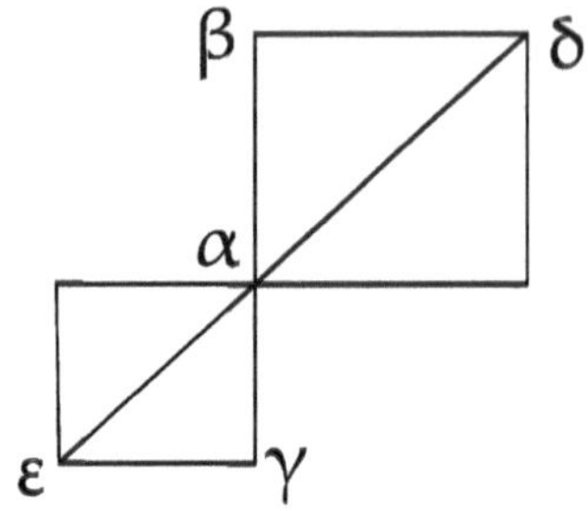

Fig. 4.4 Diagram of the parallelogram of motions as it is contained in codex P[t]

same diagonal (Fig. 4.4). When compared with the manuscript tradition, many letters appear to be missing from P[t] as only five labels are depicted in the diagram. However, the corresponding text is rather paraphrastic and offers a different version of rectilinear motion. Lines 848b13-21 in P[t] read as follows:

> Let the ratio in which the moved object is moved be that of AB to AΓ; and let the line AB be moved to Δ, and the line AΓ to E. If the ratio of the movement is that of AB to AΓ, then the line AΔ must have the same ratio to AE. For the small quadrilateral is in the same ratio [to the larger], so that their diagonal ΔE is the same.[69]

As we notice from this passage, the scribe and/or illustrator of P[t] had at least a basic mathematical understanding. He attempted to construct a mathematically sensible diagram based on these lines, while taking account of all letter labels mentioned in the text. He compares the sides of two different-sized squares with their diagonals: as the side of the one square is to the side of the other, so is the ratio of their diagonals. This interpretation becomes problematic in the following passage, where it is stated by the author that the motion along the diagonal consists of two different motions that correspond to the motions according to the sides of the quadrilateral. Hence it follows that the lines AB and AΓ constitute the sides of one and the same figure. Both the text and diagram to the following description of circular motion are very incomplete in P[t], but again we notice that the scribe was looking for the best possible diagrammatical solution to the text he was confronted with.

Bibliography

Andrist, Patrick. 2007. *Les manuscrits grecs conservés à la Bibliothèque de la Bourgeoisie de Berne – Burgerbibliothek Bern: catalogue et histoire de la collection*. Zurich: Urs Graf.

Berryman, Sylvia. 2009. *The mechanical hypothesis in ancient Greek natural philosophy*. Cambridge: Cambridge University Press.

Drabkin, Israel E. 1950. Aristotle's wheel: Notes on the history of a paradox. *Osiris* 9: 162–198.

[69] ἔστω γὰρ ὁ λόγος οὗ φέρεται τὸ φερόμενον, ὃν ἔχει ἡ AB πρὸς τὴν AΓ· καὶ ἡ μὲν AB φερέσθω πρὸς τὸ Δ, ἡ δὲ AΓ πρὸς τὸ E. οὐκ οὖν ἐπὶ τῆς φορᾶς ὁ λόγος ἦν ὃν εἶχεν ἡ AB πρὸς τὴν AΓ, ἀνάγκη καὶ τὴν AΔ πρὸς τὴν AE τοῦτον ἔχειν τὸν λόγον. ὅμοιον ἄρα ἐστι τῷ λόγῳ τὸ μικρὸν τετράπλευρον, ὥστε ἡ αὐτὴ διάμετρος αὐτῶν ἡ ΔE.

I translate εἰ οὖν as emended by van Cappelle.

Drachmann, Aage Gerhardt. 1963. *The mechanical technology of Greek and Roman antiquity.* Copenhagen: Munksgaard.

Eastwood, Bruce. 2007. *Ordering the heavens: Roman astronomy and cosmology in the Carolingian Renaissance.* Leiden: Brill.

Fowler, David H. 1987. *The mathematics of Plato's academy: A new reconstruction.* Oxford: Clarendon Press.

Harlfinger, Dieter. 1971. *Die Textgeschichte der pseudo-aristotelischen Schrift ΠΕΡΙ ΑΤΟΜΩΝ ΓΡΑΜΜΩΝ: Ein kodikologisch-kulturgeschichtlicher Beitrag zur Klärung der Überlieferungsverhältnisse im Corpus Aristotelicum.* Amsterdam: Hakkert.

Heath, Thomas L. 1949. *Mathematics in Aristotle.* Oxford: Clarendon Press.

Heiberg, Johan Ludvig. 1904. Mathematisches zu Aristoteles. In *Abhandlungen zur Geschichte der mathematischen Wissenschaften*: 3–49. Leipzig: Teubner.

Hunger, Herbert. 1989. *Schreiben und Lesen in Byzanz: Die byzantinische Buchkultur.* Munich: Beck.

Jones, Alexander. 1986. *Pappus of Alexandria: Book 7 of the collection*, 2 vols. New York: Springer.

Krafft, Fritz. 1970. *Dynamische und statische Betrachtungsweise in der antiken Mechanik.* Wiesbaden: Franz Steiner Verlag.

Netz, Reviel. 1999. *The shaping of deduction in Greek mathematics: A study in cognitive history.* Cambridge: Cambridge University Press.

Netz, Reviel. 2004. *The works of Archimedes: Translated into English, together with Eutocius' commentaries, with commentary, and critical edition of the diagrams.* Cambridge: Cambridge University Press.

Netz, Reviel. 2011. The bibliosphere of ancient science (outside of Alexandria). *NTM Zeitschrift für Geschichte der Wissenschaften, Technik und Medizin* 19(3): 239–269.

Neugebauer, Otto. 1975. *A history of ancient mathematical astronomy: In three parts.* Berlin: Springer.

Nix, Leo Michael Ludwig and Wilhelm Schmidt (eds.). 1900. *Herons von Alexandria Mechanik und Katoptrik.* Leipzig: Teubner.

Saito, Ken. 2006. A preliminary study in the critical assessment of diagrams in Greek mathematical works. *Sciamus* 7: 81–144.

Saito, Ken and Nathan Sidoli. 2012. Diagrams and arguments in ancient Greek mathematics: Lessons drawn from comparisons of the manuscript diagrams with those in modern critical editions. In *The history of mathematical proof in ancient traditions*, ed. K. Chemla: 135–162. Cambridge: Cambridge University Press.

Stornajolo, Cosimo. 1895. *Codices urbinates Graeci Bibliothecae Vaticanae.* Rome: Ex Typographeo Vaticano.

Stückelberger, Alfred. 1993. Aristoteles illustratus: Anschauungshilfsmittel in der Schule des Peripatos. *Museum Helveticum* 50: 131–143.

Taub, Liba. 2003. *Ancient meteorology.* London: Routledge.

Vilain, Christiane. 2008. Circular and rectilinear motion in the *Mechanica* and in the 16th century. In *Mechanics and natural philosophy before the scientific revolution*, eds. W.R. Laird and S. Roux, 149–172. Dordrecht: Springer.

Weitzmann, Kurt. 1947 (repr. 1970). *Illustrations in roll and codex: A study of the origin and method of text illustration.* Princeton: Princeton University Press.

Chapter 5
Diagrammatic Reasoning

Chapter 4 offered a preliminary critical examination of the diagrams in the Greek manuscripts of the Aristotelian *Mechanics*.[1] In the present chapter, I will contrast the manuscript diagrams with the figures contained in the modern critical editions of the treatise and show that it was a mistake on the part of the modern editors not to examine the manuscript diagrams. The diagrams in the manuscripts can be attributed to Byzantine times and are a valuable addition to our knowledge of the Byzantine reception of the treatise. The text of the *Mechanics* that the Byzantine scholar had in front of him was in certain aspects different from that of the modern critical editions. The main reason for this is that all of the critical editions contain traces of the early fourteenth century paraphrase of the *Mechanics* by Georgius Pachymeres. The examination of the manuscript diagrams allows us to compare the diagrams with the variants in the codices, enabling us to decide whether a certain variant is authentic or due to an emendation by Pachymeres. On this basis we can reconstruct a more reliable text. The recovery of authentic readings from the archetype is not, however, the only significant use to which the diagrams in the Greek codices can be put. They are also an important tool for the verification of the *stemma codicum* of the treatise, as will be illustrated in Sect. 5.2.

Furthermore, the first problem of the *Mechanics* on rectilinear and circular motion will be analysed in detail, since this problem is essential for the argument of the text. The interpretation of this problem in the Byzantine period by means of the accompanying diagrams will be contrasted with the readings and modifications proposed by modern editors of the treatise. We will see that the diagrams in the manuscripts and those in the modern editions are very different from one another, and that the manuscript figures offer a more coherent interpretation of rectilinear and circular motion. First, however, I will begin by examining some of the principal characteristics of the figures in the modern critical editions of the *Mechanics*, which differentiate them from those in the manuscripts.

[1] Parts of this chapter have already been published in van Leeuwen (2014).

J. van Leeuwen, *The Aristotelian Mechanics*, Boston Studies in the Philosophy and History of Science 316, DOI 10.1007/978-3-319-25925-3_5

5.1 Diagrams in Modern Critical Editions

Whereas some of the editions of the *Mechanics* contain diagrams, as for instance the nineteenth-century editions by Johannes van Cappelle (1812)[2] and Otto Apelt (1888), other editors, for example Immanuel Bekker (1831) and Maria Elisabetta Bottecchia (1982), did not include any figures in their editions. This is remarkable, especially considering that the author constantly uses letter labels to draw the reader's attention to an accompanying diagram. Furthermore, several passages in the text are almost incomprehensible without a corresponding diagram. The omission of diagrams in certain editions suggests, firstly, that these editors took no interest in the pictorial aspects of the treatise. Secondly, the diagrams that are included in the critical editions are unrelated to those in the manuscripts. The editors did not consult the diagrams contained in the manuscripts, but simply inserted figures of their own that interpret the mechanical principles in modern terms. We notice that the diagrams in the manuscripts are accorded a different status than the manuscript text; whereas for the text, all variants are accurately noted in the critical apparatus, in the case of the diagrams, the reader is not informed whether a certain diagram is also contained in the manuscripts nor what it looks like.

The diagrams in the modern critical editions of the *Mechanics* are very different from the ones contained in the Greek codices. Even in the case of relatively simple figures, where the diagram can easily be constructed on the basis of the text, editors did not include a diagram similar to the representations in the manuscripts. For example, the wedge in the diagram for problem 17 of the text has a different orientation in the modern editions. If we compare the diagram contained in the editions with my reconstruction of the archetype (Fig. 5.1), we notice that the wedge splits the object from above in all editions, whereas the codices depict a wedge that penetrates the object from the left side. This might appear to the modern reader as a minor deviation, but it actually instructs us on the Byzantine conception of a wedge.

The diagram for problem 5 on the working of a ship's rudder also shows that the modern critical editions have a different perspective. Both diagrams in Fig. 5.2 depict the rudder from above, with the difference that the figure in the editions is turned 90° and the labelling is mirrored. Along with the frequently differing orientation of the figures in the editions, modern editors tend to place diagram labels in a clockwise direction, as opposed to the counterclockwise orientation often encountered in the Greek manuscripts. This is obvious in the diagram of the parallelogram of motions included in problem 1 of the text.

The diagrams in the Greek manuscripts show a certain indifference to metrical accuracy, as we have seen in the previous chapter. Modern pictures, however, are metrically correct representations of the mechanical objects. For instance, in problem 23 on the motions of the extreme points in a rhombus, the angle at A is described as an obtuse angle, and the angle at B as an acute angle. In the diagram contained in the Hett edition we can actually distinguish these angles from one another, while in the manuscripts it is irrelevant whether one angle appears greater than the other or both are equal (Fig. 5.3).

[2] Johannes van Cappelle was the first modern editor to include figures in his critical edition of the *Mechanics*. Later editions generally contain figures based on the figures by van Cappelle.

Fig. 5.1 (a) Edition by Hett (b) Reconstruction of the archetype

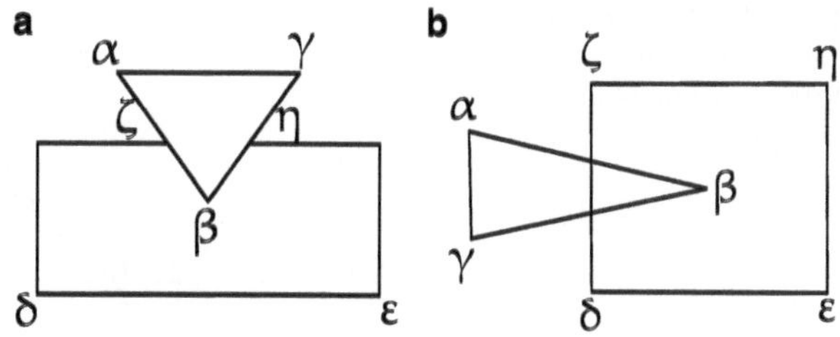

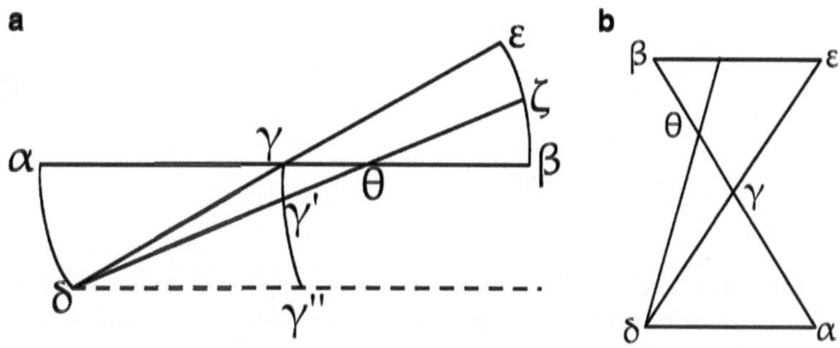

Fig. 5.2 (**a**) Edition by van Cappelle (**b**) Reconstruction of the archetype

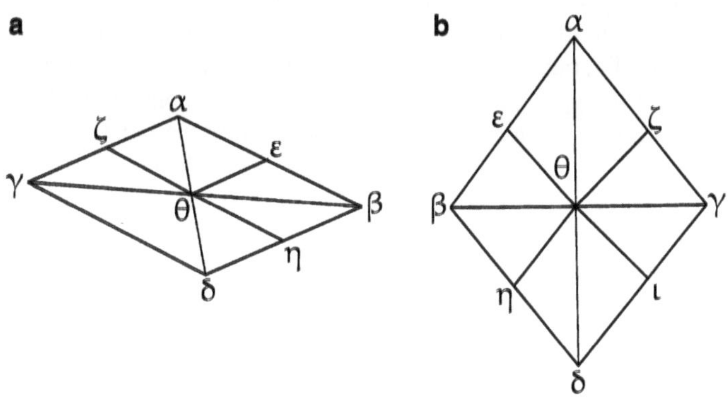

Fig. 5.3 (**a**) Edition by Hett (**b**) Reconstruction of the archetype

Another example can be found in problem 3 of the text on the working of a lever. The diagrams in the manuscripts do not consider the fact that the force is further away from the centre than the weight and depict the fulcrum at the centre of the bar (Fig. 5.4).[3] The diagrams in the editions, however, present a metrically correct picture, in which the fulcrum is moved to a position away from the centre.

[3] Only the figures in *Vat.gr.* 253 (L) and Pachymeres depict the force somewhat further from the fulcrum than the weight. This shift is, however, very minimal and might also be caused by the fact that the diagrams were inaccurately drawn freehand.

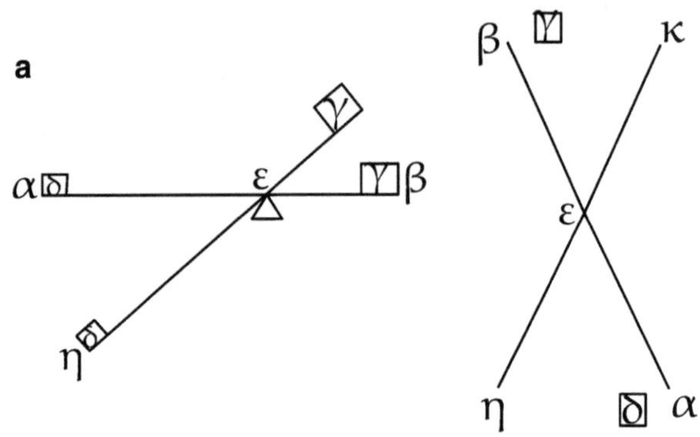

Fig. 5.4 (**a**) Edition by van Cappelle (**b**) Reconstruction of the archetype

Sometimes the labels in the editions do not correspond to those in the codices. These diagrams are of particular interest, since they might signal a different reading of the text in the critical editions; in such cases, the manuscript diagrams are helpful in confirming an authentic reading from the archetype. In the manuscript figures for the second problem, for instance, we see two letters that do not occur in the diagrams from the modern critical editions (Fig. 5.5).

These letters, P in the first diagram, and X in the second, are contained in both the text and the diagrams from the manuscripts. They are not necessary for an understanding of the principle, and for that reason van Cappelle, the first editor of the *Mechanics*, probably emended the text and omitted the labels from his figure. All later editors of the text adopted van Cappelle's emendations and, since they did not examine the diagrams from the manuscripts themselves, no one noticed that the readings from the text, P and X, were actually confirmed by the diagrams. These variants offer the authentic readings from the archetype and should therefore be included in a critical edition.

There are other reasons why the labelling of the figures in the editions is not in agreement with the manuscript diagrams. Sometimes additional labels were inserted by modern editors in order to make a certain mechanical problem more comprehensible. If we look again at Fig. 5.2 of the rudder, we notice some additional labelling in the editions: the letters Γ′ and Γ″ were added to indicate the changed positions of the pin of the rudder. These labels were probably added by van Cappelle to help the reader to understand the principle, and are not required by the text. However, additional labelling and drawing in the modern editions does not always involve a clarification of a certain mechanical principle. In problem 25, for example, on different ways of cording beds, the figures contained in the editions are the editor's own interpretations and do nothing to enhance our understanding of the Greek text. Part of the text to this problem is incomprehensible, and therefore it is understandable

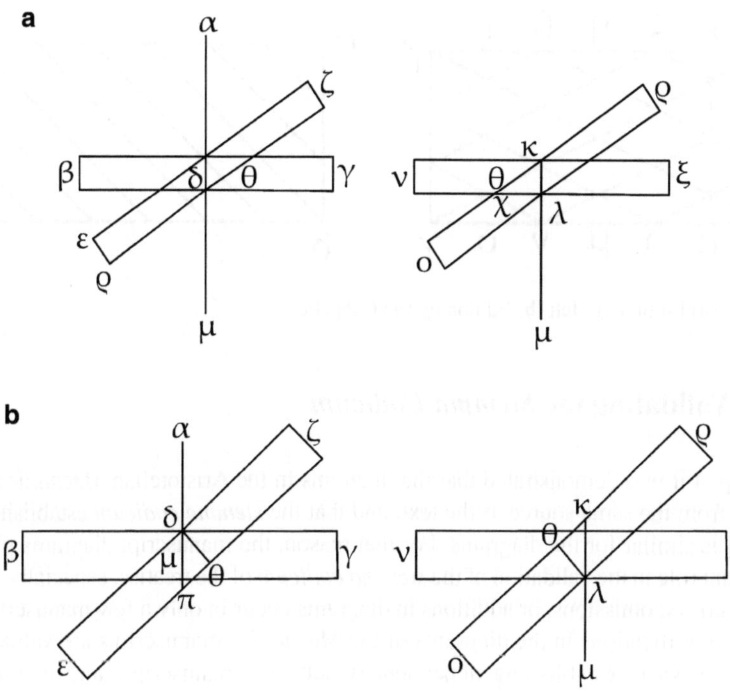

Fig. 5.5 (**a**) Reconstruction of the archetype (**b**) Edition by van Cappelle

that the editors attempted to give their own coherent reading of the text.[4] Their figures, however, are at least partly opposed to the text. Figure 5.6a illustrates Hett's interpretation of the diagonal cording of the bed. If the author of the *Mechanics* means a cord from Α to Γ and Β to Δ and then other cords parallel to these diagonals, Hett notes, he will be left with holes on the longer side that are unemployed (e.g. ΕΚ).[5] Hett must not have read the text carefully, since lines 857a1-2 point out that the number of holes on the longer side is equal to the number on the shorter side. If we halve the holes on the longer side in Hett's diagram, there are actually no holes left unemployed. The diagram in the edition by Van Cappelle also contains twice the number of holes on the longer side of the bed (Fig. 5.6b). Furthermore, the ropes are arranged not diagonally, but from side to side. These diagrams do not explain the principle as it was meant by the author; rather, they mislead the reader in his attempt to understand the text.

---

[4] Some of the editors omit the diagram to this case, e.g. Apelt (1888), 130 who notes: *figuram in re incerta non addidimus*.

[5] See Hett (1936), 399.

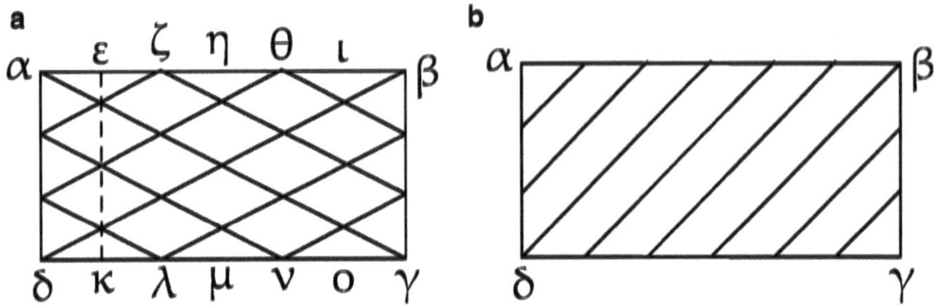

Fig. 5.6 (**a**) Edition by Hett (**b**) Edition by van Cappelle

5.2 Validating the *Stemma Codicum*

In Chap. 4 it was demonstrated that the diagrams in the Aristotelian *Mechanics* were copied from the same source as the text, and that the *stemma codicum* established for the text is similar for the diagrams. For that reason, the manuscript diagrams play an important role in the validation of the *stemma codicum* of the treatise, especially where certain errors, omissions, or additions in diagrams occur in only a few manuscripts.

The modifications in the diagrams of the *Mechanics* manuscripts are valuable in various ways in establishing dependences between manuscripts and manuscript groups. Firstly, the position of the diagrams gives us an important indication of the relations between manuscripts, for example, in those cases where a diagram is contained at the same incorrect position in the text for a complete branch of the *stemma*, where a certain diagram is missing from one group of manuscripts only, or where the special positioning of a diagram hints at a connection between manuscripts, for instance when it is integrated in the text while all of the other diagrams are marginal. Secondly, and especially useful for asserting dependences between individual manuscripts, are the errors that are frequently made in the labelling of the diagrams: letters that occur twice (or more) in a single diagram, the use of letters that are not prescribed by the text, letters that are included at the wrong position in the diagram, or letters that are missing from a diagram.

These characteristics can be illustrated by the relationship between the manuscripts *Marc.gr.* IV.57 (U^m) and *Laur.Plut.* 28.45 (F^2). Both codices can be dated to the middle of the fifteenth century, and as my analysis of the manuscript tradition has shown, F^2 is a copy of U^m. This dependence is confirmed by the figures in both manuscripts. The pages in U^m have been trimmed, and consequently individual letters and whole parts of certain diagrams have disappeared. This is the case, for example, in problem 24 of the *Mechanics*, which asks why it is that in two concentric circles, the larger circle traces the same path as the smaller circle, and vice versa (Fig. 5.7).

Due to the fact that the bottom of the page in U^m is trimmed, essential information is missing from the diagram: part of the circles, the paths traced out by the circles, and the corresponding labels. The copyist of F^2 did not know what to do with the incomplete diagram he found in his source. As we can see from Fig. 5.7c, he completed both circles, but did not add the paths traced out by the circles, or any of the letters that are missing, something he could easily have done based on his text. There are several other errors in these diagrams that point to a relation between both manuscripts: the smaller circle is almost the same size as the larger one, the horizontal diameter is not drawn through to the left side of the larger circle, and the letter E is contained twice in both diagrams.

Both manuscripts contain only one figure that is integrated in the text (Fig. 5.8). This diagram adjoins problem 1, where the motions of radii in two circles of different sizes are explained, a question that will be discussed in more detail below. In comparison with my reconstruction of the archetype in Fig. 5.8a, both U^m and F^2 are very erroneous; several lines are missing from the diagram, the labelling contains numerous errors and is incomplete, and the rectangle ΔΨΡΓ is not drawn tangentially to the outer circle. Upon examining the two manuscripts, we notice that many letters from U^m are hard to read, and that the copyist of F^2 clearly had difficulties in deciphering his source. The letter K, for example, is contained at three different places in F^2, but it is only correct in one position. The copyist introduced several new errors: the letters Δ and Θ are missing, and the horizontal diameter has not been drawn through to point B. The diagram is integrated in the left part of the text of both manuscripts, and, as is apparent from Figs. 5.8b and 5.9, the letter Γ, which belongs to the diagram in U^m, stands amidst the text of line 849a25.[6] The copyist of F^2 consequently made an obvious error. He understood that the letter Γ was part of the text and gave the diameter of the larger circle in his text the labels BEΓ instead of BE, whereas the Γ is not present in his figure (Fig. 5.10).

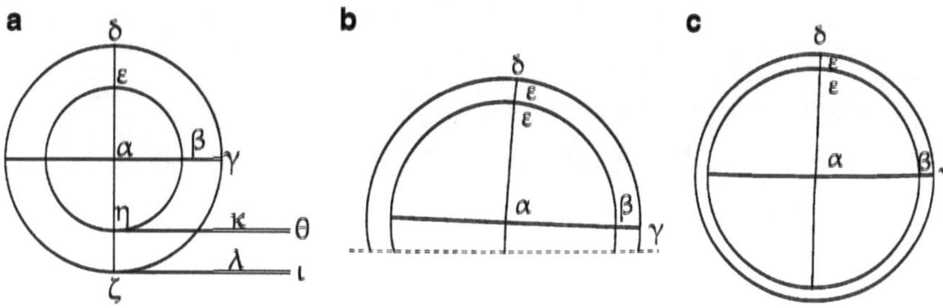

Fig. 5.7 (**a**) Reconstruction of the archetype (**b**) Codex U^m (**c**) Codex F^2

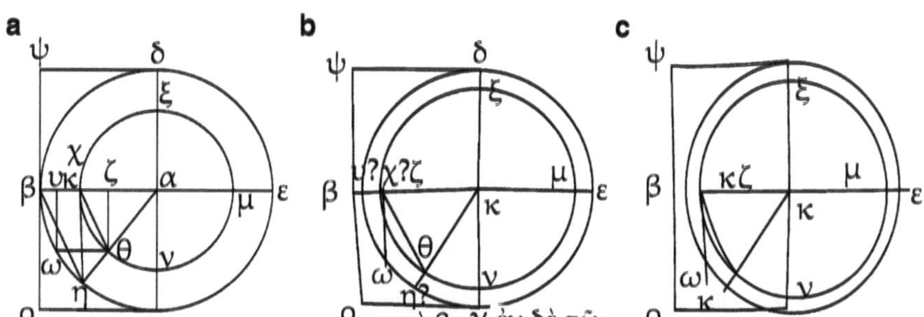

Fig. 5.8 (**a**) Reconstruction of the archetype (**b**) Codex U^m (**c**) Codex F^2

[6] Lines 849a23-25 read: καὶ ἐκβεβλήσθωσαν αἱ διάμετροι, ἐν μὲν τῷ μεγάλῳ, ἐφ' ὧν ΓΔ καὶ ΒΕ, ἐν δὲ τῷ ἐλάττονι αἱ ΜΧ ΝΞ.

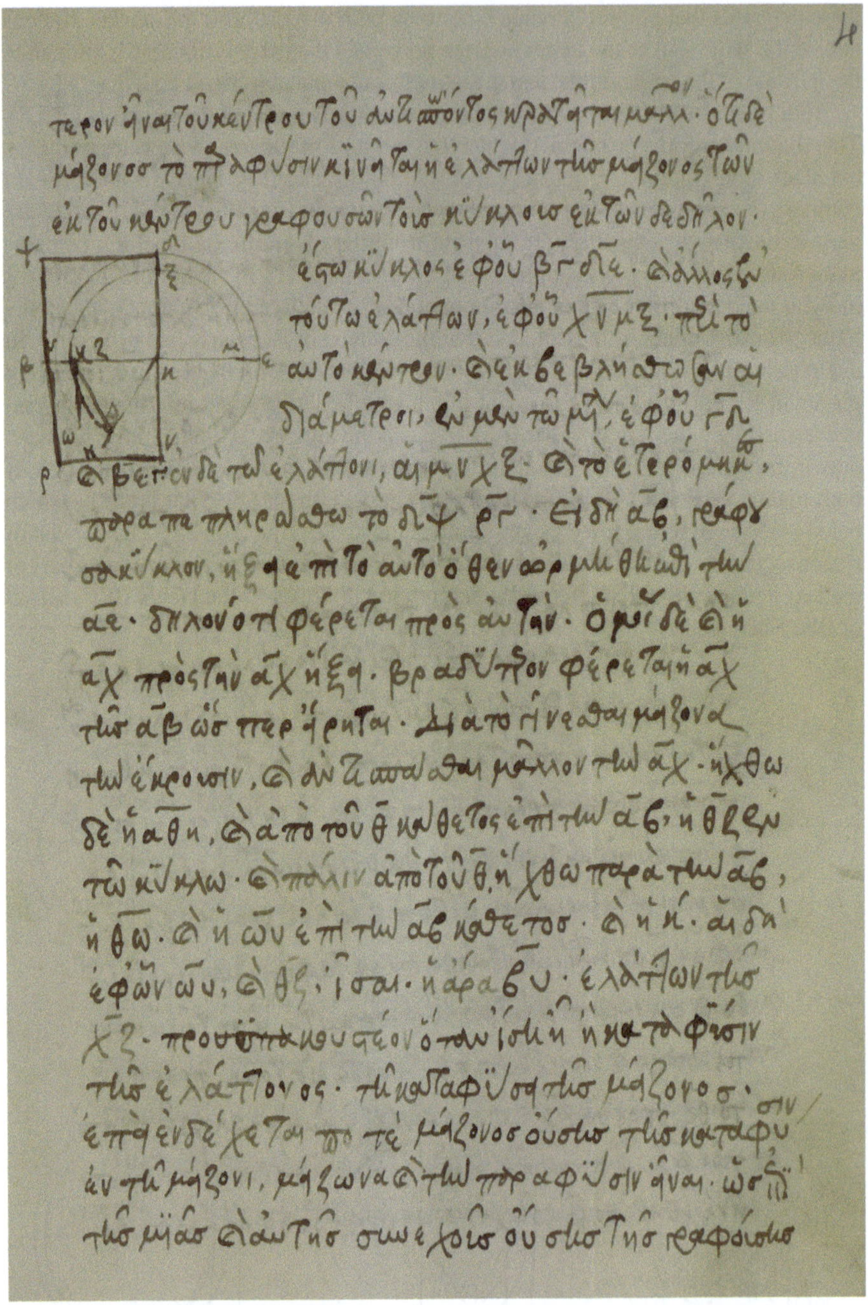

Fig. 5.9 Cod. Marc.Gr. IV.57 (U^m), f.4r. Diagram depicting circular motion to problem 1 of the *Mechanics*. Su concessione del Ministero dei Beni e delle Attività Culturali e del Turismo – Biblioteca Nazionale Marciana. Divieto di riproduzione

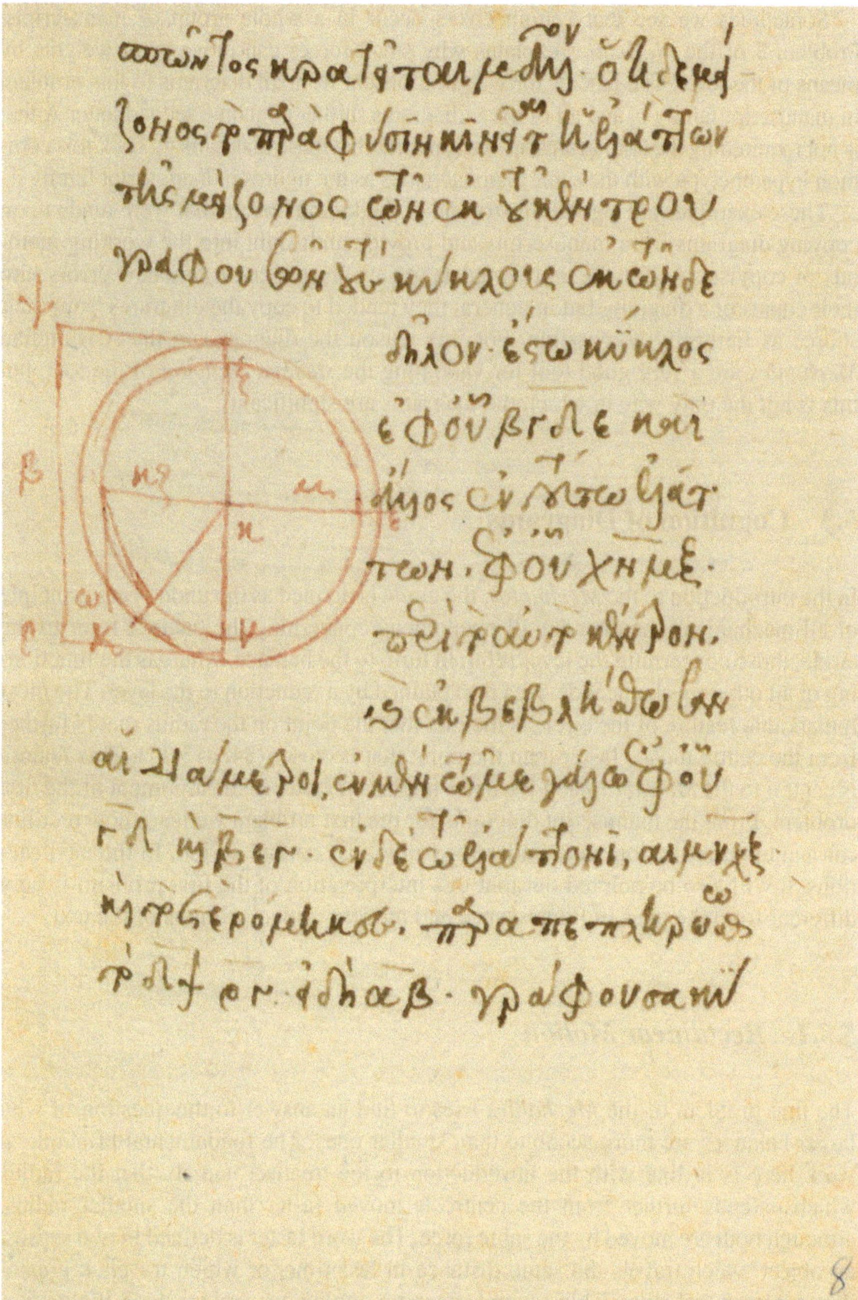

Fig. 5.10 Firenze, Biblioteca Medicea Laurenziana, Ms. Plut. 28.45 (F^2), f.8r. Diagram depicting circular motion to problem 1 of the *Mechanics*. È vietata ogni ulteriore riproduzione con qualsiasi mezzo

Sometimes we see that certain errors occur in a whole group of manuscripts. Problem 3 of the *Mechanics* explains why small forces can move great weights by means of the lever. The label Γ for weight is absent from all diagrams to this problem in manuscript family *c*, and all of these diagrams further contain a second letter A that is not required by the text. From this we can infer that these diagrams go back to a common hyparchetype with the same characteristics as the figures in manuscript family *c*.

These examples are a good illustration of the kind of errors that were made upon copying diagrams from manuscripts and provide an insight into the working methods of copyists. As for the text, copyists always introduced a few new errors into their copies of a diagram, but in general they tended to copy the diagrams from their source as faithfully as possible. For this reason the diagrams in the Aristotelian *Mechanics* are a very good tool for validating the *stemma codicum* of the text; but this is not the only way in which the diagrams are significant.

5.3 Cognition of Diagrams

In the introduction to the *Mechanics*, the circle is defined as the underlying principle of all mechanical phenomena: all phenomena concerning the balance refer to the circle, those concerning the lever refer, in turn, to the balance, whereas the functioning of all other mechanical devices is explained by a reduction to the lever. The most remarkable feature of the circle is the fact that the point on the radius that is further from the centre moves faster than the point that is closer (848a15-17). This feature is central to the *Mechanics* and forms an important factor in the argument of the first problem. From the manuscript diagrams for the first problem we learn how rectilinear and circular motion were understood in the Byzantine period. In the next sections, it will also be pointed out that this interpretation of the first problem is very different from the ones in Renaissance and modern critical editions of the text.

5.3.1 Rectilinear Motion

The first problem of the *Mechanics* tries to find an answer to the question of why larger balances are more accurate than smaller ones. The fundamental principle at work here is in line with the introduction to the treatise, namely that the radius which extends further from the centre is moved faster than the smaller radius, although both are moved by the same force. The word faster is defined in two senses: an object which travels the same distance in less time, or which travels a greater distance in equal time. This second meaning applies to circles, since the greater radius describes a greater circle in equal time. For the outer circle is greater than the inner.[7] To this statement in line 848b9 of the text, we find the first diagram to

[7] ὁ γὰρ ἐκτὸς μείζων τοῦ ἐντός.

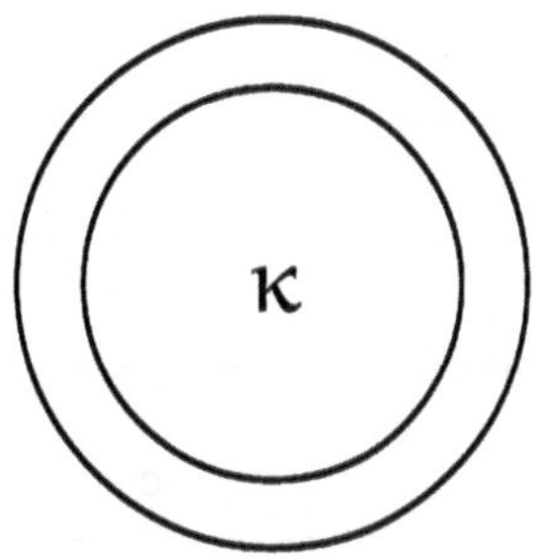

Fig. 5.11 Reconstruction of the archetype

problem 1 in the manuscripts, the reconstruction of which is shown in Fig. 5.11. We do not find a reference in the text to this diagram, and neither do we require any clarification of the text at this point. The figure was probably inserted in the Byzantine period for didactical reasons. The scribe of the archetype might have added it to aid a better understanding of the text, perhaps together with a scholium to this passage that has not been preserved.

The reason that the point on the greater radius moves faster than the point on the smaller one is that the radius describing the circle undergoes two movements. The author first applies this division of motion into two components to the case of rectilinear motion. In rectilinear motion, the components of motion move in a fixed proportion, and therefore, the motion will necessarily occur in a straight line. Rectilinear motion is illustrated by the parallelogram of movements (lines 848b13-23):

> Let the ratio in which the moved object is moved be that of AB to AΓ; and let A be moved to B, and let the line AB be moved down to ΘΓ; now let A be brought to Δ, and AB to E. If the ratio of the movement is that of AB to AΓ, then the line AΔ must have the same ratio to AE. For the small quadrilateral is in the same ratio [to the larger], so that their diagonal is the same, and A will be at Z. It can be shown in the same way, wherever the movement be interrupted; for it will always be on the diagonal.[8]

If the motion of a body starting in A is represented by the ratio of AB to AΓ, it will always be on the diagonal of the parallelogram. Point A moves towards B,

[8] ἔστω γὰρ ὁ λόγος ὃν φέρεται τὸ φερόμενον, ὃν ἔχει ἡ AB πρὸς τὴν AΓ· καὶ τὸ μὲν Δ φερέσθω πρὸς τὸ B, ἡ δὲ AB ὑποφερέσθω πρὸς τὴν ΘΓ· ἐνηνέχθω δὲ τὸ μὲν A πρὸς τὸ Δ, ἡ δὲ ἐφ' ᾗ AB πρὸς τὸ E. εἰ οὖν ἐπὶ τῆς φορᾶς ὁ λόγος ἦν ὃν ἡ AB ἔχει πρὸς τὴν AΓ, ἀνάγκη καὶ τὴν AΔ πρὸς AE τοῦτον ἔχειν τὸν λόγον. ὅμοιον ἄρα ἐστὶ τῷ λόγῳ τὸ μικρὸν τετράπλευρον, ὥστε ἡ αὐτὴ διάμετρος αὐτῶν, καὶ τὸ Δ ἔσται πρὸς τὸ Z. τὸν αὐτὸν δὴ τρόπον δειχθήσεται κἂν ὁπουοῦν διαληφθῇ ἡ φορά· αἰεὶ γὰρ ἔσται ἐπὶ τῆς διαμέτρου.

AΓ editions.

ΘΓ family *a*; BΓ *cett.*; HΓ editions.

οὐκ οὖν manuscripts.

AΔ πρὸς AE H[a]; AΔ πρὸς AZ L; AΔ πρὸς δὲ family *b* and *c*; AΔ πρὸς τὴν AE Pachymeres and editions.

add. τῷ μεγάλῳ B[e]; add. τῷ μείζονι editions.

add. καὶ Pachymeres and editions.

Δ family *b* and *c*.

om. τὸ editions.

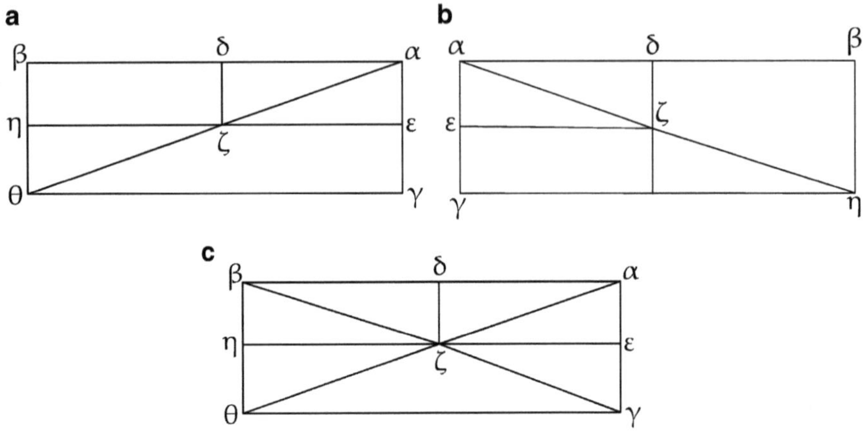

Fig. 5.12 (**a**) Reconstruction of the archetype (**b**) Edition by van Cappelle (**c**) Reconstruction of the hyparchetype of family *c*

while the line AB moves towards ΘΓ. Figure 5.12 illustrates several of the diagrams of the parallelogram of movements in the manuscripts and the critical editions of the text. The orientation of the diagrams contained in the manuscripts is different from the representation in the printed editions. In all manuscript versions the motion proceeds from the upper right to the bottom left of the parallelogram, the labelling progressing in a counterclockwise direction. The diagram from the editions, however, places the letters clockwise, which results in a motion in another direction.

There are some further disagreements in the labelling of the different diagrams. The ancient copyist labels the diagram in accordance with the manuscript readings, only adding the letter H, since this label is not mentioned in the manuscript text. He found one free spot for a letter in his diagram, and probably wanted to avoid any gaps in the alphabetical order of the lettering.[9] The modern editors by contrast did not find the correct variant ΘΓ in line 848b16, since they did not consult any manuscript from family *a* on this passage. Given that they did not pay any attention to the diagrams either, they only had the reading BΓ from the other manuscript families, which was clearly a corruption. Therefore, they had to change the text and emend BΓ to HΓ, and accordingly placed an H at the bottom right of their figure. From analysing the diagrams, however, we learn that emending the text is unnecessary, for we have found the authentic reading of the archetype, ΘΓ, in family *a*. The only question remaining is whether the presence of the letter H in the Byzantine

[9] This need to complete the alphabetical order of the labels occurs more often in the *Mechanics*, for example in problem 23. Here the parallelogram of motions is applied to the case of a rhombus, where one diagonal is longer than the other. The copyist adds the letter I, not mentioned in any of the manuscripts, to his figure in order to have all endpoints of the diagonals labelled. Since this is the last letter in the alphabetical order, there are no shifts in lettering compared with the diagram in the printed editions. Unlike the figure of the parallelogram, where one letter is missing from the alphabetical order, it is impossible to say anything definite about the original intentions of the author.

diagram can be attributed to ancient practice. We cannot give a definite answer to this, but I believe there is at least a chance that this label was intended by the author. Generally in ancient scientific texts, the labels are ordered alphabetically and it would be a little unusual for the H here to be missing. There is a further possibility that letters that occur in the diagram are not contained in the text.[10] The copyist of the archetype might have made the right choice in adding this letter to the figure, thus bringing us somewhat closer to the authentic text.

Figure 5.12c shows that not all manuscript diagrams are similar. All manuscripts in family *c* contain a second diagonal in the parallelogram. This diagonal does not have any special function in the figure; rather, it distracts the reader from the direction of the motion that takes place from the upper right to the bottom left of the parallelogram. For this reason, and due to the fact that the second diagonal is only included in one manuscript family, it is probably not authentic. I believe that the copyist of the hyparchetype added this diagonal erroneously and that it was copied by all later scribes, who remained faithful to their exemplar.

5.3.2 Circular Motion

In the next step of the argument, these two components of motion are applied to the case of circular motion. In contrast to rectilinear motion, here the proportion between the two motions is not fixed, and therefore the movement will occur in a curve instead of a straight line. The manuscripts at lines 849a2-6 read:

> Let there be a circle, ΑΒΓ, and let the endpoint Β be moved to Δ; it comes at some time to Γ. If then it were moved in the ratio of ΒΕ to ΕΓ, it would move along the diagonal ΒΓ. But now, as it is moved in no such ratio, it moves along the circumference ΒΕΓ.[11]

Figure 5.13 illustrates the diversity of the diagrams that are constructed based on these lines. The endpoint Β of the diameter in the circle ΑΒΓ is displaced to Δ, and arrives at some point at Γ. The position of points Β, Δ, and Γ in the diagram reconstructed from the archetype is slightly awkward: how should we imagine point Β moving in the direction of Δ but coming to Γ, with all three points lying in a straight line? If it is moved in the ratio of ΒΕ to ΕΓ, it would move along the diameter ΒΓ. This is evident from the reconstructed diagram, where Ε is a flexible point that can take up a position everywhere along the curve. Since the endpoint Β, however, is moved in no such proportion, it will move along the curve ΒΕΓ. The diagram reconstructed from the archetype is a perfect attempt to reconcile all aspects of the text: points Α, Β, and Γ are all positioned on the circle, the ratio of the motions is that of

[10] See Netz (2004), 170–171.

[11] ἔστω κύκλος ὁ ΑΒΓ, τὸ δ’ ἄκρον τὸ ἐφ’ οὗ Β φερέσθω ἐπὶ τὸ Δ· ἀφικνεῖται δέ ποτε ἐπὶ τὸ Γ. εἰ μὲν οὖν ἐν τῷ λόγῳ ἐφέρετο ὃν ἔχει ἡ ΒΕ ΕΓ, ἐφέρετο ἂν τὴν διάμετρον τὴν ἐφ’ ᾗ ΒΓ. νῦν δέ, ἐπείπερ ἐν οὐδενὶ λόγῳ, ἐπὶ τὴν περιφέρειαν φέρεται τὴν ἐφ’ ᾗ ΒΕΓ.

ἡ ΒΔ πρὸς τὴν ΔΓ Pachymeres and editions.

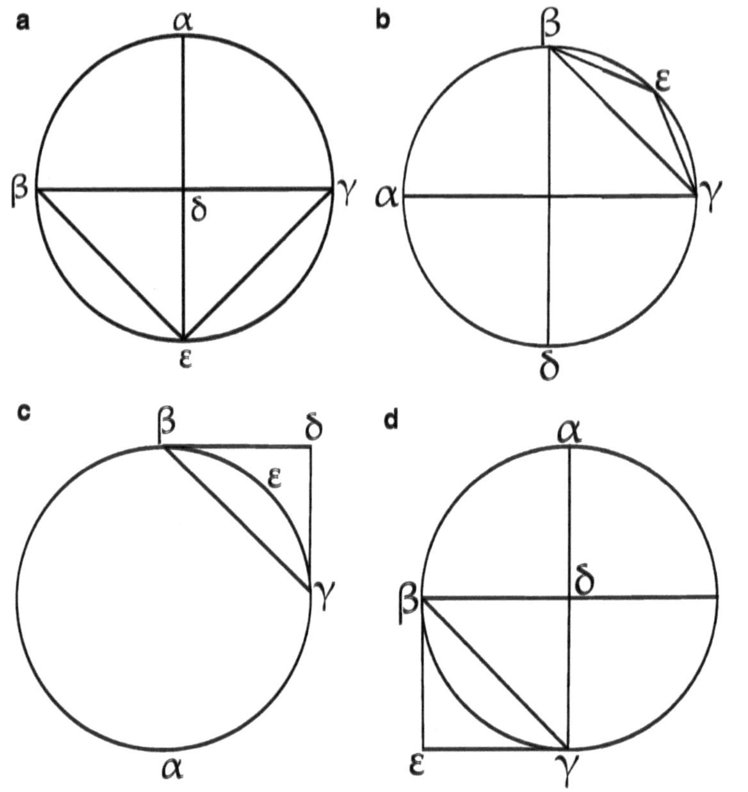

Fig. 5.13 (**a**) Reconstruction of the archetype (**b**) Renaissance interpretation (Niccolò Leonico Tomeo) (**c**) Edition by van Cappelle (**d**) My construction

BE to EΓ, and if such a motion existed point B would move along BΓ, which is at the same time diameter of the circle and diagonal of the parallelogram. The motion along the curve BEΓ takes place at the bottom left segment of the circle in a downward direction, which is consistent with the initial question on balances at the beginning of problem 1 of the *Mechanics*.

Another interpretation of the text can be found in the Latin translation of the *Mechanics* by the Renaissance scholar Niccolò Leonico Tomeo (Fig. 5.13b).[12] In his diagram point B is placed at the top of the circle and moves in a clockwise direction towards Γ in the ratio of BE to EΓ. Point Δ is moved from the centre of the circle to a position on the curve, where we can picture one end of the diameter B to pass through point Γ on its way to Δ.

[12] Leonico Tomeo (1525), 26v-27r translates the text as follows: *Sit circulus ABCD: extremum autem ubi est B feratur ad ipsum D; pervenit sane aliquando ad ipsum C. Si quidem igitur in proportione feratur, quam habet BE, EC: fertur utique secundum diametrum ubi BC. Nunc autem quoniam in nulla proportione, in circumferentia certe fertur ubi BEC.* Similar diagrams can be found in other Renaissance works; see e.g. Piccolomini (1565), 13r or Blancanus (1615), 151.

Both the reconstruction of the archetype and Leonico Tomeo's diagram, however, do not take into account the direction of the two motions of which circular motion is made up. From lines 849a14-16 of the text it is evident that one of the components of circular motion is directed towards the centre, whereas the other is tangential. This notion of tangential and radial motion is considered in the diagram from the critical editions of the text (Fig. 5.13c). Here, another difficulty arises, namely in the ratio of the motions. As we can see in the diagram, the ratio of BE to EΓ has been replaced by a different proportion that is only possible with an emendation of the text in line 849a4: ἡ BΔ πρὸς τὴν ΔΓ.[13] This emendation was made by Pachymeres and adopted in all printed editions, without anyone noticing that this might not have been the authentic manuscript reading. Unfortunately, we do not know how Pachymeres understood circular motion, since he did not include a diagram here.

All critical editions further depict point B at the top of the circle, thereby translating ἄκρον in line 849a2 with "summit". The use of the word ἄκρον is somewhat unusual in relation to circles, but from other occurrences in problem 1 of the text we can infer its meaning.[14] Here in the *Mechanics*, as is often the case in ancient mathematical texts, ἄκρον is used to denote an endpoint of a diameter,[15] and it is understood as such by the scribe of the archetype and in Renaissance interpretations.[16] The exact position of point B is not specified by the text; for this reason it is not necessary to place it at the top of the circle, as the editions suggest.

From among these different representations of circular motion, the interpretation from the Byzantine period is the only one that remains completely faithful to the manuscript readings, attempting to combine all different aspects of the text in one diagram. However, since the two components of motion here are not related to the directions towards the tangent and the centre of the circle, this diagram is not likely to represent the authentic diagram intended by the author. Because of the fact that many scribes had difficulties with this diagram,[17] and because the manuscripts read at 849a4 ἡ BE EΓ where we would have expected ἡ BE πρὸς τὴν EΓ, the text is likely to be corrupt to some degree. Still, instead of emending the text as the modern editors did, I have attempted to construct a figure that integrates all aspects of the text (Fig. 5.13d).

I depict point B at the left side of the circle, since this is where it was also placed in the archetype, and gives the most comprehensible interpretation of circular motion in the context of problem 1. In this problem the accuracy of larger and smaller balances is explained by the motion of the different radii, which is conse-

[13] See n.11 for this emendation of the Greek.

[14] See, among other instances, its occurrence in lines 849a12, 849a14 and 849b2.

[15] See Mugler (1958), 47.

[16] E.g. Leonico Tomeo (1525), 27r translates ἄκρον as *extremum*; Blancanus (1615), 151 contains *extremum diametri*.

[17] These difficulties are visible from the letters that are missing from different manuscript diagrams, but it is most conspicuous in the incomplete diagrams in the independent manuscripts, *Par.gr.* 2507 (P[1]) and *Par.gr.* 2115 (Par).

quentially a vertical and downward motion. The diagram takes into account both the tangential and radial components of motion required by the text: point B is placed at the left side of the figure, and moves towards Δ, which is again placed at the centre of the circle as it is in the manuscript diagrams, but arrives in Γ. Now we have the same situation as in the parallelogram of movements, and the ratio of the motions is represented by BE to EΓ. However, since it is moved in no such ratio, it will move along the curve BEΓ. The difficulty with this diagram is that point E in this figure cannot be on the tangent to B and, on the curve simultaneously.[18] We might consider this diagram to represent the impossible case: if the ratios of the sides of the parallelogram held, point B would move along the diagonal of the figure. Since, however, in circular motion the two components of motion are in no proportion, the situation displayed in the diagram is nonexistent. As a reader of the text, we can picture the actual motion taking place from B to Γ along the arc.

This reconstruction offers only one possible interpretation of circular motion. I will by no means argue that this is the figure intended by the author of the text. There are difficulties with all figures and the text was probably corrupted to some degree. We should perhaps acknowledge the fact that we cannot make a perfect reconstruction of the authentic text and diagram for this passage. The least to be done in a new critical edition of the *Mechanics* is to offer the reconstructed text and the diagram from the archetype side by side. As my examination of the manuscript diagrams has shown, all diagrams confirm the manuscript reading in line 849a4: ἡ BE EΓ. Thus, Pachymeres' emendation of this line, adopted in all modern editions and translations, does not represent the only comprehensible interpretation of the text. We learn much more from the manuscript diagram and by comparing it with the other figures, than we do by looking at a figure in the modern editions that is remote from ancient tradition.

5.3.3 Circular Motion Continued

The analysis of circular motion is continued by the argument that where two objects are moved by the same force and one is more interfered with than the other, the one that suffers more interference will move more slowly. This seems to happen in the case of circles with longer and shorter radii. A point on a shorter diameter is attracted more to the centre, and moves more slowly. The fact that a point closer to the centre is deflected more than a point further away is explained at lines 849a21-849b19:

> Let there be a circle BΓΔE, and XNMΞ another smaller one inside it, about the same centre; and let the diameters be drawn, ΓΔ and BE in the large circle, and MX and NΞ in the small; and let the rectangle ΔΨPΓ be completed. If the line AB describing the circle comes back to the same position from which it started, i.e. to AB, it is clear that it moves towards itself. In the same way AX will come back to AX. But AX moves more slowly than AB, as has

[18] Although one might consider the possibility that the label E occurs twice at different positions in the same figure, which would, however, be somewhat exceptional in ancient diagrams.

> been said, because the interference is greater and AX is more held back. Let AΘH be drawn, and from Θ a perpendicular ΘZ on AB within the circle, and again from Θ let ΘΩ be drawn parallel to AB, and ΩY and HK perpendiculars on AB. Now the lines ΩY and ΘZ are equal. BY then is less than XZ; for in unequal circles equal straight lines drawn perpendicular to the diameter cut off smaller parts of the diameter in the greater circles, and ΩY is equal to ΘZ. In the same time then in which AΘ travels along XΘ the endpoint of BA has moved along a greater arc than BΩ in the greater circle. For the natural movement is equal, but the unnatural less; and BY is less than ZX. But they need to be in proportion, the natural motion to the natural motion, and the unnatural to the unnatural. Now it has moved along the arc HB, which is greater than ΩB. It must have completed HB in this time; for that will be its position, when the proportion between the natural and unnatural motion holds in both ways. If, then, the natural movement is greater in the greater circle, the unnatural movement would only agree in the case that B is moved along BH in the same time as the point X [is moved along XΘ]. For in that case point B naturally moves to H; and unnaturally to K; for HK is the perpendicular from H. Then HK is to KB, as ΘZ is to ZX. This will be obvious, if B and X be joined to H and Θ. But if B is moved along an arc either greater or less than HB, it will not be the same, nor will the natural movement be proportional to the unnatural in the two circles.[19]

The motions of two radii, AB in the larger circle and AX in the smaller one, are analysed. If the longer radius AB moves along the curve, point B moves faster than point X, since B covers a greater distance than X (BH > XΘ), although they are moved by the same force. This is the case, since for the same natural motion (ΘZ = ΩY), the unnatural motion is larger in the smaller circle (XZ > BY); point X suffers greater interference than point B, and travels more slowly. In the time it takes

[19] ἔστω κύκλος ἐφ’ οὗ ΒΓΔΕ, καὶ ἄλλος ἐν τούτῳ ἐλάττων, ἐφ’ οὗ ΧΝΜΞ, περὶ τὸ αὐτὸ κέντρον· καὶ ἐκβεβλήσθωσαν αἱ διάμετροι, ἐν μὲν τῷ μεγάλῳ, ἐφ’ ὧν ΓΔ καὶ ΒΕ, ἐν δὲ τῷ ἐλάττονι αἱ ΜΧΝΞ· καὶ τὸ ἑτερόμηκες παραπεπληρώσθω, τὸ ΔΨΡΓ. εἰ δὴ ἡ ΑΒ γράφουσα κύκλον ἥξει ἐπὶ τὸ αὐτὸ ὅθεν ὡρμήθη ἐπὶ τὴν ΑΒ, δῆλον ὅτι φέρεται πρὸς αὐτήν. ὁμοίως δὲ καὶ ἡ ΑΧ πρὸς τὴν ΑΧ ἥξει. βραδύτερον δὲ φέρεται ἡ ΑΧ τῆς ΑΒ, ὥσπερ εἴρηται, διὰ τὸ γίνεσθαι μείζονα τὴν ἔκκρουσιν καὶ ἀντισπᾶσθαι μᾶλλον τὴν ΑΧ. ἤχθω δὲ ἡ ΑΘΗ, καὶ ἀπὸ τοῦ Θ κάθετος ἐπὶ τὴν ΑΒ ἡ ΘΖ ἐν τῷ κύκλῳ, καὶ πάλιν ἀπὸ τοῦ Θ ἤχθω παρὰ τὴν ΑΒ ἡ ΘΩ, καὶ ἡ ΩΥ ἐπὶ τὴν ΑΒ κάθετον, καὶ ἡ ΗΚ. αἱ δὴ ἐφ’ ὧν ΩΥ καὶ ΘΖ ἴσαι. ἡ ἄρα ΒΥ ἐλάττων τῆς ΧΖ· αἱ γὰρ ἴσαι εὐθεῖαι ἐπ’ ἀνίσους κύκλους ἐμβληθεῖσαι πρὸς ὀρθὰς τῇ διαμέτρῳ ἔλαττον τμῆμα ἀποτέμνουσι τῆς διαμέτρου ἐν τοῖς μείζοσι κύκλοις, ἔστι δὲ ἡ ΩΥ ἴση τῇ ΘΖ. ἐν ὅσῳ δὴ χρόνῳ ἡ ΑΘ τὴν ΧΘ ἐνηνέχθη, ἐν τοσούτῳ χρόνῳ ἐν τῷ κύκλῳ τῷ μείζονι μείζονα τῆς ΒΩ ἐνήνεκται τὸ ἄκρον τῆς ΒΑ. ἡ μὲν γὰρ κατὰ φύσιν φορὰ ἴση, ἡ δὲ παρὰ φύσιν ἐλάττων· ἡ δὲ ΒΥ τῆς ΖΧ. δεῖ δὲ ἀνάλογον εἶναι, ὡς τὸ κατὰ φύσιν πρὸς τὸ κατὰ φύσιν, τὸ παρὰ φύσιν πρὸς τὸ παρὰ φύσιν. μείζονα ἄρα περιφέρειαν διελήλυθε τὴν ΗΒ τῆς ΩΒ. ἀνάγκη δὲ τὴν ΗΒ ἐν τούτῳ τῷ χρόνῳ διεληλυθέναι· ἐνταῦθα γὰρ ἔσται, ὅταν ἀνάλογον ἀμφοτέρως συμβαίνῃ τὸ παρὰ φύσιν πρὸς τὸ κατὰ φύσιν. εἰ δὴ μεῖζόν ἐστι τὸ κατὰ φύσιν ἐν τῇ μείζονι, καὶ τὸ παρὰ φύσιν μᾶλλον ἂν ἐνταῦθα συμπίπτοι μοναχῶς, ὥστε τὸ Β ἐνηνέχθαι ἂν τὴν ΒΗ ἐν τῷ ἐφ’ οὗ Χ σημεῖον. ἐνταῦθα γὰρ κατὰ φύσιν μὲν γίνεται τὸ Β σημεῖον ἐς τὸ Η· παρὰ φύσιν δὲ ἐς τὸ Κ· ἔστι γὰρ ἡ ΗΚ ἀπὸ τοῦ Η κάθετος. ἔστι δὲ ὡς τὸ ΗΚ πρὸς τὸ ΚΒ, τὸ ΘΖ πρὸς τὸ ΖΧ. φανερὸν δὲ ἐὰν ἐπιζευχθῶσιν ἀπὸ τῶν ΒΧ ἐπὶ τὰ ΗΘ. εἰ δὲ ἐλάττων ἢ μείζων τῆς ΗΒ ἔσται, ἣν ἠνέχθη τὸ Β, οὐχ ὁμοίως ἔσται οὐδὲ ἀνάλογον ἐν ἀμφοῖν τὸ κατὰ φύσιν πρὸς τὸ παρὰ φύσιν.

add. τὸ A Pachymeres and editions.

AB family *a*; AE *cett.*

ἐνταῦθα γὰρ κατὰ φύσιν μὲν γίνεται τὸ Β σημεῖον ἐς τὸ Η· παρὰ φύσιν δὲ ἐς τὸ Κ· ἔστι γὰρ ἡ ΗΚ ἀπὸ τοῦ Η κάθετος Par; ἐνταῦθα γὰρ κατὰ φύσιν μὲν γίνεται τῷ Β σημείῳ τὸ κέντρον, ἔστι γὰρ αὐτὴ ἀπὸ τοῦ Η κάθετος, παρὰ φύσιν δὲ ἐς τὸ ΚΒ *cett.*

X to move to Θ, B moves further than Ω, namely to H (since BK:KH = XZ:ZΘ). This analysis enables the author to answer his initial question of why larger balances are more accurate than smaller ones. The cord of the balance acts as the centre, while both arms of the balance are the radii from the centre. The longer the arms are, the more they will move, even under a very small weight, of which the displacement might not even be visible in a smaller balance.

Figure 5.14 compares the diagram in the Byzantine manuscript tradition with its modern counterpart in the critical editions. The manuscript versions of the diagrams to this problem contain numerous errors. On the one hand, this is due to the complexity of the diagram; here, a diagram is required for a proper understanding of the principle in question. On the other hand, the fact that the diagrams were drawn freehand, especially in a diagram where many different letters are packed so closely together, increases the number of errors. The letter A as the centre of both circles is not specified by the text in the manuscripts at line 849a23. Since it receives increased specification from line 849a26 onwards, the copyist was able to insert the letter into his diagram. From this passage we learn that the radius AB will come back to the same position from which it started, namely AB, and that the same is true for the shorter radius AX, which will return to AX. This practice whereby labels are unspecified at the beginning of a description, but receive increased or complete specification in the course of an argument, is common to Greek mathematics.[20] Therefore, the emendation by Pachymeres at 849a23 add. τὸ A, that was adopted in all critical editions, is an unnecessary addition and needs to be removed in a new critical edition of the *Mechanics*.

The reconstructed diagram from the archetype differs again in many respects from the diagram in the printed editions. It contains, for instance, two additional lines, ΘK and HB. These lines are required by the text at 849b14-16: "Then HK is to KB, as ΘZ is to ZX. This will be obvious, if B and X be joined to H and Θ."[21] From the similar triangles which then appear in the diagram, we recognize that the ratios of the lines are the same. The first editor to include diagrams in his critical edition was van Capelle. He probably overlooked this passage in the text, and therefore omitted lines ΘK and HB from his diagram. Since none of the subsequent editors checked the manuscript diagrams themselves, they too failed to notice that these lines were missing.

From Fig. 5.14 we further see that the orientation of the diagram in the editions is very different from that of the archetype.[22] Whereas the motion in the archetype takes place at the bottom left segment of the circle and is thus depicted as a downward motion, the motion in the figure from the modern editions is demonstrated by a lateral motion. As I have already explained regarding Fig. 5.13 on circular motion, it is much more comprehensible to depict the motion here as a vertical and downward motion than a horizontal and lateral one. Problem 1 describes the motions in

[20] See Netz (1999), 21 and Sect. 4.2.2 where it is described as a characteristic of the *Mechanics* as well.

[21] ἔστι δὲ ὡς τὸ HK πρὸς τὸ KB, τὸ ΘZ πρὸς τὸ ZX. φανερὸν δὲ ἐὰν ἐπιζευχθῶσιν ἀπὸ τῶν BX ἐπὶ τὰ HΘ.

[22] Note that the letter B is again depicted at the top of the diagram in the critical editions, just as in the first diagram on circular motion.

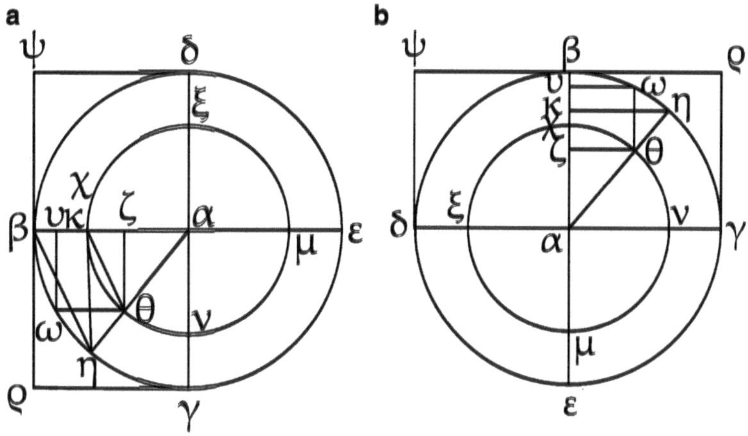

Fig. 5.14 (**a**) Reconstruction of the archetype (**b**) Edition by van Cappelle

larger and smaller balances; line BE should therefore be compared to a balance beam. From the diagram in the archetype, it is evident that the principle was understood as such by the scribe in the Byzantine period.

Bibliography

Apelt, Otto (ed.). 1888. *Aristotelis quae feruntur De Plantis, De Mirabilibus Auscultationibus, Mechanica, De Lineis Insecabilibus, Ventorum Situs et Nomina, De Melisso Xenophane Gorgia*. Leipzig: Teubner.

Bekker, Immanuel (ed.). 1831. *Aristotelis Opera II*. Berlin: Reimer.

Blancanus, Josephus. 1615. *Aristotelis Loca Mathematica: Ex universis ipsius Operibus collecta et explicata*. Bologna: Cochius.

Bottecchia, Maria Elisabetta (ed.). 1982. *Aristotele: MHXANIKA, Tradizione manoscritta, testo critico, scolii*. Padua: Antenore.

Hett, Walter S. (ed.). 1936. *Aristotle: Minor works*. Cambridge, MA: Harvard University Press.

Leonico Tomeo, Niccolò. 1525. *Nicolai Leonici Thomaei Opuscula nuper in lucem aedita quorum nomina proxima habentur pagella*. Venice: Bernardino Vitali.

Mugler, Charles. 1958. *Dictionnaire historique de la terminologie géométrique des Grecs*. Paris: Klincksieck.

Netz, Reviel. 1999. *The shaping of deduction in Greek mathematics: A study in cognitive history*. Cambridge: Cambridge University Press.

Netz, Reviel. 2004. The limits of text in Greek mathematics. In *History of science, history of text*, ed. K. Chemla, 161–176. Dordrecht: Springer.

Piccolomini, Alessandro. 1565. *In Mechanicas Quaestiones Aristotelis, Paraphrasis paulo quidem plenior*, 2nd ed. Venice: Traianus Curtius.

Van Cappelle, Johannes Petrus (ed.). 1812. *Aristotelis Quaestiones Mechanicae*. Amsterdam: Den Hengst.

Van Leeuwen, Joyce. 2014. Thinking and learning from diagrams in the Aristotelian *Mechanics*. *Nuncius* 29: 53–87.

Chapter 6
Early Modern Transformations of Mechanics

While the text of the *Mechanics* had been apparently unknown in the Middle Ages, it was rediscovered in the early modern period and intensively studied in numerous translations, paraphrases and commentaries.[1] The first Latin translation of the treatise was made by Vittore Fausto in the year 1517, followed in 1525 by the work of the next translator Niccolò Leonico Tomeo. It was Leonico Tomeo's edition that became the standard translation in the sixteenth century.[2] His text and Alessandro Piccolomini's Latin paraphrase (1547) have been very influential for the spread of mechanical knowledge in this century and formed the basis for many later commentaries on the text. Bernardino Baldi, for example, refers to both texts in his 1621 published *In Mechanica Aristotelis Problemata Exercitationes*.[3] This chapter will focus primarily on the Latin translations of the *Mechanics* from the first half of the sixteenth century. These works by the humanists Fausto, Leonico Tomeo and Piccolomini are characterized by a philological concern for establishing a reliable text, and reveal which Greek texts underpinned the reception of the *Mechanics* in the early modern period. Leonico Tomeo's *Quaestiones Mechanicae* (*Mechanical Questions*) represent an exceptional case in this context. We possess both a handwritten Greek manuscript and a Latin printed edition in the hand of Leonico Tomeo, which enables us to observe closely the transmission processes of the text. Apart from the humanists' search for sound Greek manuscripts, the text itself was interpreted according to the altered scope of Renaissance mechanics. The diagrams contained in the translations and commentaries are especially relevant, since they reflect the focus of the individual interpreters. Therefore, some of the problems will be analysed in more detail by means of the diagrams in order to describe the processes of transformation and appropriation of mechanical knowledge in the early modern period. An evaluation of the conception of the mechanical discipline in antiquity and the early modern period will help to gain a better understanding of these processes.

[1] Parts of this chapter have appeared in van Leeuwen (2014).

[2] See Rose and Drake (1971), 80.

[3] See Nenci (2011) for a discussion of Baldi's commentary on the *Mechanics*, including a facsimile of this important work.

J. van Leeuwen, *The Aristotelian Mechanics*, Boston Studies in the Philosophy and History of Science 316, DOI 10.1007/978-3-319-25925-3_6

6.1 The Mechanical Discipline

The introduction to the *Mechanics* contains several remarks on the scope of mechanics in antiquity. These passages will be compared with Aristotle's division of the sciences and analysed against the background of the conception of the mechanical discipline in the early modern period. The very beginning of the text instructs us on the relationship between art and nature in ancient mechanics (847a11-13): "We wonder at, firstly, those things that occur according to nature, of which the cause is unknown, and secondly at those things which are produced unnaturally by art for the benefit of mankind."[4] Of particular interest is the question of how to understand the motion *para phusin* that is produced by mechanical devices. In modern translations of the *Mechanics*, it is sometimes translated by "contrary to nature",[5] which seems to suggest that in antiquity forced mechanical motions were conceived of as acting entirely against nature. As we will see, this neither complies with the Aristotelian view, nor was it interpreted as such by the early modern commentators. However, in the literature the concept of mechanics in antiquity is often contrasted with later views on the subject. The ancients would have held that mechanics works against nature and is therefore excluded from natural philosophy.[6] Mechanical devices were designed to outwit nature and can be considered an "Überlistung der Natur", as Fritz Krafft defines it.[7] For that reason, the study of mechanics cannot yield any knowledge of nature. This line of thought stems from a misapplication of Galileo Galilei's rejection of the idea that results from mechanics cannot be applied to natural philosophy.[8] Galileo challenged the contemporary view that mechanics would break nature's laws, by showing that mechanical devices depend upon the principles of natural philosophy. Apparently, as Sylvia Berryman argues, this idea was held by some engineers in Galileo's day[9]; it is, however, very different from the ancient understanding of mechanics.

That Aristotle did not regard mechanics as being excluded from natural philosophy, is illustrated by the example of machines. In chapter 7 of his *Movement of Animals*, he compares the motions of animals with those of automated puppets. In this case, machines can serve as models to understand the processes in the natural world. Art and nature were thus not regarded by Aristotle as two distinct fields, in which the knowledge gained in the one could not be transferred to the other. In order to explain Aristotle's view on the relationship between art and nature, we need to distinguish between two different meanings of *para phusin*. Mechanical motions can be *para phusin* in the sense that they go against the nature of the moved object,

[4] θαυμάζεται τῶν μὲν κατὰ φύσιν συμβαινόντων, ὅσων ἀγνοεῖται τὸ αἴτιον, τῶν δὲ παρὰ φύσιν, ὅσα γίνεται διὰ τέχνην πρὸς τὸ συμφέρον τοῖς ἀνθρώποις.

[5] See e.g. Hett (1936), 331.

[6] See Krafft (1970), 156.

[7] See Krafft (1970), 157.

[8] See Berryman (2009), 43.

[9] See Berryman (2009), 44.

for example when a lever moves a heavy body in an upward direction. This is visible from *Physics* VIII.4, where Aristotle states that a lever causes an unnatural motion (lines 255a21-22): "Some of them are capable of causing motion unnaturally (e.g. the lever is not naturally capable of moving a weight), […]."[10] This does, however, not imply that mechanical motions are against nature per se. In the Aristotelian corpus we find many parallels between art and nature, as, for example, the following passage from *Physics* II.8 (lines 199a12-17):

> Thus, if a house were a natural product, the process would pass through the same stages that it in fact passes through when it is produced by art; and if natural products could also be produced by art, they would move along the same line that the natural process actually takes. We may therefore say that the earlier stages are for the purpose of leading to the later. Indeed, as a general proposition, the arts either, on the basis of nature, carry things further than nature can, or they imitate nature.[11]

This passage emphasizes the close analogy between art and nature, upon which art imitates nature, but it also demonstrates that art can improve on nature. From this, we can derive the second meaning of *para phusin*, as something that goes beyond nature. One should not think of beyond nature to mean the supernatural, but it should be understood in the sense of bringing about effects that would not be possible without the intervention of art, as Mark Schiefsky has argued.[12] Art imitates nature and is therewith able to complete what nature alone cannot. The second meaning of the term *para phusin* is very clear from the introduction to the text, which states that some things cannot be accomplished by nature alone; in such cases, the intervention of mechanics is required, and mechanical devices are applied in order to improve on nature. Examples of such devices are the lesser that masters the greater, or things possessing little weight that move heavy weights (847a21-23).

The view of mechanics as something that is beyond rather than against nature, is supported by the Latin tradition. We notice that the term *para phusin* was not translated with *contra naturam*. Fausto in the first Latin translation of the text renders the term as *extra rerum naturam* or *praeter naturae lege*, to be understood as being outside or beyond nature. Leonico Tomeo in the comments to his translation offers the same interpretation, while stating that art transgresses the laws of nature.[13] He consequently translates *para phusin* by *praeter naturam*. Piccolomini offers the same translation while following a similar line of thought, in that he states that mechanics imitates and helps nature, and can produce versatile effects when compared with the simplicity of nature.[14]

The author of the *Mechanics* goes on to define the place of mechanics amongst the other sciences. The problems of mechanics should be regarded as a mixed science, for "these are not wholly identical with nor entirely separate from physical problems, but they are common to both mathematical and physical speculations; for

[10] Transl. Hardie and Gaye in Barnes (1984), 426.

[11] Transl. Hardie and Gaye in Barnes (1984), 340.

[12] See Schiefsky (2007), 72.

[13] See Leonico Tomeo (1525), 23r.

[14] See Piccolomini (1565), 7v.

the *how* is demonstrated by mathematics, but the *about* by physics" (847a24-28).[15] The relationship between mechanics and mathematical science coheres with Aristotle's classification of the mechanical discipline in the corpus. In the *Physics* II.2, Aristotle defines optics, harmonics and astronomy as the more physical of the mathematical sciences (lines 194a7-8): "Similar evidence is supplied by the more natural of the branches of mathematics, such as optics, harmonics, and astronomy."[16] Even though Aristotle does not mention mechanics in this context, from other passages in the corpus we can infer that he would probably also think of mechanics as a mathematical science.[17] In the *Posterior Analytics*, Aristotle further specifies the relations in these mathematical sciences, when he speaks of a subordination to pure mathematics (lines 78b35-39): "And such are those which are related to each other in such a way that the one is under the other, e.g. optics to geometry, and mechanics to solid geometry, and harmonics to arithmetic, and star-gazing to astronomy."[18] The next few lines give a clarification of how we should understand this subordination (79a2-3): "For here it is for the empirical scientists to know the fact and for the mathematical to know the reason why."[19] The fact is provided by the subordinate science itself, for example, optics, but the demonstration is given by the superior science, which, in the case of optics, is geometry. The discipline of mechanics makes use of principles that are common to solid geometry, and can as such be regarded as subordinate to this science.

The early modern scholars also comment upon the position of mechanics amongst the other sciences. Lines 847a27-28 of the text expounding the relation of mechanics to mathematics and physics were omitted from Leonico Tomeo's Latin translation. A remark by Leonico Tomeo in the right margin of f.99r in codex *Bern.* 402 (B^e) shows that he considered these lines a scholium, and thus intentionally omitted them from the later edition. However, in his notes he explains that the mechanical problems are common to both physics and mathematics. The subject matter is natural, since mechanical devices are made of natural materials such as iron or wood, whereas the controlling principles of these devices are mathematical.[20] When abstracted from the natural material, the underlying mathematical principles of mechanics will be disclosed. Piccolomini in his paraphrase takes a similar view of the subject, while explaining the intermediate status of mechanics. Its principles are derived from mathematics, and to this end mechanics is subordinate to

[15] ἔστι δὲ ταῦτα τοῖς φυσικοῖς προβλήμασιν οὔτε ταὐτὰ πάμπαν οὔτε κεχωρισμένα λίαν, ἀλλὰ κοινὰ τῶν τε μαθηματικῶν θεωρημάτων καὶ τῶν φυσικῶν· τὸ μὲν γὰρ ὥς διὰ τῶν μαθηματικῶν δῆλον, τὸ δὲ περὶ ὃ διὰ τῶν φυσικῶν.

[16] Transl. Hardie and Gaye in Barnes (1984), 331.

[17] See for example *Metaph.* 1078a14-17 and *APo.* 76a22-25 on the close connection between optics, harmonics and mechanics.

[18] Transl. Barnes in Barnes (1984), 128.

[19] Transl. Barnes in Barnes (1984), 128. The terms ὅτι and διότι from the *APo.* may correspond to περὶ and ὥς from the introduction of our text.

[20] See Leonico Tomeo (1525), 23v.

geometry.[21] At several points, Piccolomini mentions the connection of mechanics with optics, astronomy and harmonics in the works of Aristotle.[22] We see that Piccolomini accepts Aristotle's division of the sciences and also defines mechanics as being more mathematical than physical, since, most importantly, its mode of demonstration is mathematical.

In this context, the early modern commentators also discuss the issue of whether mechanics should be considered an art or a science. In the introductory lines of the text, mechanics is specified as an art or *technē* as it concerns practical applications in the physical world. Apart from dealing with physical objects, we have seen that mechanics can be considered mathematical in that the working of mechanical devices is demonstrated by mathematics. This is clear from the argumentative structure of the text, in which all mechanical phenomena ultimately refer to the geometrical properties of the circle. From this theoretical understanding it follows that mechanics can be identified as a science or *epistēmē* as well. This definition of the mechanical discipline as both *epistēmē* and *technē* seems somewhat problematic for Aristotle. In the *Nicomachean Ethics* Book VI, Aristotle makes a clear distinction between an *epistēmē* as an ἕξις ἀποδεικτική, a demonstrative *habitus*, and a *technē* as an ἕξις μετὰ λόγου ἀληθοῦς ποιητική, a productive *habitus* which involves reasoning.[23] However, this distinction between the theoretical and the practical is not consistently applied throughout Aristotle's works.[24] The definition of mechanics in the introduction to the treatise might be comparable to the way in which Aristotle describes medicine as both a science and an art.[25]

It is exactly this ambiguity in Aristotle's works that was pointed out by Piccolomini in his paraphrase. Piccolomini mentions that on numerous occasions, not only in the *Mechanics*, but also in his *Generation of Animals* and *Metaphysics*, Aristotle uses the noun art while referring to science.[26] By highlighting this incongruity in the Aristotelian corpus, we notice that Piccolomini acknowledged the authority of Aristotle and made a great effort to reconcile the text of the *Mechanics* with the other works by Aristotle. For Piccolomini and other commentators in the early modern period, mechanics could be considered as both an art and a science. It is an art to the extent that it involves the construction of machines, but a science pertaining to the geometrical causes and principles of these mechanical devices. According to Piccolomini, it is again the mathematical mode of demonstration that

[21] See Piccolomini (1565), 8r.

[22] See Piccolomini (1565), 5r.

[23] See *EN* 1139b31-32 and 1140a10.

[24] See Micheli (1995), 31 n.41, where he points to the ambiguity in Aristotle between the terms *technē* and *epistēmē*; sometimes they are distinct from one another, whereas at other times they are treated as synonyms.

[25] See e.g. *APo.* 79a13-14, where medicine is defined as a science (*epistēmē*); and *Metaphysics* 1032b9-10, in which a physician is described as being responsible for producing health, and medicine is defined as a *technē*.

[26] See Piccolomini (1565), 8r.

determines that mechanics is first and foremost a science.[27] That mechanics was primarily deemed to be a science in the early modern period indicates its elevated status in comparison to the Middle Ages, when it was identified with the mechanical or sellularian arts that employed mathematics.[28] In the early modern period, the theoretical and practical aspects of mechanics were further elaborated, and we will see in the next sections that these developments led to a new kind of diagrams.

6.2 Latin Translations and Their Sources

The first Latin translation of the *Mechanics* was made by the Italian humanist Vittore Fausto and published in Paris in the year 1517. However, Fausto's aspirations went beyond the theoretical level and he displayed a particular interest in shipbuilding, as is evident from his experiences at the Venetian Arsenal.[29] The edition consists of a preface and a Latin translation of the text accompanied by diagrams. Fausto does not make mention of the Greek source on which he based his translation, but Rose and Drake consider it likely that it was the Aldine *editio princeps* printed in Venice between 1495 and 1498.[30] A comparison between the *editio princeps* and Fausto's translation for the most part confirms this assumption; however, the Aldine cannot have been the only source for Fausto's text. On the first pages of his translation we find two remarks from Georgius Pachymeres' paraphrase of the *Mechanics* that are not present in the Aldine. These phrases at lines 848a17 and 848a37 from the introduction of the text have been rendered literally in Latin by Fausto. This connection with the paraphrase by Pachymeres is also apparent at lines 854b30-31, a passage omitted from the Aldine, but present in both Pachymeres and Fausto.[31] Further smaller similarities in the labelling of the diagrams confirm a relationship between the two. At the bottom left of the diagram of the parallelogram of motions we find in Fausto the label *t*. Since in other questions, such as problem 23 or 25 of the text, Fausto consequently translates the Greek letter Θ by *t*, it is likely that this label was contained here in his source as well. However, the Aldine contains the erroneous label B, whereas only manuscript family *a* and Pachymeres contain the correct reading Θ as set out in the previous chapter. In this case, an influence from the paraphrase by Pachymeres is more likely than that Fausto would have emended the text on his own account and inserted the letter *t*, which forms a clear

[27] See Piccolomini (1565), 8rv.

[28] See Laird (1986), 46–47.

[29] See Valleriani (2010), 118–119.

[30] See Rose and Drake (1971), 78.

[31] Despite the similarities between Fausto and Pachymeres, Fausto's translation cannot have been completely based on a copy of Pachymeres' paraphrase. Many passages that are omitted from Pachymeres, e.g. at 847a27-28, 850a31, 850b7-8 and 853a17-18, are present in the Aldine and Fausto. Moreover, if such a reliance on Pachymeres existed, we would expect to see more influences from it in Fausto's translation, amongst others, in the diagrams.

break in the otherwise alphabetical ordering of the letter labels. Because of the fact that the majority of the readings in Fausto's translation agree with the *editio princips*, whereas influences from Pachymeres are less frequent, it seems likely that Fausto made use of a copy of the Aldine that contained annotations derived from Pachymeres' paraphrase. If Fausto had consulted a manuscript of Pachymeres himself, we would have expected to see more substantial influences, for instance in the number of phrases that were adopted from it and in the diagrams. Fausto's diagrams appear to be unrelated to any of the manuscript diagrams, and are likely to represent his own representations of the mechanical questions. These diagrams display a careful reading and understanding of the text, as will be shown below.

Despite the fact that Fausto's Latin translation made the text of the *Mechanics* available to a larger audience for the first time, it was not his work, but that by the subsequent translator that was most influential during the early modern period. Niccolò Leonico Tomeo, a humanist and professor of Aristotelian philosophy in Padua, made translations and commentaries on several works by Aristotle. His Latin translation of the *Mechanics* with his diagrams and comments, the *Mechanical Questions*, appeared in the year 1525. The translation was reprinted many times, however, only the 1525 (Venice) and 1530 (Paris) editions include his diagrams and comments.[32] Leonico Tomeo's text was used by many successive commentators and became the standard Latin translation in the sixteenth century. We have already seen that Leonico Tomeo was the copyist of the Greek codex B^e that, together with some influences from the paraphrase by Pachymeres, provided the basis for his Latin translation. The manuscript contains later emendations and annotations in Leonico Tomeo's hand, many of which were adopted in his translation. At f.99r, for instance, Leonico Tomeo added a marginal note indicating that he believed lines 847a27-28 to be a scholium; accordingly, these lines have been omitted from the translation. When looking at the case of circular motion in the first problem of the text, Leonico Tomeo added the label Δ to the circle originally defined by the letters ΑΒΓ only. The Latin translation thereupon labels the circle ABCD. In the second diagram to circular motion, we have seen that the centre of the circles was specified by Pachymeres at line 849a23 as τὸ A. This emendation was added later by Leonico Tomeo in B^e and translated as such in the *Mechanical Questions*. However, not every annotation in codex B^e was adopted in the Latin translation, as, for example, passage 856b28-29 that was added in the left margin of f.111v yet is missing from the Latin edition. Moreover, Leonico Tomeo was not always accurate in his Latin translation as is visible from several omissions (for example, at lines 850a37-39) and incorrect letter labels. The relationship between the diagrams in Leonico Tomeo's manuscript and the printed edition will be explored in the next section.

Another Italian humanist, Alessandro Piccolomini, was responsible for making a Latin paraphrase of the *Mechanics*, first published in Rome in 1547. An Italian translation of this paraphrase was produced in 1582 by Oreste Vannocci Biringucci. Throughout the paraphrase, we notice Piccolomini's philological interests and his concern for establishing a correct text. In the preface, he mentions the obscurity of

[32] See Rose and Drake (1971), 79.

the text to the improvement of which, with the exception of Leonico Tomeo who had made some minor notes, no one had contributed anything until his day.[33] In his search for sound texts, Piccolomini consulted Greek manuscripts in libraries in Venice, Padua, Bologna and Florence.[34] Due to the strongly paraphrastic character of Piccolomini's text, it is difficult to connect this activity with any of the extant manuscripts based on the readings. Piccolomini may have used codex *Marc.gr.* 214 (H^a) in the Biblioteca Marciana in Venice, as Rose and Drake presume.[35] Given that in his comments to problem 25 Piccolomini refers to an old manuscript in the Marciana,[36] codex H^a is certainly the best candidate. The other manuscripts at his time in the Marciana, *Marc.gr.* 215 (N^a), *Marc.gr.* 216 (O^a) and *Marc.gr.* 200 (Q), all date from the period 1450 to 1470, and are thus at least one and a half centuries younger than H^a. In his analysis of problem 20, Piccolomini refers to a manuscript in the Biblioteca Medicea Laurenziana in Florence.[37] It can probably be identified with *Laur.Plut.* 28.45 (F^2), the only manuscript then in the Laurenziana.[38] Codex *Laur.Acq.* 65 (F^1), while copied in Florence, was not acquired by the Laurenziana until the year 1815.[39] Even though we cannot make more precise statements about the manuscripts examined by Piccolomini, it can be inferred that they were most likely all contained in manuscript families *a* and *c*. These would have offered Piccolomini valuable variants when compared with Leonico Tomeo's translation, that is, in the first place, connected with family *b*. Piccolomini does not seem to have been influenced by any of the extant diagrams. Rather, he included his own diagrams and added many new ones to the text that give evidence of his interest in contemporary devices and machines. It is this focus on the practical applications of mechanics that becomes preeminent in most later commentaries on the *Mechanics*.

6.3 Niccolò Leonico Tomeo's *Mechanical Questions*

The Latin translation of the *Mechanics* by Niccolò Leonico Tomeo played an important role in the spread of mechanical knowledge in the sixteenth century and provided the basis for many later commentaries on the text. Leonico Tomeo's primary concern was of a philological nature, as is evident from his emendations and marginal notes in B^e indicating whether a specific passage should be considered a scholium. His notes to each of the problems in his *Mechanical Questions*, however, only marginally involve difficulties concerning the mechanical and mathematical issues of the text. He was criticized for this by later scholars, amongst others by Bernardino

[33] See Piccolomini (1565), 6r.
[34] See Piccolomini (1565), 6v.
[35] See Rose and Drake (1971), 84.
[36] See Piccolomini (1565), 53r.
[37] See Piccolomini (1565), 42v.
[38] See Rose and Drake (1971), 84.
[39] See Moraux (1976), 335.

Baldi, who claimed in the preface to his commentary on the *Mechanics* that Leonico Tomeo added very short marginal notes of negligible meaning.[40] Although Baldi might have been overstating his position here, it is true that Leonico Tomeo's comments are for the most part re-phrasings, whereas commentators such as Baldi explained and developed the mathematical foundations of the text and supplied the problems with alternative demonstrations. All the more interesting, therefore, are the mechanical diagrams drawn by Leonico Tomeo which constitute commentaries on the text.

These diagrams are an excellent example illustrating what could occur in consecutive stages of the transmission process. First, we will take a closer look at the diagrams contained in codex B^{e}, describe their different status when compared with the Greek tradition, and investigate whether influences from the Byzantine paraphrase by Pachymeres can be discerned. In the next step, the diagrams in B^{e} will be compared with those contained in Leonico Tomeo's Latin translation in order to analyse to what extent the diagrams from the manuscript correspond to those in the printed edition.[41] Finally, we will see that the fates of Leonico Tomeo and Pachymeres are again closely interwoven, as the diagrams from Leonico Tomeo's Latin translation were copied into the Latin translation of Pachymeres' paraphrase published by Philipp Bech in Basel in 1560. This transfer of the diagrams from one text to the other has major implications for the relationship between the text and the diagrams.

The diagrams in B^{e} were added by Leonico Tomeo at a later point after completion of the manuscript. Since neither the exemplar of B^{e} contained any diagrams, nor do there appear to be any similarities between the diagrams in B^{e} and the Greek tradition, it is clear that Leonico Tomeo was himself responsible for drawing the diagrams. The two most pervasive characteristics of the diagram tradition, indifference to metrical accuracy and overspecification, are absent from codex B^{e}. In general, Leonico Tomeo paid close attention to the metrical precision of the diagrams. In the introduction to the *Mechanics*, the author describes a mechanical device consisting of three circles in contact with each other. Leonico Tomeo accordingly drew a series of contiguous circles, an aspect entirely ignored by the Byzantine scribes. The reconstructed diagram from the archetype shows that the manuscripts depict three separate circles. Another example can be found in the third problem on the working of the lever. The author of the text states that the force is further from the centre of the lever than the weight. This detail was overlooked in the manuscript tradition where the fulcrum is positioned at the exact centre of the lever. Leonico Tomeo, however, attempted to present a geometrically exact picture corresponding to the description in the text and placed the force at a further distance from the

[40] See Baldi (1621), praefatio 6.

[41] Codex *Reg.lat.* 1291 in the Vatican Library forms an additional connecting link between B^{e} and the printed edition of the *Mechanical Questions*. This manuscript contains two versions of the Latin translation by the hand of Leonico Tomeo, the second of which is the author's ready-for-press manuscript, as noted by Nobis (1969), 328.

centre of the lever in his diagram. The second feature of overspecification can be illustrated by the diagram of the parallelogram of motions to the first problem of the *Mechanics*. The sides of the parallelogram stand for the ratio of two motions; if the motion of a point starting at one corner of the parallelogram is represented by the ratio of the sides, the point will always move along the diagonal of the figure. The author of the text expresses the idea that the motion of the point can be interrupted at any arbitrary moment, and that one will always find that the point is on the diagonal of the parallelogram. In order to prove this, the author takes two points, Δ and E, on the sides of the parallelogram in the same ratio as that of the sides to one another, for which he shows that they meet on the diagonal. The author does not define the specific position of the points on the sides; they can be placed anywhere as long as the ratio of the sides holds. In the Byzantine manuscript figures, we see that points Δ and E are drawn at the exact middle of the sides of the parallelogram at a position that shows greater regularity than is demanded by the argument of the text, thus representing a case of overspecification. In Leonico Tomeo's diagram in B^e, points Δ and E are depicted, but so are a number of other points, each of them representing the ratio of the sides of the parallelogram. Instead of taking very specific points for Δ and E at the middle of the sides of the parallelogram, Leonico Tomeo demonstrates that any two points may be chosen as long as they represent the ratio of the sides. Thus, the diagram in B^e presents the level of generality discussed in the text. This diagram thereby achieves precisely the opposite from overspecification, in that it depicts the greatest possible regularity.

In all cases where a diagram is prescribed by the author of the text, Leonico Tomeo included one in B^e. Sometimes, he added a different number of diagrams, as, for example, to the second problem of the *Mechanics* on the suspension of balances. Whereas only two diagrams are required by the text, Leonico Tomeo inserted five diagrams of balances in B^e. Another example can be found in problem 5 on the working of the rudder; one diagram would be sufficient to illustrate the device described in the text, but three diagrams are contained in the manuscript from Bern. From these additional diagrams, we notice that Leonico Tomeo was experimenting with how to best represent a certain mechanical principle. Only once did he add a diagram not required by the text, that is a diagram of a steelyard to problem 20. This diagram is devoid of any letter labels and is merely an illustration of this type of balance. Apart from the inclusion of additional diagrams, we also notice smaller differences between the Byzantine manuscript tradition and B^e in the orientation of diagrams or the positioning of letter labels. In the diagram of the lever to problem 3, for instance, Leonico Tomeo places the weight Γ on the side of the lever that is identified with an A, and the moving force Δ on the other side B, whereas in the manuscript tradition it is precisely the other way around. Both diagrams are in fact a correct representation of the lever principle, since it is not spelled out by the text which sides of the lever weight and moving force need to be positioned on. The comparison has shown that Leonico Tomeo drew the diagrams in B^e independently of the manuscript tradition. This is supported by the fact that B^e does not contain any of the diagrams from the Byzantine archetype that are not required by the text, for example, the diagram to the first problem showing two concentric circles with the letter *kappa* inside to denote their centre.

We now need to examine whether there are connections between the paraphrase by Pachymeres and codex B^e at the level of the diagrams. Leonico Tomeo inserted all prescribed diagrams, including those absent from Pachymeres. This resulted in some very idiosyncratic interpretations, for instance the diagram of circular motion to problem 1 of the text described in the previous chapter. There are many further differences between B^e and Pachymeres, especially in those cases in which Pachymeres' diagrams are in agreement with the diagram tradition, whereas B^e contains variations such as five different diagrams of balances to problem 2 of the text. I could find only two instances in which there might be an influence from Pachymeres on the diagrams in B^e. The first pertains to the diagram of the parallelogram of motions, where the diagram in Pachymeres presents a similar general argument as was made by Leonico Tomeo. The second case is concerned with problem 25 on the stringing of beds. Two diagrams are defined by the text, but Pachymeres contains three diagrams in line with the manuscript tradition, as my reconstructions from the archetype show (see Fig. 6.7). The exact same diagrams are found in B^e which hints at a possible connection. This becomes especially relevant when we take a closer look at one of these diagrams depicting the different strings of a bed from one side to the other. Here, the metrical inaccuracy is preserved, since the letter B is not placed at the middle of the long side of the rectangle as required by the text. While Leonico Tomeo otherwise paid close attention to metrical precision in his diagrams, he supposedly adopted this diagram from Pachymeres. Overall, this results in an independent picture of the diagrams in B^e; Leonico Tomeo neither had a manuscript from the tradition at his disposal while drawing the diagrams, nor does he seem to have been inspired by any of the diagrams from Pachymeres other than these two possible connections.

Next we will analyse what happened with the diagrams from B^e in Leonico Tomeo's Latin translation. Several diagrams from B^e were adopted in the exact same form in the *Mechanical Questions*. The diagram of a steelyard that was added by Leonico Tomeo to problem 20 of the text was copied in the printed edition. Another example involves the diagram to circular motion from the first problem of the text that is identical in the Latin translation, including Leonico Tomeo's renaming of the figure in ABCD. However, there are also a number of changes when we compare the manuscript with the printed edition, the most prominent of which I will describe. We notice a high degree of errors in the Latin letter labels: amongst others, erroneous labels in the text, or labels mentioned in the text but not present in the corresponding diagram. For the most part, Leonico Tomeo retained the alphabetical order of the letters, with the exception of the diagram of the parallelogram of motions in which the label Θ is translated by the letter M, thus deviating from the alphabetical ordering. From several variant diagrams in B^e, we notice that Leonico Tomeo was sometimes looking for the best way to represent a principle described in the text. The five diagrams of balances in B^e appear to be such attempts, two of which were adopted in the printed edition. Problem 25 offers a somewhat different picture; of the three diagrams contained in B^e, two were copied in the Latin translation. However, next to these incomplete diagrams, two additional ones were added illustrating the fully corded beds. Other alterations in the diagrams in the *Mechanical Questions* are of a stylistic nature. Diagrams are provided with more details and pictorial elements. For example, instead of depicting the balances to the second problem by lines only, in the Latin translation the scales of the balances are also drawn in a three-dimensional representation. Likewise, the diagram to problem 17 of the text, on the working of a wedge, is provided with a three-dimensional effect. In these cases, the geometrical abstractions of the mechanical objects were turned into pictures. These diagrams foreshadow a different type of diagrams, next to the strictly geometrical diagrams of the manuscript tradition, that becomes more pronounced in later Renaissance commentaries on the *Mechanics*.

In transmission processes it could happen that diagrams were transferred from one text to the other. This also occurred with the diagrams from Leonico Tomeo's Latin translation that were copied in the exact same form in the Latin translation of the paraphrase by Pachymeres printed by Philipp Bech. This transfer has some serious consequences for the relationship between the text and the diagrams, since some of Leonico Tomeo's diagrams do not coincide with the text by Pachymeres. This can be exemplified by the previously mentioned label M at the bottom left of the diagram of the parallelogram. Pachymeres' Latin text contains the letter G here, which leads to an incongruity between the text and the corresponding diagram. A rather drastic example of such a discrepancy is represented in the diagram to circular motion. The ratio of the motions in Leonico Tomeo's diagram is depicted by BE to EC; as stated in the previous chapter, Pachymeres emended the text and substituted for it a different proportion, that of BD to DC. In the Latin translation of Pachymeres we now have two different ratios in the text and the diagram, which makes the case of circular motion quite incomprehensible when attempting to understand the text with the help of the corresponding diagram.

6.4 Practical Knowledge

Whereas the Latin translations of the *Mechanics* by Fausto and Leonico Tomeo are closely related with the Greek text, most early modern commentators further extend the mechanical knowledge that is presented in the text. We see that alternative solutions to the problems were proposed, but also that new questions were added to the ones considered by the author. These questions have arisen from progress in certain fields of practical knowledge, most importantly in architecture and construction. The interest in practical activities is evident from the early modern translations of and commentaries on the *Mechanics*, some of which were already produced by practitioners, as, for example, the 1573 Italian translation and commentary by the engineer Antonio Guarino. Piccolomini in his paraphrase strengthens the connection to practical activities as he often refers to contemporary machines that were in use on the building site. For instance, with respect to problem 18 of the text on the working of pulleys, Piccolomini includes a digression on various devices reliant on the same principle. He seems particularly interested in a machine used by architects for the lifting and drawing of heavy objects, of which he gives a detailed description with an accompanying diagram (Fig. 6.1). This machine operates and can be explained by the principle of the pulley known from the text. From this and other

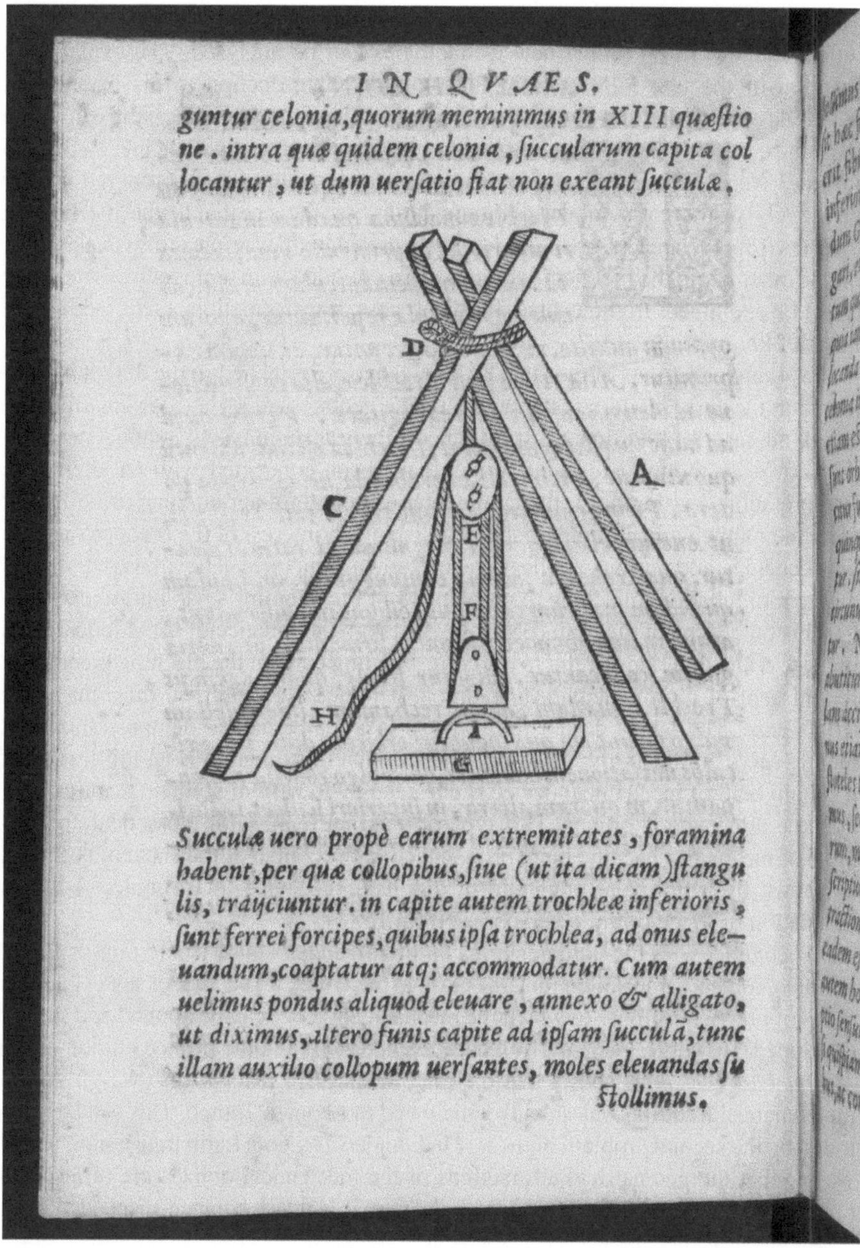

I N Q V AE S.

guntur celonia, quorum meminimus in XIII quæstio ne. intra quæ quidem celonia, ſuccularum capita col locantur, ut dum uerſatio fiat non exeant ſucculæ.

Succulæ uero propè earum extremitates, foramina habent, per quæ collopibus, ſiue (ut ita dicam) ſtangu lis, traijciuntur. in capite autem trochleæ inferioris, ſunt ferrei forcipes, quibus ipſa trochlea, ad onus ele- uandum, coaptatur atq; accommodatur. Cum autem uelimus pondus aliquod eleuare, annexo & alligato, ut diximus, altero funis capite ad ipſam ſuccula͂, tunc illam auxilio collopum uerſantes, moles eleuandas ſu ſtollimus.

Fig. 6.1 Alessandro Piccolomini's Latin paraphrase of the *Mechanics* (1565), 40v. Diagram of a hauling machine to problem 18 of the text. By courtesy of the Library of the Max Planck Institute for the History of Science

examples, we notice that the text of the Aristotelian *Mechanics* provides the theoretical framework by which new technologies can be analysed. The explanatory function of the text follows from Piccolomini's introduction to his paraphrase, where he asserts that the work of Aristotle examines the true causes of admirable machines, not only those that already exist, but also those that will be invented in the future.[42] We see that practical and theoretical concerns have been brought together effectively in the early modern commentaries: on the one hand, the mechanical theories from the ancient text were freely expanded with examples from experience; on the other hand, the practical activities were provided with a model by which they could be explained. In the course of this process, transformations occurred in both fields of knowledge.[43]

The transformations of mechanical knowledge are reflected in the diagrams contained in the early modern commentaries. The geometrical style diagrams from the Greek manuscript tradition are still found in the translations and commentaries on the text, even though occasionally they have been connected with an alternative interpretation. Furthermore, the commentaries contain diagrams that are different from the mathematical tradition and that display an interest in the practical aspects of mechanics and various applications of known principles. The above-mentioned diagram of a hauling machine to problem 18 in Piccolomini's paraphrase exemplifies this other type of diagram (Fig. 6.1). Although the practice of using letter labels in the text and diagram to refer to the corresponding parts of the machine is certainly reminiscent of the mathematical tradition, it is not a geometrical diagram in the strict sense. The labels do not fulfill a role in a mathematical demonstration, but merely serve an illustrative function. The fact that this diagram is an illustration of a certain kind of mechanical device is enforced by the use of stylistic features such as three-dimensional effects. This type of diagram representing practical applications of mechanics became more pronounced in the course of the sixteenth century, as shown by a full-page miniature of this hauling device in Biringucci's Italian translation of Piccolomini's paraphrase from 1582.[44]

By contrast with the commentaries on the *Mechanics* by Piccolomini or Baldi, Leonico Tomeo's translation remains close to the Greek text and does not expand on fields of practical activity to which the mechanical theory can be applied and further developed. However, a comparison between codex B[e] and the printed edition of the *Mechanical Questions* shows that the kind of diagram that moves away from the mathematical tradition was already anticipated in Leonico Tomeo. This can be illustrated by the second problem of the text in Leonico Tomeo's Latin translation, where we see that the geometrical abstractions of the mechanical objects are turned into pictures. Problem 2 of the *Mechanics* asks why it is that a balance suspended from

[42] See Piccolomini (1565), 6r.

[43] See Valleriani (2009), 185. In this context, several studies have been published on long-term developments of mechanical knowledge, which in some cases show a continuous tradition from antiquity until the eighteenth century; see Damerow and Renn (2010) and (2012). Valleriani (2010) focuses more specifically on the early modern interplay between practical and theoretical knowledge.

[44] See Biringucci (1582), 72.

above returns to its original position after the weight is removed, while a balance supported from below does not. The author begins with an explanation of the first case. In a balance suspended from above, the longer part of the beam is above the perpendicular; for this reason the balance returns to its original position after removal of the weight. Lines 850a11-19 of the text describe the diagram to the first case:

> Let ΒΓ be the straight beam, and ΑΔ the cord. When this is produced downwards, it will be the perpendicular ΑΔΜ. If the weight is placed upon it at B, B will then be at E, and Γ at Z, so that the line dividing the beam in two, which was first ΔΜ of the perpendicular itself, will be ΔΘ when the weight is placed; so that the part of the beam EZ which is outside the perpendicular AM will be greater than half the beam by ΦΠ. If then the weight is taken away from E, Z must move down; for P is smaller. Therefore, if the cord is above, the beam rises again.[45]

Figure 6.2 shows the diagram as it is contained in the Greek manuscripts. The straight beam ΒΓ is bisected by the perpendicular ΑΔΜ; when the weight is placed, the beam takes up the position EZ where a new perpendicular ΑΔΘ bisects the beam. This new perpendicular does not cut the beam into two equal parts, but the raised side of the beam is heavier, which is the reason why the balance moves back after removal of the weight. The line ΑΔΘ is missing from the diagram, but it can be supplied by the reader in order to understand the principle.

When the balance is supported from below, the contrary takes place, as is explained by the description of the second diagram at lines 850a24-29:

> Let ΝΞ be the straight beam, and ΚΛΜ the perpendicular. ΝΞ is divided in two. If the weight is placed upon it at N, N will then be at O, and Ξ at P, and ΚΛ at ΛΘ, so that KO is greater than ΛΡ by ΘΚΛ. Now when the weight is taken away the beam must stay there; for the excess over half the beam at X lies upon it as a weight.[46]

Again, in the balance supported from below, the perpendicular to the moved balance will not divide the beam into two equal parts. Now the lowered side of the beam is heavier and prevents the balance from returning to its original position after the weight is removed. Figure 6.3 illustrates the manuscript diagram of the balance supported from below. We see that the perpendicular to the moved balance ΛΘ is missing from the diagram. If it were drawn in the diagram, the triangle ΘΚΛ would

---

[45] ἔστω ζυγὸν ὀρθὸν ἐφ' οὗ ΒΓ, σπαρτίον δὲ τὸ ΑΔ. ἐκβαλλόμενον δὴ τοῦτο κάτω κάθετος ἔσται ἐφ' ἧς ἡ ΑΔΜ. ἐὰν οὖν ἐπὶ τὸ Β ἡ ῥοπὴ ἐπιτεθῇ, ἔσται τὸ μὲν Β οὗ τὸ Ε, τὸ δὲ Γ οὗ τὸ Ζ, ὥστε ἡ δίχα διαιροῦσα τὸ ζυγὸν πρῶτον μὲν ἦν ἡ ΔΜ τῆς καθέτου αὐτῆς, ἐπικειμένης δὲ τῆς ῥοπῆς ἔσται ἡ ΔΘ· ὥστε τοῦ ζυγοῦ ἐφ' ᾧ ΕΖ τὸ ἔξω τῆς καθέτου τῆς ἐφ' ἧς ΑΜ, τοῦ ἐν ᾧ ΦΠ, μείζω τοῦ ἡμίσεος. ἐὰν οὖν ἀφαιρεθῇ τὸ βάρος ἀπὸ τοῦ Ε, ἀνάγκη κάτω φέρεσθαι τὸ Ζ· ἔλαττον γάρ ἐστι τὸ Ρ. ἐὰν μὲν οὖν ἄνω τὸ σπαρτίον ἔχῃ, πάλιν διὰ τοῦτο ἀναφέρεται τὸ ζυγόν.

AM family *a*; AB *cett.*

ΘΠ van Cappelle.

E editions.

[46] ἔστω ζυγὸν τὸ ἐφ' οὗ ΝΞ, τὸ ὀρθόν, κάθετος δὲ ἡ ΚΛΜ. δίχα δὴ διαιρεῖται τὸ ΝΞ. ἐπιτεθέντος δὲ βάρους ἐπὶ τὸ Ν, ἔσται τὸ μὲν Ν οὗ τὸ Ο, τὸ δὲ Ξ οὗ τὸ Ρ, ἡ δὲ ΚΛ οὗ τὸ ΔΘ, ὥστε μεῖζόν ἐστι τὸ ΚΟ τοῦ ΛΡ τῷ ΘΚΛ. καὶ ἀφαιρεθέντος οὖν τοῦ βάρους ἀνάγκη μένειν· ἐπίκειται γὰρ ὥσπερ βάρος ἡ ὑπεροχὴ ἡ τοῦ ἡμίσεος τοῦ ἐν ᾧ τὸ Χ.

ΑΘ family *a*; ΛΘ *cett.*

K editions.

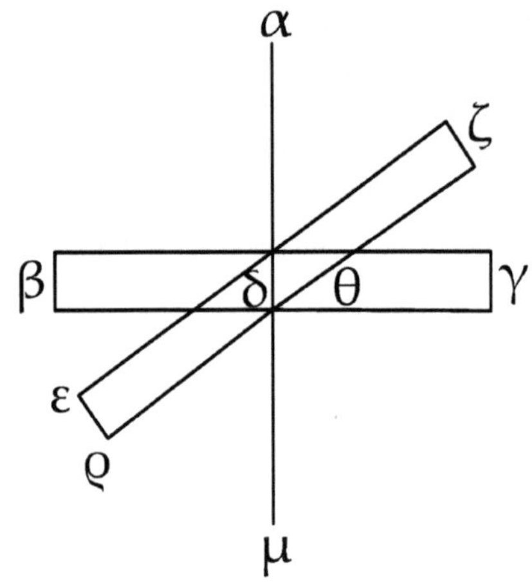

Fig. 6.2 Reconstruction of the archetype

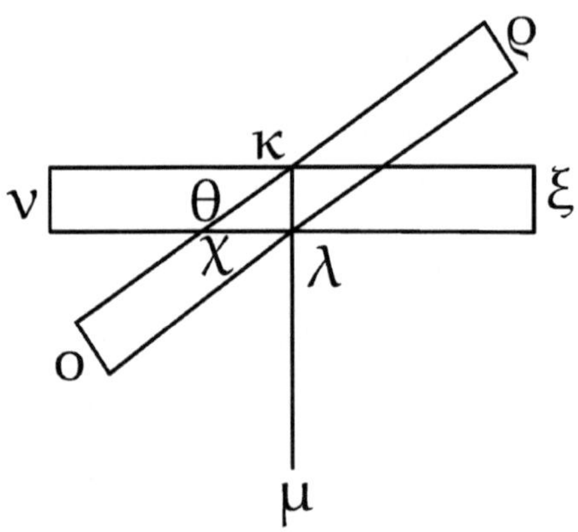

Fig. 6.3 Reconstruction of the archetype

become visible, depicting the area by which the lowered side of the beam is heavier than the raised side.

My reconstructed diagrams show that the balances in the archetype were most likely depicted by double lines. There are, however, some manuscript diagrams that represent the balances by single lines, which complicates the question of how the authentic diagram looked. In order to resolve this difficulty, we may apply the principle of *lectio difficilior potior*. The balance depicted by double lines probably represents the diagram from the archetype, on the basis of which some later scribes simplified the figure.

Let us now take a look at Leonico Tomeo's diagrams in B^{e}. Differing from the Greek text that demands only two diagrams, this codex contains five diagrams of balances in the left margin and at the bottom of the page (Fig. 6.4). Three of them depict the first case of a balance suspended from above and two of them the second case of a balance supported from below. We notice that Leonico Tomeo, as in the

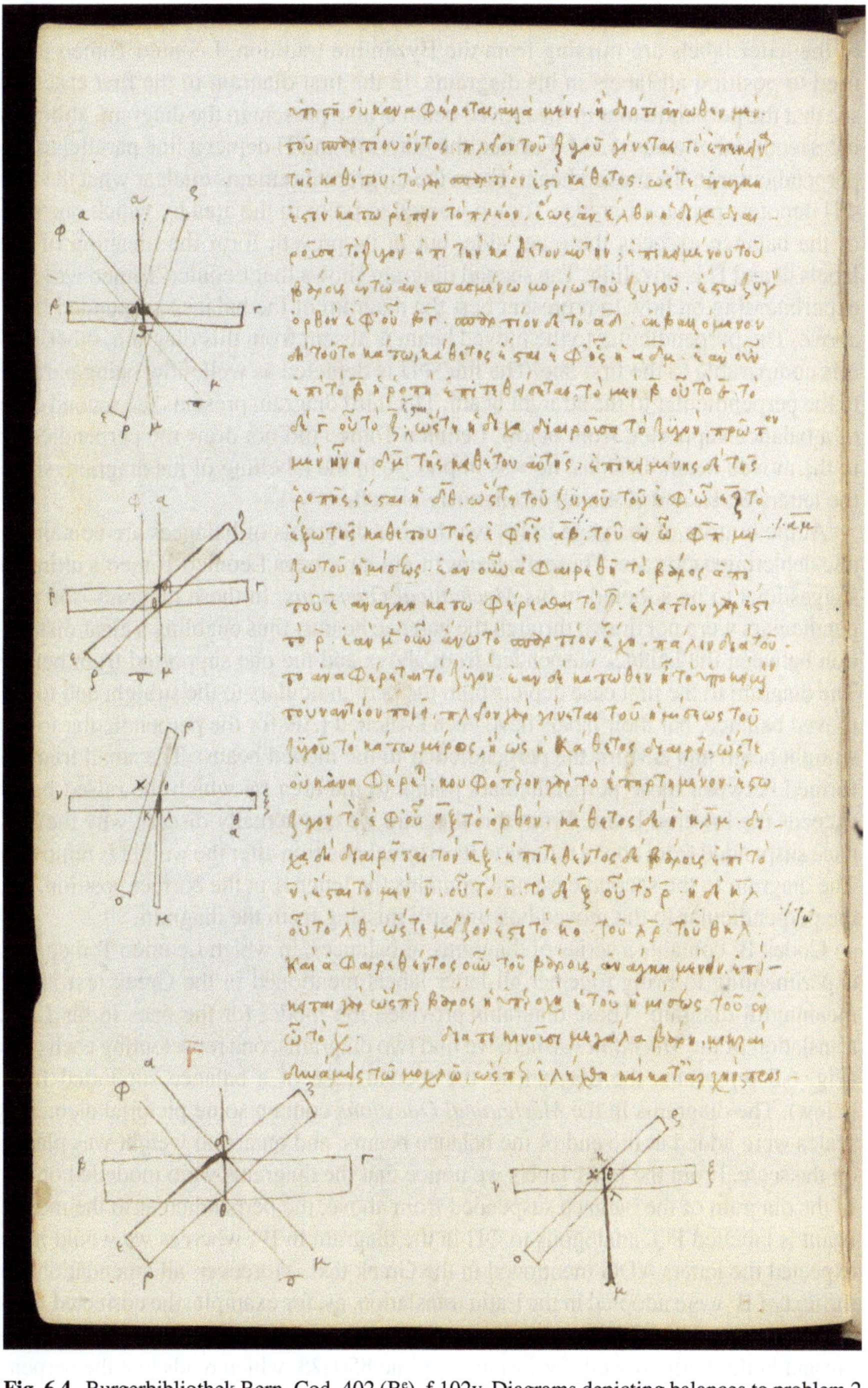

Fig. 6.4 Burgerbibliothek Bern, Cod. 402 (B^e), f.102v. Diagrams depicting balances to problem 2 of the *Mechanics*

manuscript diagrams, depicted all balances by means of double lines. Whereas some of the letter labels are missing from the Byzantine tradition, Leonico Tomeo in B[e] tried to position all labels in his diagrams. In the first diagram to the first case, we see that the perpendicular to the moved beam is also present in the diagram, although incorrectly labelled as AΔM. Further, the letters Φ and Π depict a line parallel to the perpendicular to the moved beam. From the diagram it remains unclear what the line ΦΠ denotes; perhaps Leonico Tomeo wanted to point to the area by which one side of the balance exceeds the other side, but in its present form the meaning of the labels Φ and Π is puzzling. The second diagram shows that Leonico Tomeo was still experimenting on how to represent best the diagram of the balance suspended from above. The perpendicular to the moved beam is absent from this diagram, otherwise it is comparable to the first one. The line ΦΠ is depicted as well, now being parallel to the perpendicular to the straight beam. The third diagram presents the second case of a balance supported from below. Leonico Tomeo did not draw the perpendicular to the moved beam; besides, he was imprecise in the labelling of his diagram, since the letters Θ, K, and X are all erroneously placed.

At the bottom of the page in B[e], two further diagrams of balances are contained, one depicting each case. These diagrams might have been Leonico Tomeo's ultimate suggestions to be adopted in his *Mechanical Questions*. In these diagrams the perpendiculars were not drawn through the balance beams, thus enabling a clear distinction between the balance suspended from above and the one supported from below. The diagram to the first case depicts both the perpendiculars to the straight and to the moved balance, but their labels have been switched (ΔΘ for the perpendicular to the straight beam and ΔM for the perpendicular to the moved beam). The small triangle formed between these perpendiculars points to the area by which the raised beam exceeds the lowered beam. From this diagram we can actually discern why the balance suspended from above returns to its original position after the weight is removed. The diagram to the second case now contains the letter K at the correct position, but the perpendicular to the moved beam is still missing from the diagram.

Codex B[e] contains a series of diagrams of balances, in which Leonico Tomeo was experimenting to bring together all letter labels mentioned in the Greek text into a meaningful diagram. These diagrams provided the model for the ones in his Latin translation of the *Mechanics*, where we find two diagrams, one representing each case (Fig. 6.5 represents the diagram to the second case of a balance supported from below). The diagrams in the *Mechanical Questions* contain some pictorial elements: scales were added at the end of the balance beams, and an actual weight was placed on the scale. From the letter labels we notice that the diagrams were modelled on B[e]. In the diagram of the balance suspended from above, the perpendicular to the moved beam is labelled PQ, analogous to ΦΠ in the diagram in B[e], whereas we would have expected the letters ADG mentioned in the Greek text. Moreover, all emendations in the text of B[e] were adopted in the Latin translation, as, for example, the corrected variant AM at line 850a17 instead of the erroneous reading AB. Some errors from B[e] are present in the Latin as well, for instance, at line 850a25, which reads that the perpendicular KLM bisects the beam KG (instead of the correct NG). In the Latin translation not all the letters mentioned in the text are present in the diagrams. In the diagram of

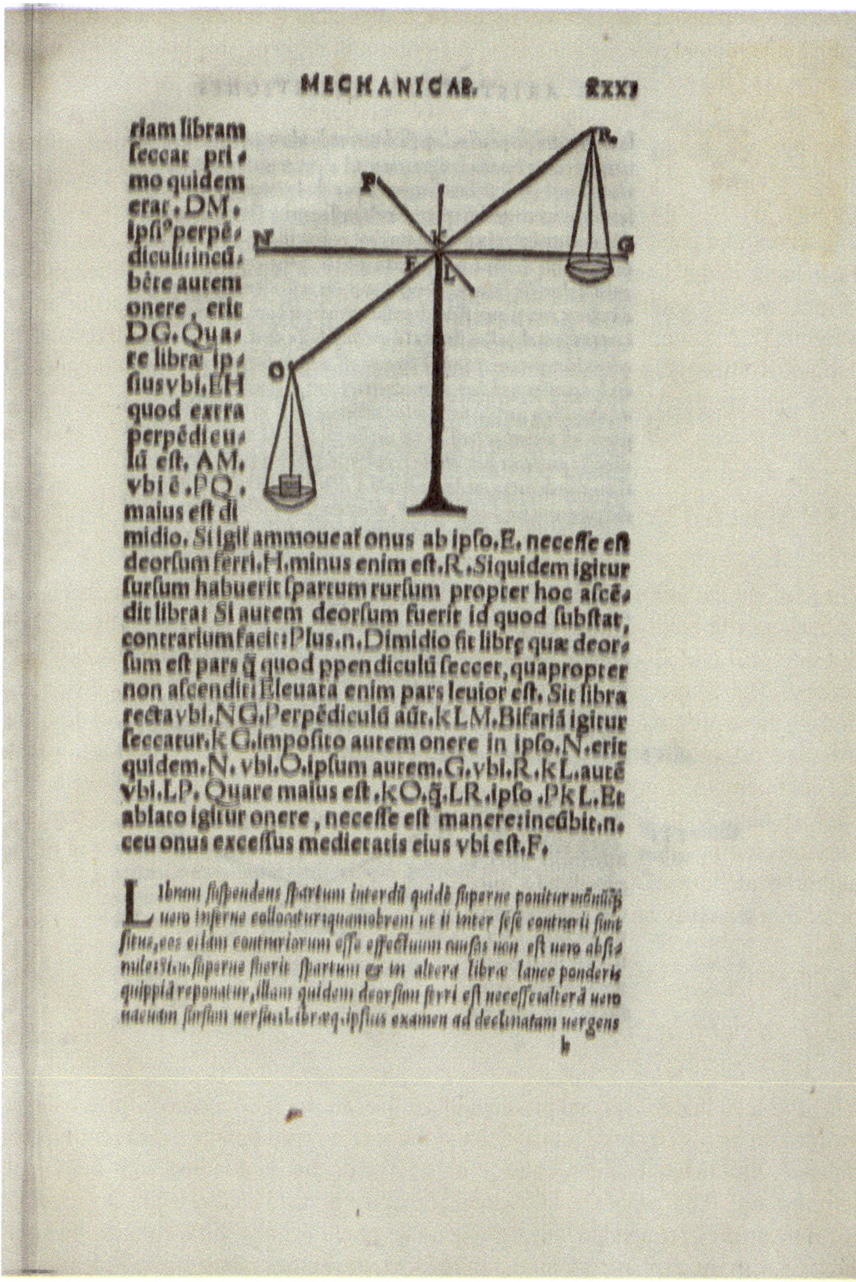

MECHANICAE. XXXI

riam libram seccat primo quidem erat.DM. ipsius perpendiculi linea. habere autem onere, erit DG. Quare librae ipsius vbi.EH quod extra perpendiculum est. AM. vbi est.PQ. maius est dimidio. Si igitur ammoueatur onus ab ipso.E. necesse est deorsum ferri.H.minus enim est.R. Siquidem igitur sursum habuerit spartum rursum propter hoc ascendit libra. Si autem deorsum fuerit id quod substat, contrarium facit. Plus.n. Dimidio fit librae quae deorsum est pars quam quod perpendiculum seccet, quapropter non ascendit. Eleuata enim pars leuior est. Sit libra recta vbi.NG.Perpendiculum autem.KLM.Bifariam igitur seccatur.KG.imposito autem onere in ipso.N.erit quidem.N. vbi.O.ipsum autem.G.vbi.R.KL.autem vbi.LP. Quare maius est.KO.quam.LR.ipso.PKL.Et ablato igitur onere, necesse est manere: incumbit.n. ceu onus excessus medietatis eius vbi est.F.

Libram suspendens spartum interdum quidem superne ponitur nonnumquam uero inferne collocatur quamobrem ut ii inter sese contrarii sunt situs, eos etiam contrariorum esse effectuum causas non est uero absimile. Si.n. superne fuerit spartum et in altera librae lance ponderis quippiam reponatur, illam quidem deorsum ferri est necesse alteram uero ... sursum uersus. Librae q. ipsius examen ad declinatam uergens

h

Fig. 6.5 Niccolò Leonico Tomeo's *Mechanical Questions* (1525), 31r. Diagram of a balance to problem 2 of the text. By courtesy of the Library of the Max Planck Institute for the History of Science

a balance suspended from above, for example, the label R is missing, and in the diagram to the second case the letter F is only present in the text. This kind of error might be related to the different status of the diagrams in the *Mechanical Questions*.

Whereas in codex B^e there is a specific relationship between the text and the diagrams, we do not find such an argument from the picture in the printed edition. In the *Mechanical Questions*, the physical properties of the balance have been ignored. While depicting the balances by single instead of double lines, we are not able to see at a glance that one side of the balance is heavier than the other. When the balance in the diagram has an actual thickness as in B^e or the other Greek manuscripts, we can infer from the diagram which side of the balance carries more weight. This example illustrates the different role of some of the diagrams in Leonico Tomeo's Latin translation. These diagrams have lost their significance as proofs and are now reduced to pictures merely illustrating the text.

6.5 Selected Problems

The mechanical problems were commented upon in the early modern period in the framework of practical knowledge. Some of these questions, such as problem 20 on the steelyard and 25 on the stringing of beds, have raised much discussion in this context. The commentators digress from the Greek text and give alternative solutions to these problems. By looking at their successive interpretations, I will describe the transformations these problems underwent in the sixteenth century. However, the commentators' notes do not only relate to the mechanical devices described in the text, but also to the textual difficulties in these problems. The many corrupt and obscure readings in questions 20 and 25 complicated a proper understanding of the author's intentions. As a result of this, the approaches to finding a solution are diverse in the early modern commentaries.

6.5.1 Problem 20

We have seen that the second problem of the *Mechanics* is concerned with a balance with equal arms (ζυγόν). Question 20 of the text now addresses a different type of balance, the unequal-armed balance or steelyard, that is denoted with the Greek term φάλαγξ. This problem asks why it is that a steelyard can weigh heavy loads of meat by a small suspended weight, when the whole is only half a balance. In compliance with the explanatory model of the text, the author raises a subsequent question in which the steelyard is referred to another mechanical principle. His explanation is twofold, in that the steelyard works as a balance and a lever at the same time. First, the steelyard can be conceived of as a balance with equal arms, in which the cord is the centre of the balance and the weights are placed in the scales on either side of the balance. In the case of the steelyard there is only a scale pan on one side of the balance, the other side has a counterweight attached to the beam. The position of the cord determines the weight in the scale pan. There are many different balances, the author continues, as many as there are different cords, i.e. suspension points of the balance. Second, at the end of the problem the author briefly compares the steelyard to the working of a lever. The cord can be identified with the fulcrum of the lever, and, the closer the cord is to the scale pan, the more easily the appended weight will be able to move the weight in the scale pan on the opposite side.

The balance with unequal arms in problem 20 of the text is characterized by a moveable suspension point and a counterweight attached to the beam. This kind of unequal-armed balance is also known as the Bismar type.[47] It is distinguished from another type of unequal-armed balance, the Roman steelyard, that has a fixed suspension point but a moveable counterpoise. By the first century BC, the Roman steelyard was widely used.[48] As we will see below, many of the commentators on the *Mechanics* include a discussion of the Roman steelyard in their notes, since this was the most commonly used type in the early modern period.

The Renaissance translators and commentators Fausto, Leonico Tomeo, Piccolomini and Baldi agree in their translations of the Greek terms for balances. A φάλαγξ is consistently translated by the Latin word *statera*, and ζυγόν by *libra*. Even though no diagram to this problem is prescribed by the author of the text, most commentators include one in their explanation of the unequal-armed balance. Leonico Tomeo already inserted a diagram to problem 20 in codex B^e, that was adopted in the exact same form in his Latin translation of the text. This diagram is devoid of any letter labels and does not reveal the exact working of the device. It remains unclear from Leonico Tomeo's illustration that this type of balance in fact has a varying suspension point. The *Mechanical Questions* contain an accurate rendition of the Greek text, except for one passage (853b38-854a3) that is omitted from the Latin translation. When we look at B^e we notice that these lines were left out by Leonico Tomeo intentionally, as he remarks in the left margin of f.107v that he considers this passage as a scholium.

At the beginning of problem 20 in his paraphrase of the *Mechanics*, Piccolomini makes mention of the corrupt state of the text.[49] He informs us about a manuscript he found in the Laurenziana library in Florence that is less erroneous than the others. This manuscript is likely identical with codex F^2 in manuscript family *c*, as explained above. Family *c* is distinguished from the other families by the presence of scholia. The text of problem 20 contains two scholia at lines 853b25 and 853b37 that are integrated in

[47] See Jenemann (1989), 321 and Damerow et al. (2000), 5–6.

[48] See Jenemann (1989), 326.

[49] See Piccolomini (1565), 42v.

the text of F^2. Since family *c* in problem 20 does not contain any superior readings when compared with the other manuscripts, Piccolomini possibly had in mind these scholastic passages when referring to a more complete text in the manuscript from the Laurenziana. Next, Piccolomini states that a certain interpreter ruined the whole meaning of the question. As Leonico Tomeo's translation of the *Mechanics* appears to have been the only text known to Piccolomini, he is probably criticizing Leonico Tomeo here. Indeed, we will see that Piccolomini's explanation of the unequal-armed balance strongly differs from Leonico Tomeo's close reading of the text.

Piccolomini explains the working of the *statera* by means of a diagram. The balance is defined by AB, from one end of which A is suspended the scale C, on the other end is the appended weight F, and the suspension points are D and E.[50] This account of the unequal-armed balance appears to differ from the type described by the author of the *Mechanics*. Whereas the counterpoise in the balance described in the Greek text is fixed, Piccolomini states that the appended weight F runs about the balance beam. From his further explanation of the diagram, we see that Piccolomini has in mind the Roman steelyard with a moveable counterweight. When the balance is suspended at D, the appended weight F is at a further distance from the centre than when suspended at E, for which reason the weight in the scale C is more easily moved. Irrespective of which suspension point is chosen, the balance operates in the same way, namely the counterpoise F is moved until equilibrium is reached. Piccolomini here refers to a specific type of Roman steelyard with two suspension points. These balances were equipped with two different weight scales.[51] To make use of the other weight range, the balance was turned upside down and suspended from the second handle. This resulted in a different ratio between the arms of the balance, and the same counterpoise would be able to weigh a heavier load. Piccolomini refers to this type of Roman steelyard only, and does not mention in his paraphrase the Bismar type from the Greek text. Since he is criticizing Leonico Tomeo for misinterpreting the text, it might be that Piccolomini did not know of the unequal-armed balance with moveable suspension point. With regard to the multiple suspension points mentioned by the author of the text, Piccolomini states that in practice there are only two that are used for heavier and lighter objects to be weighed.[52] It is obvious from this that the Bismar from the Greek text was consequently interpreted by Piccolomini as the Roman steelyard. The balance with unequal arms in his paraphrase is thus a perfect example illustrative of how mechanical knowledge from the Greek text could be transformed and adapted to contemporary practices.

The next commentator on the text was well aware of the distinction between the Bismar and the Roman steelyard. Baldi begins his comments with a synopsis of the problem in which he compares the *statera* with the balance and lever, but also makes explicit the additional explanatory step of referring back to the properties of the circle. He then explains both types of unequal-armed balances, starting with the Roman

[50] See Piccolomini (1565), 43r.

[51] See Jenemann (1989), 328 for a description of this type of balance.

[52] See Piccolomini (1565), 44r.

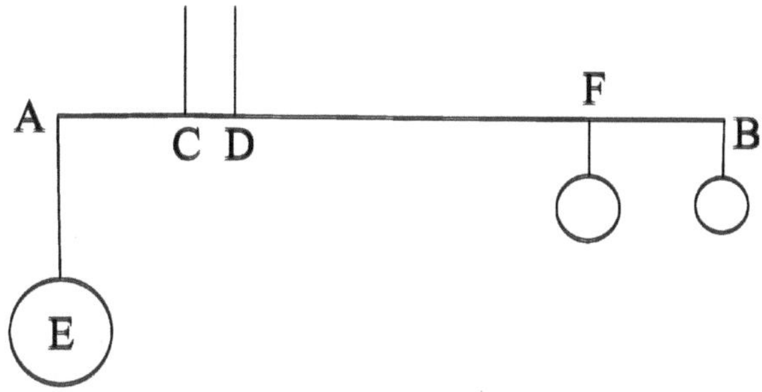

Fig. 6.6 Diagram of an unequal-armed balance as included in Bernardino Baldi's commentary on the *Mechanics*

steelyard. Baldi's diagram is similar to the one in the paraphrase by Piccolomini of a balance with two suspension points and a moveable appended weight (Fig. 6.6).

The balance is labelled AB, the suspension points are C and D, E is the weight in the scale, and F the counterweight.[53] Baldi asserts that equilibrium exists when the balance is suspended in C. This can be explained by the law of the lever: as the weight in E is to the appended weight in F, so is the length CF to the length AC. However, if the suspension point is in D, the balance tilts and the counterpoise F needs to be moved until equilibrium is restored. He continues by saying that the *statera* used in his time has two suspension points.[54] When the suspension point used is closer to the weight in the scale, the arm on the side of the counterpoise is longer, thus the same counterpoise can balance heavier weights. A second diagram illustrates that the steelyard can also be used in a different way, namely with a fixed appended weight and a moveable suspension point. Baldi informs the reader that equilibrium is reached when the suspension point of the balance coincides with the centre of gravity of the whole system. He then explains two cases in which the centre of gravity is not identical with the suspension point, from which it follows that the balance tilts in one or the other direction. Baldi's commentary to problem 20 shows that the knowledge of balances from the Greek text and practical experience can be combined in an effective way. He concludes that, although the steelyard can be used in these two different ways, in contemporary practice only the type with the moveable counterpoise is found.[55]

[53] See Baldi (1621), 133.
[54] See Baldi (1621), 134.
[55] See Baldi (1621), 135.

6.5.2 Problem 25

Question 25 is an exception among the other problems in that it does not refer to the explanatory model of the text, in which the mechanical principle under discussion is brought back to the lever, balance or circle. It deals with the cording of beds and asks two questions: why are beds made to be twice as long as they are wide? And why are they not corded diagonally but from side to side? The answer to the first question is given directly by the author, since the bed size is related to the size of the human body. The second question of why beds are corded from side to side is more complex and receives a threefold answer: the timbers are less strained in this way; the ropes are less strained; less rope is used when corded from side to side. Rather than elaborating on these points, the author of the *Mechanics* describes the details of both ways of cording beds while referring to two diagrams. The Byzantine manuscript tradition, however, contains three diagrams, two of the cording from side to side and one illustrating a diagonally corded bed (Fig. 6.7). None of these diagrams is sufficient for understanding exactly how beds were corded in antiquity. The most complete is the first diagram depicting four cords from side to side that corresponds to the description of the author at lines 856b11-18:

> For let AZHI be the bed, and let ZH be divided in two at B. The holes in ZB and ZA are equal. For the sides are also equal; for the whole ZH is double the side ZA. Now they stretch the rope as has been described from A to B, then to Γ, then to Δ, then to Θ, then to E, and so on until they return to another corner; for the ends of the rope are at two corners.[56]

The Greek manuscripts display an indifference to metrical accuracy, since the label B is not placed at the middle of ZH in any of the diagrams. Otherwise, the diagrams offer a correct representation of the text in which the cords are only partially defined. Apart from this and other incomplete descriptions, we find many incorrect letter labels in the Greek text, as well as corrupt and incomprehensible passages, such as at the end of problem 25. Here the author explains how much more rope is expended when the bed is corded diagonally than when it is corded from side to side (857a3-4): "So that the rope is so much longer than both sides are greater than the diagonal."[57] We will see that the early modern commentators were puzzled by this passage and came up with very different solutions to the bed problem.

Fausto's text offers only a Latin translation of the *Mechanics* without his comments on the single problems. However, his translation and diagrams added to problem 25, give evidence of a thorough reading and mathematical understanding of the text. This is apparent from several smaller emendations by Fausto, for instance at lines 856b31-32 which read in the Greek: "So that it is clear that there are four such lengths of rope as AB in the bed."[58] Although the first diagram from the manuscripts

[56] ἔστω γὰρ κλίνη ἡ AZHI, καὶ δίχα διῃρήσθω ἡ ZH κατὰ τὸ B. ἴσα δὴ τρυπήματά ἐστιν ἐν τῇ ZB καὶ ἐν τῇ ZA. καὶ γὰρ αἱ πλευραὶ ἴσαι εἰσίν· ἡ γὰρ ὅλη ZH διπλασία ἐστίν. ἐντείνουσι δ' ὡς γέγραπται, ἀπὸ τοῦ A ἐπὶ τὸ B, εἶτα οὗ τὸ Γ, εἶτα οὗ τὸ Δ, εἶτα οὗ τὸ Θ, εἶτα οὗ τὸ E. καὶ οὕτως ἀεί, ἕως ἂν εἰς γωνίαν καταστρέψωσιν ἄλλην· δύο γὰρ ἔχουσι γωνίαι τὰς ἀρχὰς τοῦ σπαρτίου.

[57] ὥστε καὶ τὸ σπαρτίον μεῖζον τοσούτῳ ὅσον αἱ πλευραὶ ἄμφω μείζους εἰσὶ τῆς διαμέτρου.

[58] ὥστε δῆλον ὅτι τὰ τηλικαῦτα σπαρτία ὅσον τὸ AB, τέσσαρα τοσαῦτ' ἔνεστιν ἐν τῇ κλίνῃ.

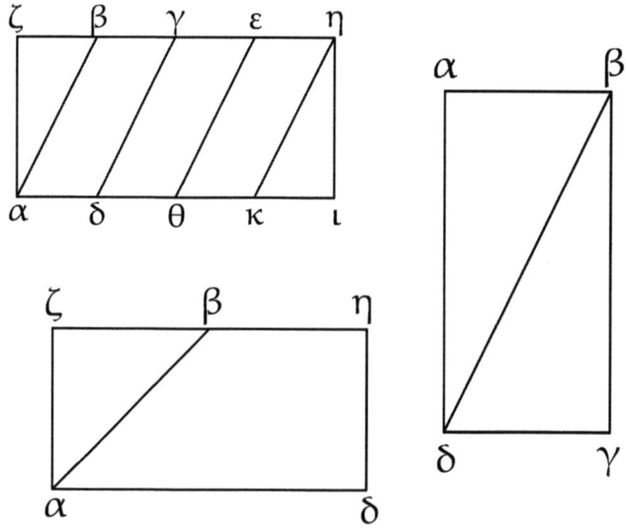

Fig. 6.7 Reconstructions of the archetype

depicts these four pieces of rope only, there is the same number of ropes again on the opposite side of the bed. This was recognized by Fausto as he emended the text and translated *in grabati dimidio*, in half of the bed. Fausto was likely responsible for drawing his own diagrams, and the two diagrams he inserted to the bed problem are very different from the ones in the manuscript tradition. These diagrams illustrate the complete side to side and diagonal cording of beds as it can be inferred from the author's description. In the first diagram, Fausto did not only depict the different cords from one side to the other, but also the shorter parts of rope along the side of the bed from one hole to the next, which shows that he conceived of the bed as being corded by one piece of rope only.

The text of problem 25 in Leonico Tomeo is less accurate than Fausto's Latin translation. Lines 856b28-29 for instance are missing from the *Mechanical Questions*. Although this passage was initially omitted in codex B^e as well, it was added by Leonico Tomeo in the left margin, but not subsequently adopted in his translation. We have already seen that the diagrams contained in B^e to problem 25 are similar to the ones from the manuscript tradition. The illustrations in the Latin translation were influenced by the diagrams in B^e, without being identical to them. Two of the manuscript diagrams depicting the partial cording from side to side were copied into the Latin translation. In addition, two diagrams were added depicting the complete side to side and diagonal cording of the beds. Apparently, the incomplete diagrams from the tradition were not considered to be sufficient in order to explain the principle and two additional ones were included. Now we are faced with a situation in which we have the diagrams from the manuscript tradition and the reinterpretations next to each other. The upper diagram in Fig. 6.8 shows the new diagram as it was constructed on the basis of the text; the lower diagram was adopted from the Greek manuscripts.

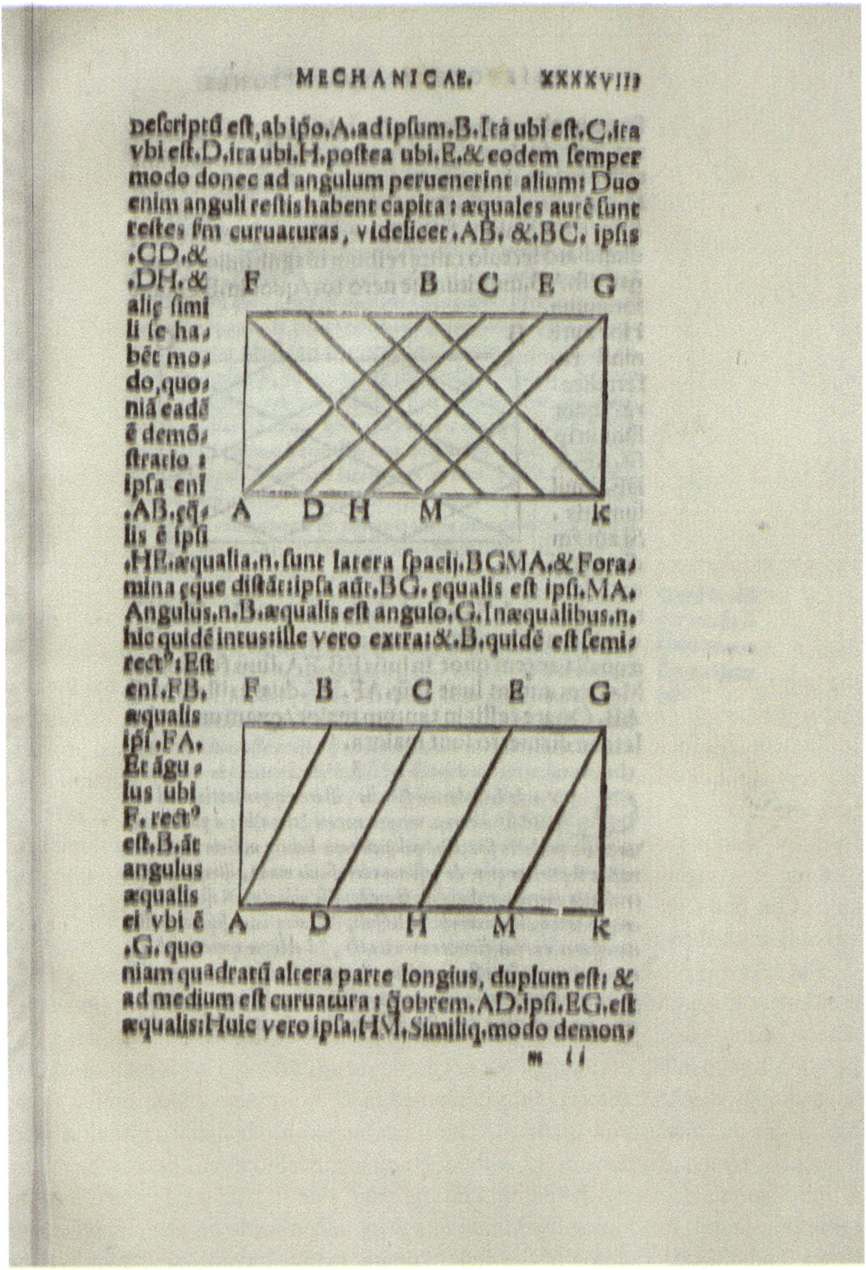

MECHANICAE. XXXXVIII

Descriptũ est, ab ipso .A. ad ipsum .B. Itã ubi est .C. ita vbi est .D. ita ubi .H. postea ubi .E. & eodem semper modo donec ad angulum peruenerint alium: Duo enim anguli restis habent capita: æquales autẽ sunt restes sm curuaturas, videlicet .AB. & .BC. ipsis .CD. & .DH. & aliæ simili se habẽt modo, quoniã eadẽ ẽ demõstratio: ipsa enĩ .AB. æqlis ẽ ipsi

F B C E G
A D H M K

.HE. æqualia .n. sunt latera spacij .BGMA. & Foramina quæ distãt ipsa autẽ .BG. æqualis est ipsi .MA. Angulus .n. .B. æqualis est angulo .G. In æqualibus .n. hic quidẽ intus: ille vero extra: & .B. quidẽ est semirect⁹: Est enĩ .FB. æqualis ipsi .FA. Et ãgulus ubi .F. rect⁹ est .B. ãt angulus æqualis ei vbi ẽ .G. quoniam quadratũ altera parte longius, duplum est: & ad medium est curuatura: quobrem .AD. ipsi .EG. est æqualis: Huic vero ipsa .HM. Similiq; modo demon

F B C E G
A D H M K

m ii

Fig. 6.8 Niccolò Leonico Tomeo's *Mechanical Questions* (1525), 48r. Diagrams depicting the cording of beds to problem 25 of the text. By courtesy of the Library of the Max Planck Institute for the History of Science

While latter still preserves the indifference to metrical accuracy that is typical for the manuscript tradition, the former presents a precise diagram in which the label B is depicted at the exact middle of FG. This merging of two sets of diagrams presents a puzzle to the reader: which of them should be taken as understanding the text? In his notes accompanying this problem, Leonico Tomeo neither mentions the obscurity of this problem, nor contributes to clarifying any of the textual difficulties.

A critical examination of problem 25 is first found in the Latin paraphrase by Piccolomini. Piccolomini refers to the corrupt state of the text, and mentions that in his search for sound readings, he came across a codex in the Marciana library in Venice.[59] Even though this manuscript could not unravel all textual problems, it shed some light on difficult passages and enabled him to discover the true sense of Aristotle's words. Piccolomini included two diagrams in his paraphrase, first of the cording from side to side, and second along the diagonal of the bed (Fig. 6.9).

These diagrams are similar to the constructions by Fausto and Leonico Tomeo, with the difference that Piccolomini added many letters to the ones mentioned by the author of the text, and included labels for every hole in the bed frame. He used these labels in his paraphrase in order to set up calculations for the amount of rope required in both ways of cording. He begins by counting the length of rope in the bed that is corded from side to side. Assuming that the bed has a length of 6 ft and a width of 3 ft, he first counts one piece of rope AB: in the triangle ABF, both AF and BF are 3 ft long, from which it follows that AB equals the square root of 18, that is 4¼. Taken together, there are eight of such pieces of rope in the bed, which yields a total length of 34 ft. Next, Piccolomini performs similar calculations for the diagonally corded bed, where we find pieces of rope of three different lengths. The shortest piece is that of MF, which equals the square root of 5, that is 2¼. As there are four pieces of rope equal to MF in the bed, this results in a length of 9 ft. The next piece of rope is EL, which equals the square root of 20, that is 4½. Again, there are four of such pieces with a total length of 18 ft. The largest piece of rope is BD, which equals the square root of 45, that is 6¾, multiplied by 2, results in a total length of 13½ ft. The total amount of rope used for the diagonal cording is thus 40½ ft. Piccolomini does not dwell on the author's cryptic remark that the bed corded along the diagonal requires as much more rope as the difference between the two sides and the diagonal. Instead, he simply concludes with the remark that Aristotle was right in his observation that the diagonally corded bed needs more rope than when corded from side to side.[60] However, Piccolomini has made an error in his calculations of the roping, since he forgot to include the shorter lengths of rope along the side of the bed from one hole to the next, that were indicated by Fausto in his diagram. We will see that Piccolomini's calculations were corrected by the next commentator on the text.

In his commentary, Baldi presents a comprehensive discussion of problem 25. He refers to the obscurity and corruption of the text and praises Piccolomini for making

[59] See Piccolomini (1565), 53v.

[60] See Piccolomini (1565), 56v.

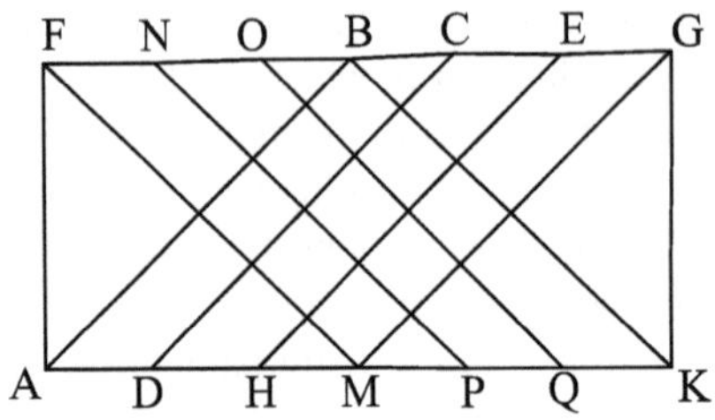

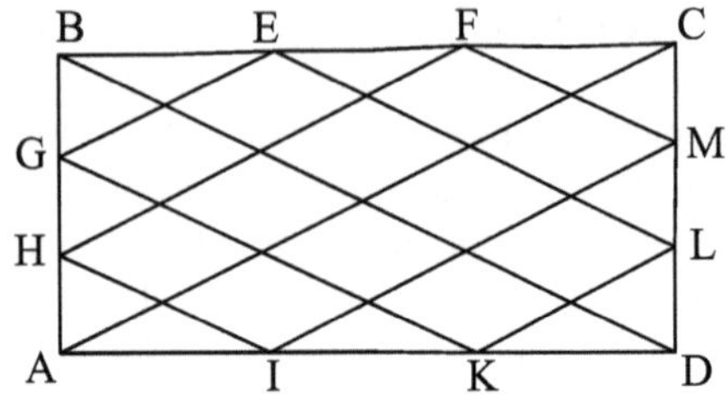

Fig. 6.9 Diagrams of the different cording of beds as included in Alessandro Piccolomini's Latin paraphrase of the *Mechanics*

good progress on the bed problem.[61] On the first question of why the length of beds is twice the width, Baldi remarks that beds in his time have a different proportion. The size of a bed is 6x4 ft, rather than 6x3 ft from the Greek text, in order that two people can comfortably sleep in it. With respect to the second question of why beds are corded from side to side, Baldi mentions the threefold reasoning by the author and analyses each of the arguments. By means of a diagram, he first demonstrates that the timbers are much strained when the ropes are fixed diagonally. He compares the cording along the diagonal with a perpendicular arrangement of the ropes and shows that in the first case the ropes exert so much pressure that they will crush the bed frame. To the second reason that the ropes are less strained when corded from side to side, he notes that this is readily apparent. The longer the ropes are, the weaker they will be, and the ropes along the diagonal are clearly longer. To the third part of the question, Baldi sets out the cording from side to side with the help of a similar diagram to the one contained in the paraphrase by Piccolomini. Starting in the corner at A, the cord is attached to F, from there to I, G, H, K, C and E, thus forming four pieces of rope equal to AF and three smaller parts equal to FI. A rope of the same length is corded in a similar way from B to the hole at F. When compared with the diagram from Fausto's translation in which the bed was corded with one piece of rope only, we notice that Baldi interpreted it as consisting of two ropes. Baldi calculates the length of rope in this bed, and reaches a different conclusion from Piccolomini. According to Baldi, the square root of 18 is 4⅓; this result times 8, as there are eight of such pieces of rope in the bed, gives an outcome of 34⅔ ft. Those six smaller parts that were forgotten by Piccolomini also need to be factored in, Baldi notes, which yields a total length for the cording from side to side of 40⅔ ft. Baldi does not himself calculate the cording along the diagonal, but refers to the length of 40½ ft from Piccolomini's text. However, when the smaller parts of rope are added here as well, it is obvious that much more rope is expended in the diagonal cording. Baldi was the first commentator to remark on the varying amounts of rope used in both ways of cording, that was expressed in the Greek text by the puzzling remark concerning the difference between the two sides and the diagonal. He is frustrated with this statement by Aristotle and adds that this passage is corrupted, complicated and obscure and that

[61] See Baldi (1621), 151.

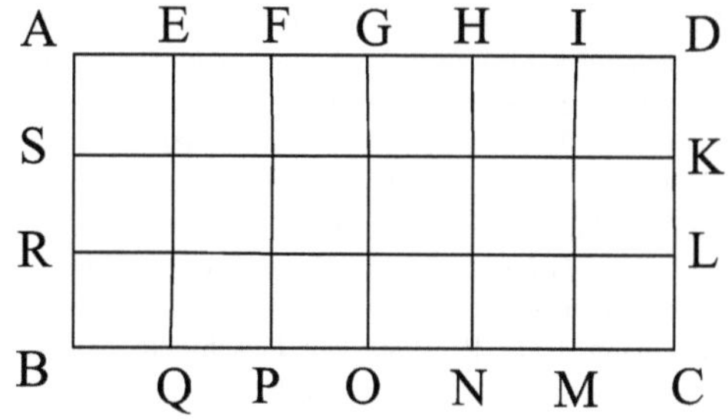

Fig. 6.10 Diagram of the alternative cording of beds as included in Bernardino Baldi's commentary on the *Mechanics*

one needs the help of a Delian swimmer.[62] Baldi decides not to linger on this problem, for which no solution will easily be found, and comes up with a rather unconventional counterproposal. While alluding to the legend of the Gordian knot, just like Alexander the Great, Baldi takes a radical approach and wonders why the ancients did not use the perpendicular arrangement of the ropes. This alternative way of fixing the ropes is illustrated by a diagram (Fig. 6.10). Baldi explains that the total length of rope is 32 ft which is clearly less than in the beds described by the author of the text. Moreover, when corded perpendicularly, less pressure is exerted on the wooden frame. By proposing this alternative solution for the question on beds, Baldi's commentary represents a decisive step toward the transformation of problem 25.

The questions raised in the *Mechanics* served as a starting point for further reflections in the early modern period. Whereas the first translations remained close to the Greek text, later commentaries display a different approach. Piccolomini's and Baldi's texts are strongly linked with contemporary practical knowledge, as is evident from their discussions of problems 20 and 25. Their diagrams give evidence of the early modern focus on practical activities and are thus a powerful tool for studying the transformation processes of mechanics in the sixteenth century.

Bibliography

Baldi, Bernardino. 1621. *In Mechanica Aristotelis Problemata Exercitationes*. Mainz: Johannes Albinus.

Barnes, Jonathan. 1984. *The complete works of Aristotle*, 2 vols. Princeton: Princeton University Press.

Bech, Philipp (ed.). 1560. *Georgii Pachymerii, in Aristotelis philosophiam epitome*. Basel: Froben.

Berryman, Sylvia. 2009. *The mechanical hypothesis in ancient Greek natural philosophy*. Cambridge: Cambridge University Press.

Biringucci, Oreste Vannocci. 1582. *Parafrasi di Monsignor Alessandro Piccolomini, sopra Le Mecaniche d'Aristotile*. Rome: Zanetti.

Damerow, Peter and Jürgen Renn. 2010. The transformation of ancient mechanics into a mechanistic worldview. In *Transformationen antiker Wissenschaften*, eds. G. Töpfer and H. Böhme, 239–263. Berlin: De Gruyter.

[62] See Baldi (1621), 153.

Damerow, Peter and Jürgen Renn. 2012. *The equilibrium controversy – Guidobaldo del Monte's critical notes on the mechanics of Jordanus and Benedetti and their historical and conceptual backgrounds*. Berlin: Edition Open Access.

Damerow, Peter, Jürgen Renn, Simone Rieger and Paul Weinig. 2000. *Mechanical knowledge and Pompeian balances*, Preprint 145. Berlin: Max Planck Institute for the History of Science.

Fausto, Vittore. 1517. *Aristotelis Mechanica Vittoris Fausti industria in pristinum habitum restituta ac latinitate donata*. Paris: Josse Badius.

Hett, Walter S. (ed.). 1936. *Aristotle: Minor works*. Cambridge, MA: Harvard University Press.

Jenemann, Hans R. 1989. Zur Geschichte der Waagen mit variablem Armlängenverhältnis im Altertum. *Trierer Zeitschrift* 52: 319–352.

Krafft, Fritz. 1970. *Dynamische und statische Betrachtungsweise in der antiken Mechanik*. Wiesbaden: Franz Steiner Verlag.

Laird, Walter Roy. 1986. The scope of Renaissance mechanics. *Osiris*, 2nd Series, vol. 2: 43–68.

Leonico Tomeo, Niccolò. 1525. *Nicolai Leonici Thomaei Opuscula nuper in lucem aedita quorum nomina proxima habentur pagella*. Venice: Bernardino Vitali.

Micheli, Gianni. 1995. *Le origini del concetto di macchina*. Florence: Olschki.

Moraux, Paul. 1976. *Aristoteles Graecus: Die griechischen Manuskripte des Aristoteles. 1.Band: Alexandrien – London*. Berlin: De Gruyter.

Nenci, Elio. 2011. *Bernardino Baldi's In Mechanica Aristotelis Problemata Exercitationes*. Berlin: Edition Open Access.

Nobis, Heribert M. 1969. Über zwei Handschriften zur frühneuzeitlichen Mechanik in italienischen Bibliotheken. In *Sudhoffs Archiv* Bd. 53, H. 3, 326–332.

Piccolomini, Alessandro. 1565. *In Mechanicas quaestiones Aristotelis, Paraphrasis paulo quidem plenior*, 2nd ed. Venice: Traianus Curtius.

Rose, Paul Lawrence and Stillman Drake. 1971. The pseudo-Aristotelian *Question of Mechanics* in Renaissance culture. *Studies in the Renaissance* 18: 65–104.

Schiefsky, Mark. 2007. Art and nature in ancient mechanics. In *The artificial and the natural: An evolving polarity*, eds. B. Bensaude-Vincent and W.R. Newman, 67–108. Cambridge, MA: MIT Press.

Valleriani, Matteo. 2009. The transformation of Aristotle's *Mechanical Questions*: A bridge between the Italian Renaissance architects and Galileo's *First New Science*. *Annals of Science* 66(2): 183–208.

Valleriani, Matteo. 2010. *Galileo engineer*. Dordrecht: Springer.

Van Cappelle, Johannes Petrus (ed.). 1812. *Aristotelis Quaestiones Mechanicae*. Amsterdam: Den Hengst.

Van Leeuwen, Joyce. 2014. Thinking and learning from diagrams in the Aristotelian *Mechanics*. *Nuncius* 29: 53–87.

Chapter 7
Conclusion

Whilst recapitulating the transmission of the Aristotelian *Mechanics*, two threads have been spun out in the present book, that of tradition and that of innovation. Whereas the first involves the copying activities by scribes in the Byzantine Period in order to preserve the ancient text for future generations, the second relates to the creative engagement with the mechanical subject matter of the text at different times. My examination of the complete Byzantine manuscript material of the *Mechanics* has demonstrated the urgent need for a new critical edition of the treatise. The newly established *stemma codicum* of the treatise points out which manuscripts should be considered in such an edition. All current editions are either based on a small part of the manuscript tradition and thus exclude important codices from the constitution of the text, or rely to a large extent on manuscripts that are contaminated. Most importantly, I have shown that all editions are influenced by the Byzantine paraphrase of the text by Georgius Pachymeres.

Eliminating these paraphrastic traces from the authentic text, would, for the most part, result in variants that are certainly different from the ones in the current critical editions, but it would not lead to a completely new interpretation of the treatise. However, some insights gained from the examination of the manuscripts could change our perspective on the text, as can be illustrated by the geometrical analysis of motion from the first problem of the *Mechanics*. In his division of motion into components, the author asserts that circular motion consists of both natural and unnatural components. According to the extant manuscripts, circular motion itself is characterized as natural, whereas its components towards the tangent and the centre of the circle are unnatural. I have argued that for the *Mechanics* natural motion should not be regarded in the Aristotelian sense of an object's motion towards its proper place. In the specific context of mechanics, it should rather be understood as a motion that is natural to the properties of a mechanical object, for instance for a circular object it is natural to move along its circumference. All modern editions, on the other hand, contain a different reading of natural and unnatural motion, one that was adopted from an emendation of the text by Johannes van

J. van Leeuwen, *The Aristotelian Mechanics*, Boston Studies in the Philosophy and History of Science 316, DOI 10.1007/978-3-319-25925-3_7

Cappelle. Upon this interpretation, the motion towards the tangent is natural and that towards the centre unnatural. Most scholars experience difficulties in interpreting the emended passage, without realizing that it is not the authentic manuscript reading. Moreover, they rely too heavily on a modern diagram of circular motion, in which the components towards the tangent and the centre of the circle have a different orientation when compared with the diagrams from the manuscript tradition.

This points to the significance of examining the diagrams contained in the manuscript traditions of ancient scientific texts. My first critical edition of the diagrams in the textual tradition of the *Mechanics* demonstrates the relevance of these diagrams for the recovery of authentic manuscript readings. In the first problem of the text, we have seen, for example, that the ratio of the motions of which circular motion is made up, was replaced by Pachymeres with an emendation that found its way into all modern critical editions of the treatise. The analysis of the manuscript diagrams, however, has shown that this emendation was not a justified correction of an otherwise corrupted passage of the text, since all diagrams confirm the original ratio of the motions from the manuscript text. A new critical edition of the *Mechanics* should therefore comprise a comprehensive analysis of both the text and the diagrams. Editions of ancient scientific texts that include the diagrams from the manuscript tradition, will provide more insight into the text and its historical context, than the modern diagrams from many editions that contain anachronistic interpretations of the scientific subjects.

Diagram studies are not only significant for a reconstruction of the text, but also shed light on processes of transfer and transformation of knowledge. At this point my research exceeds a strictly philological approach, which aims to recover the author's words and considers all later modifications to be corruptions in the transmission process. Ancient scientific texts are part of a creative tradition in which the subject matter was reworked and adapted to the specific needs of the time. The diagrams contained in these texts reflect the different contexts in which scientific knowledge was appropriated. Therefore, we should consider the diagrams in the manuscripts of the *Mechanics* with their roots in the Byzantine period as a historical document of that time. It is in the early modern period, however, that the diagrams become an especially powerful means to study processes of the transformation of knowledge. In the commentaries on the *Mechanics*, we notice a close connection between the theoretical knowledge from our treatise and the practical activities of, amongst others, Renaissance engineers and architects. The focus on practical knowledge is evident from a different type of images next to the geometrical diagrams from the manuscript tradition.

In this context, particular attention is paid to the 1525 Latin translation of the *Mechanics* by Niccolò Leonico Tomeo. A comparison of the diagrams in this translation with those from codex B^e also copied by Leonico Tomeo, illustrates altered diagrammatic practices in the transition from manuscript to print. Whereas codex B^e contains diagrams that build upon the geometrical tradition, in the Latin translation we often find a looser connection between the text and the diagrams. The fact that some of the diagrams were turned into pictures is instructive for our knowledge of the developments in mechanics in the early modern period. These images display the Renaissance interest in practical applications of mechanical knowledge.

The altered scope of the mechanical discipline in the sixteenth century is apparent from images depicting contemporary machines and other mechanical devices, but also from the audiences addressed in mechanical texts. Many of the translators and commentators were themselves versed in practical activities and spoke to practitioners in the field. In order to reach this different group of people, the vernacular was increasingly employed. At the same time, debates arose on the question of whether the vernacular was a suitable medium in which to express philosophical ideas. Alessandro Piccolomini shares some thoughts on this topic in a letter addressed to the Italian author and poet Pietro Aretino. Piccolomini advocates the importance of the vernacular and expresses his wish to translate and comment upon several ancient scientific and philosophical texts in Italian.[1] He discusses, amongst others, two options for translating vocabulary for which the Italian language lacks the appropriate terms, namely either to paraphrase those concepts or to invent new words for them.[2] In direct relation to the Aristotelian *Mechanics*, Piccolomini urged Oreste Vanucci Biringucci to translate his Latin paraphrase of the treatise into Italian. Biringucci's translation appeared in the year 1582 and was specifically designed for practitioners in mechanics. In the preface to his text, Biringucci points to the usefulness of the Italian translation, primarily for engineers and architects.[3] He mentions that Piccolomini regretted having written his paraphrase in Latin, because by this he excluded those people from using the text that would have benefited most from its subject matter. This example shows that the focus on practical knowledge in the sixteenth century goes hand in hand with the growing awareness of the importance of the vernacular. My examination of some of the diagrams in early modern mechanical texts refers to certain changes in the discipline at this time. Furthermore, including vernacular translations and commentaries on the *Mechanics* – with a focus on the technical terminology in the vernacular – in a study of the text and its transformations in the early modern period, would provide a more thorough understanding of the broader cultural contexts of mechanics.

Bibliography

Biringucci, Oreste Vannocci. 1582. *Parafrasi di Monsignor Alessandro Piccolomini, sopra Le Mechaniche d'Aristotile*. Rome: Zanetti.

Landoni, Teodorico (ed.). 1874. *Lettere scritte a Pietro Aretino*. Vol. 2, Pt. 1. Bologna: Romagnoli.

[1] See Landoni (1874), 230.

[2] See Landoni (1874), 232.

[3] See Biringucci (1582), 5.

Appendix: Reproductions and Reconstructions of the Diagrams

Introduction

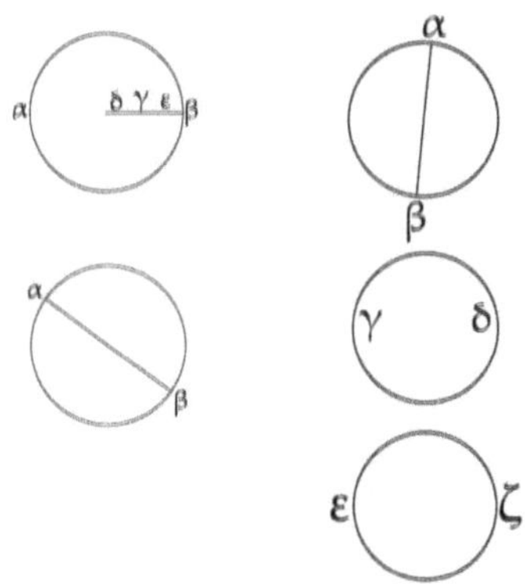

codex ω

J. van Leeuwen, *The Aristotelian Mechanics*, Boston Studies in the Philosophy and History of Science 316, DOI 10.1007/978-3-319-25925-3

Family *a*

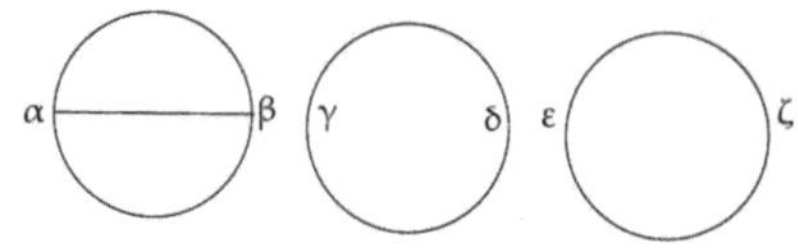

codex α

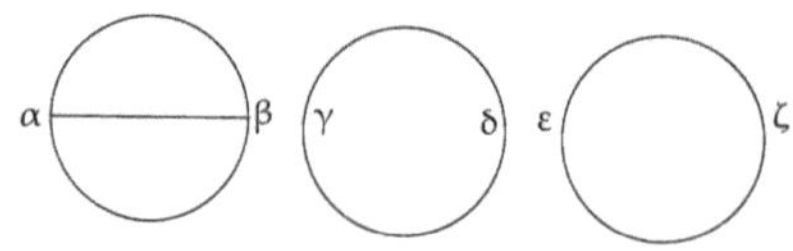

codex L

Family *b*

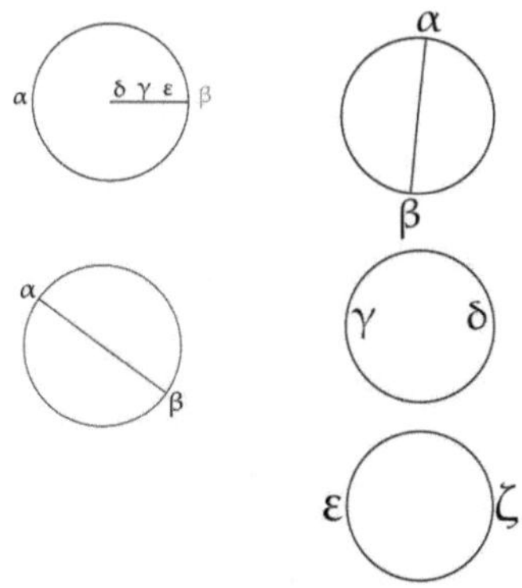

codex β

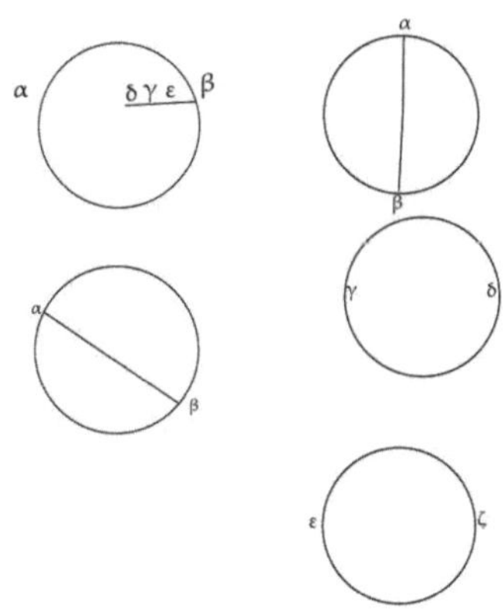

codex P

Family *c*

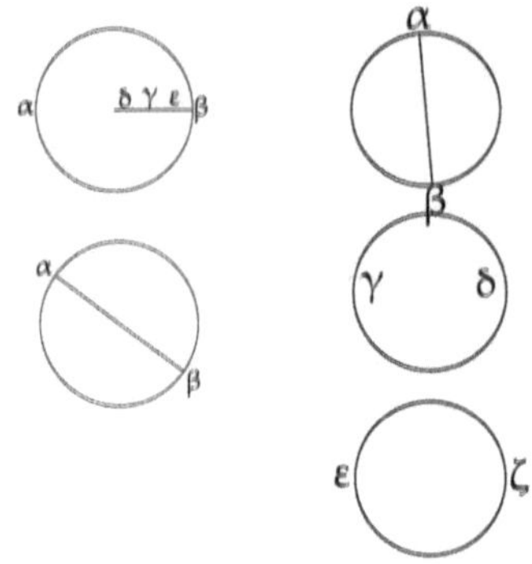

codex γ

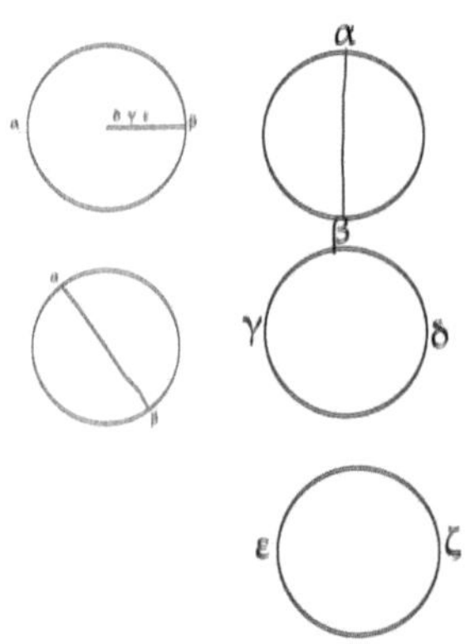

codex W^a

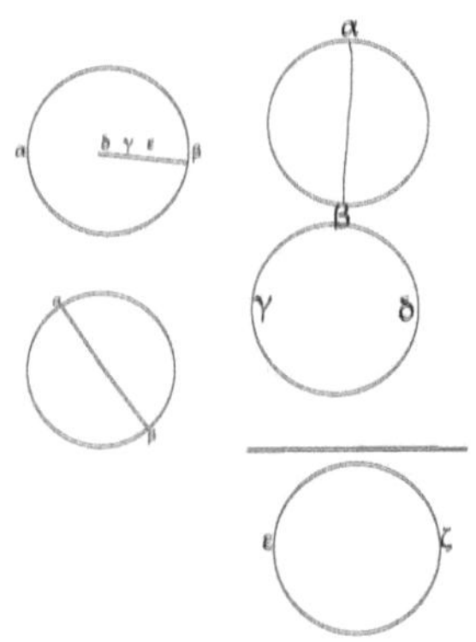

codex V^1

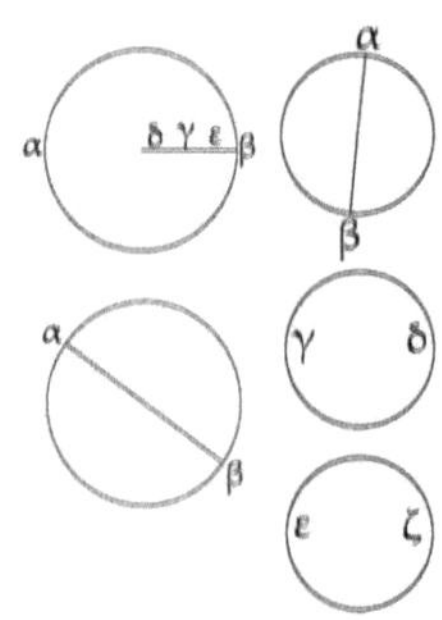

codex δ

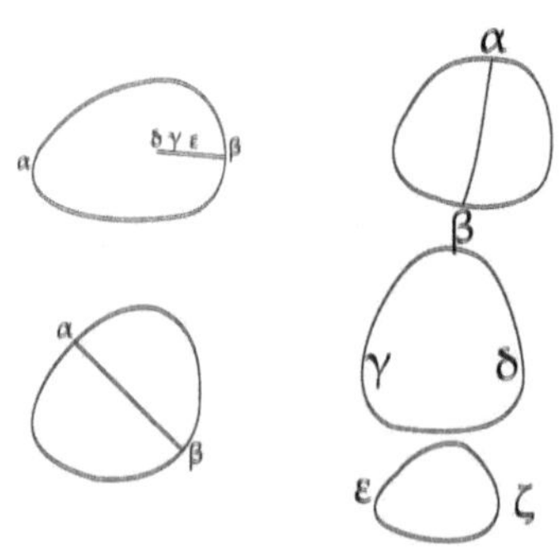

codex M^1

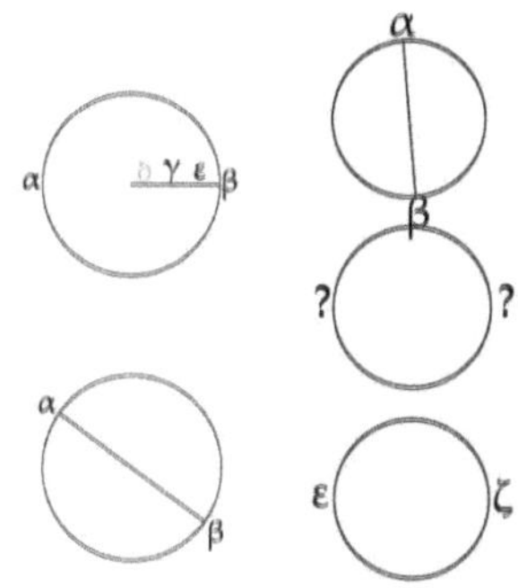

codex ε

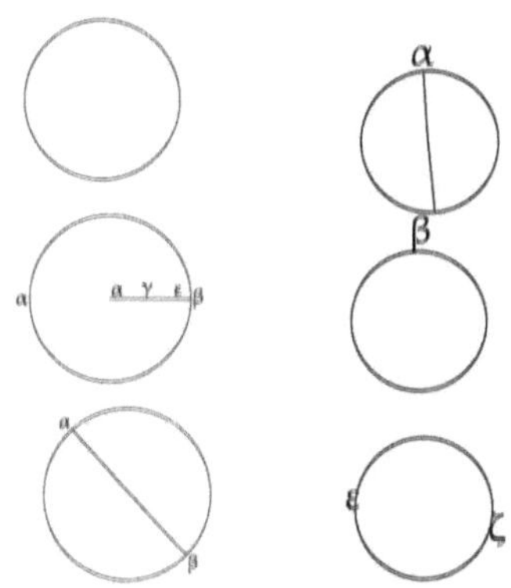

codex N^h

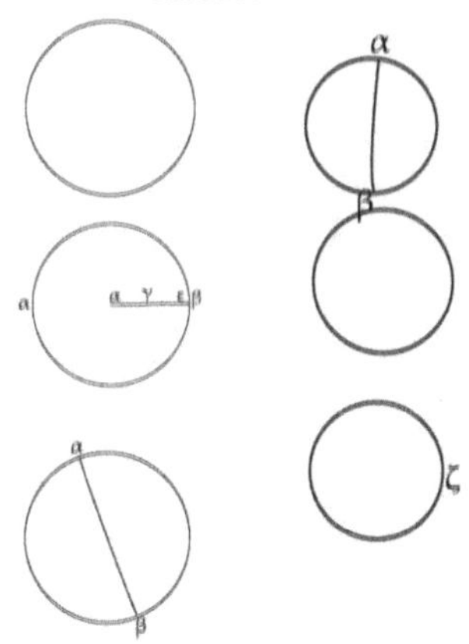

codex F^1

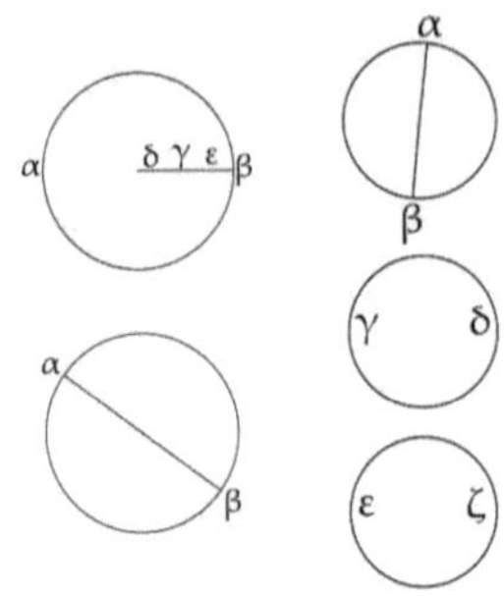

codex ζ

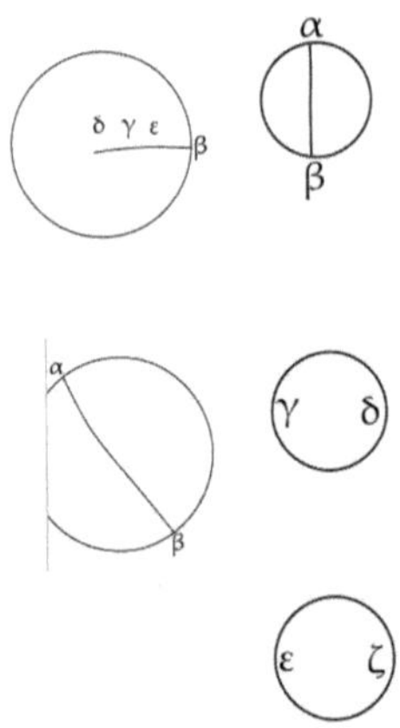

codex U^m

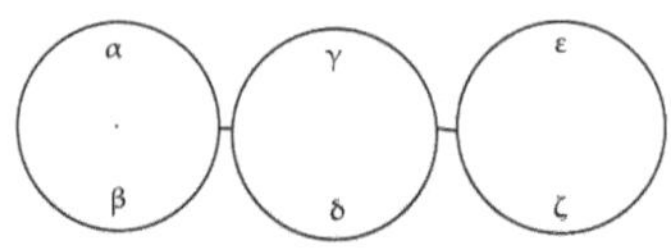

codex N^b

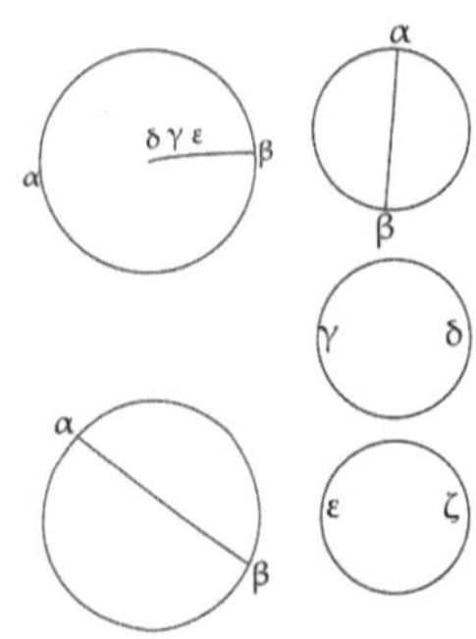

codex q

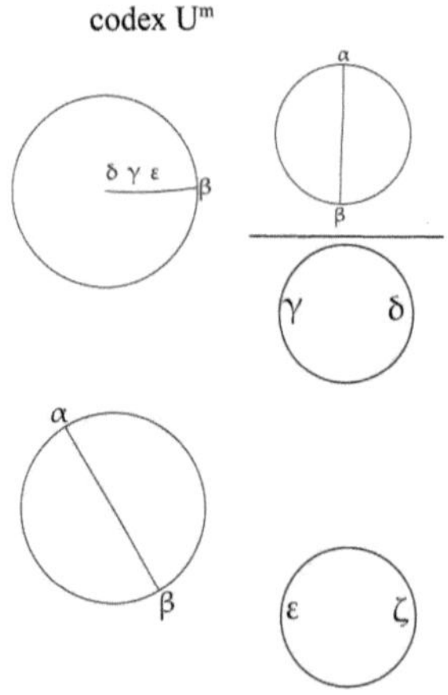

codex F^2

Problem 1.1

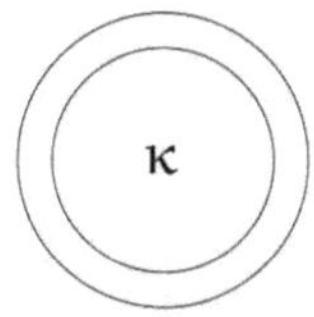

codex ω

Family *b*

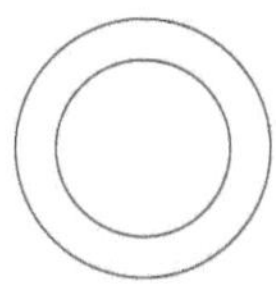

codex β

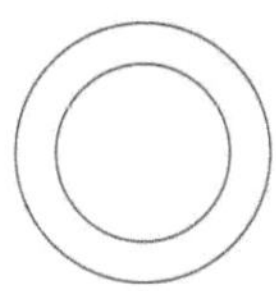

codex P

Family *c*

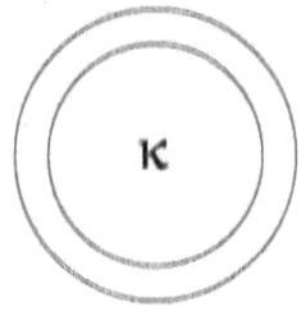

codex γ

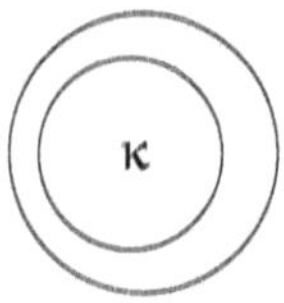

codex W^{a}

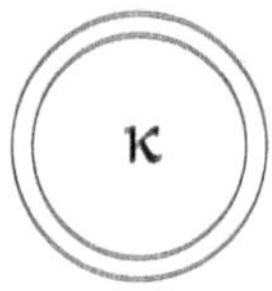

codex V^{1}

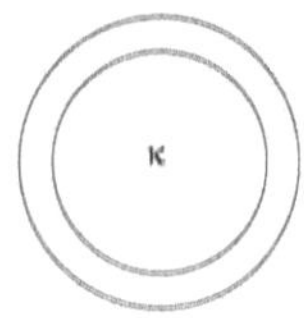

codex δ

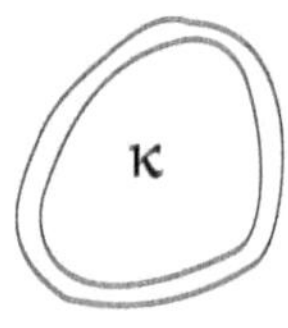

codex M^{1}

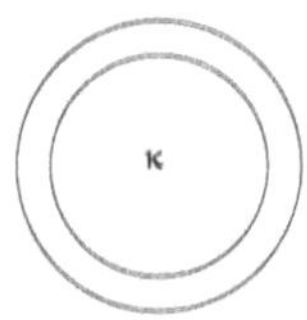

codex ε

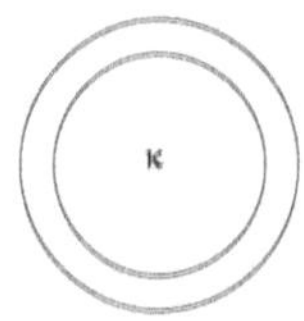

codex ζ

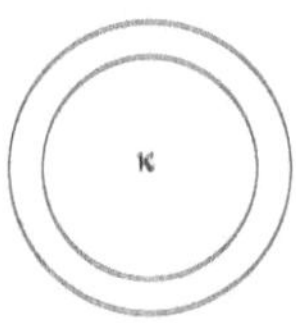

codex N^{h}

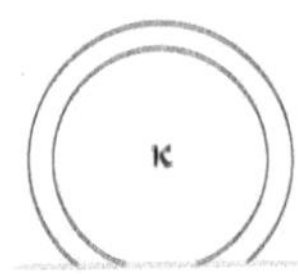

codex U^{m}

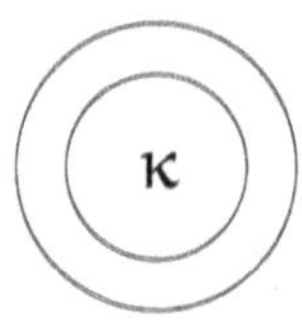

codex N^{b}

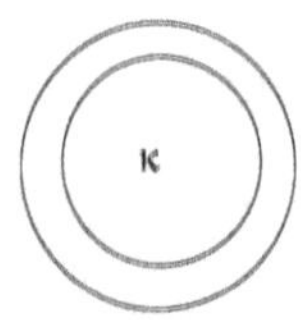

codex q

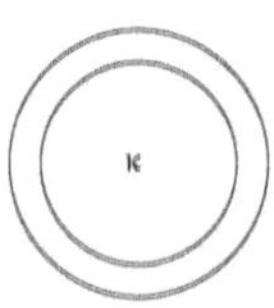

codex F^{1}

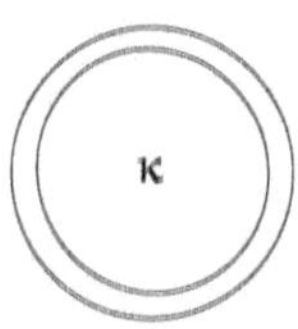

codex F^{2}

Problem 1.2

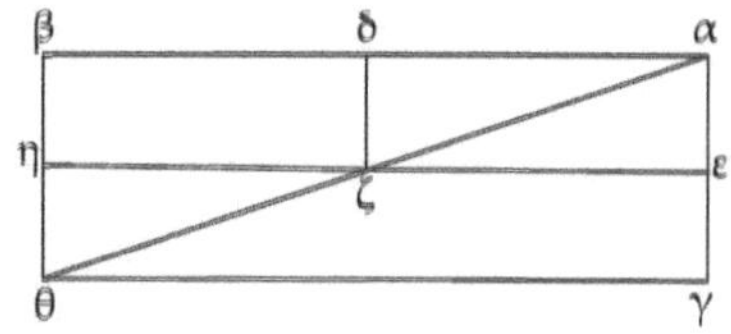

codex ω

Family *a*

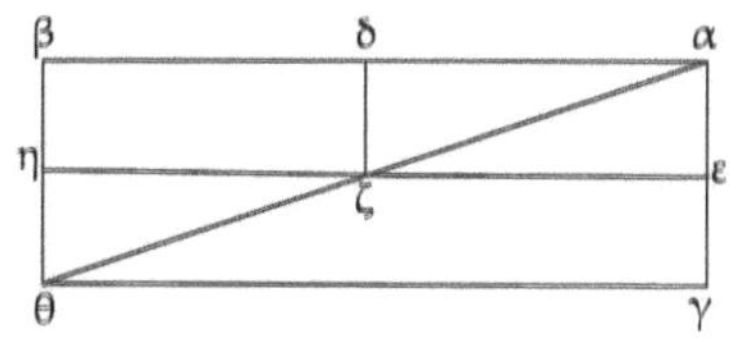

codex α

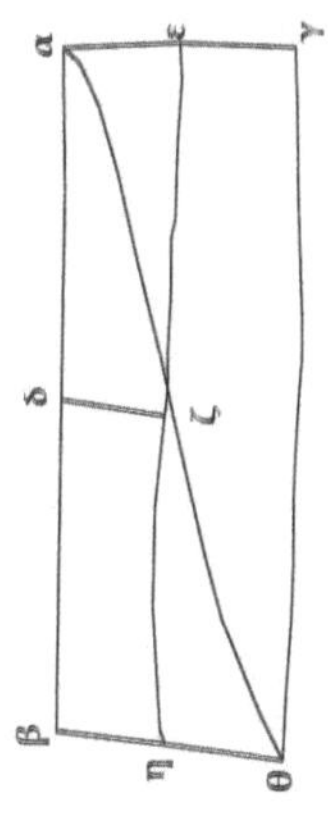

codex L

Family *b*

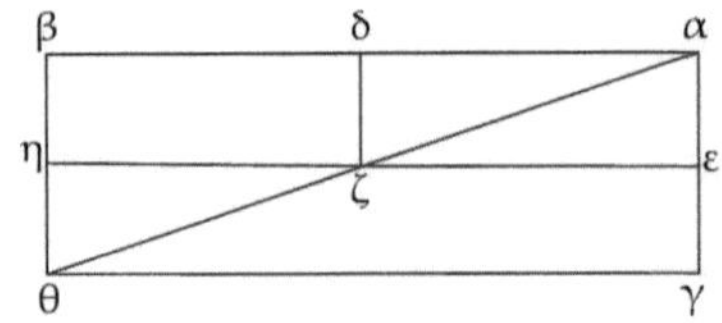

codex β

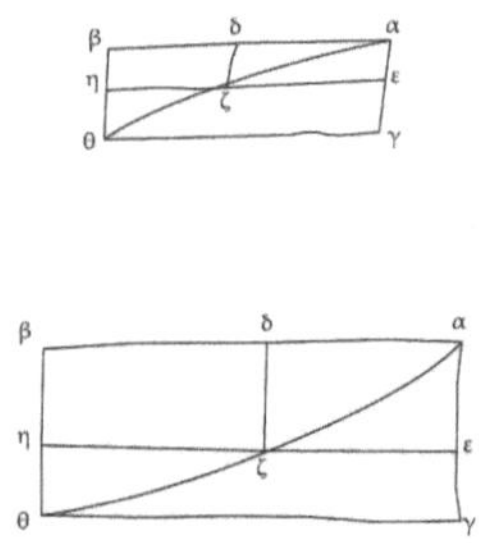

codex P

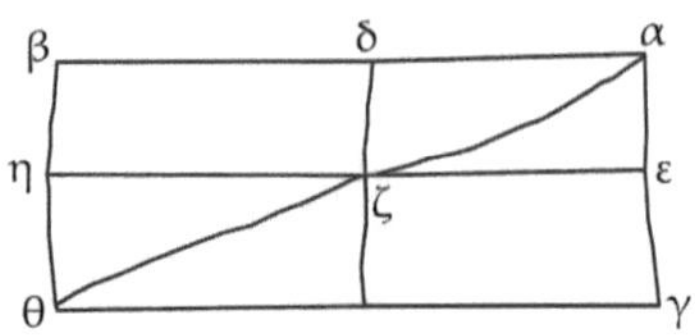

codex M[v]

Family *c*

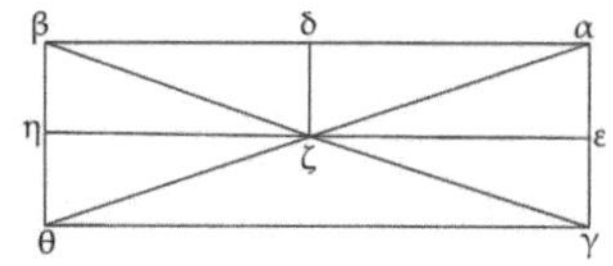

codex γ

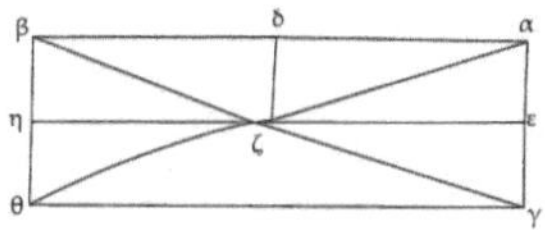

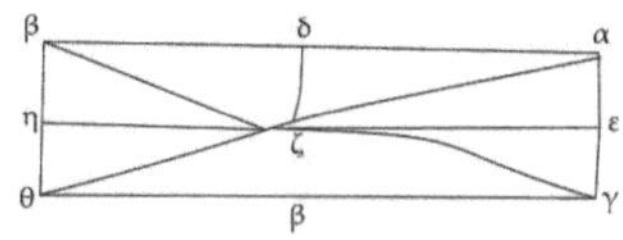

codex W^a

codex V^1

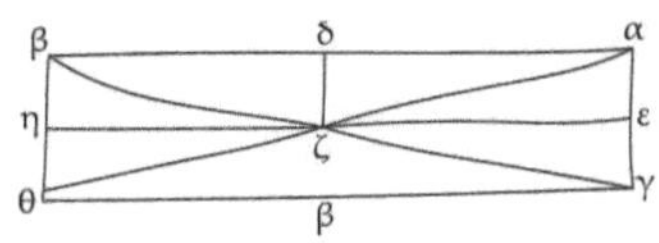

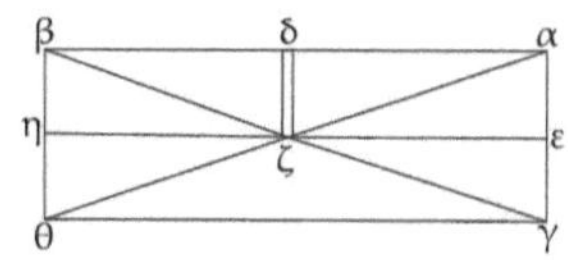

codex M^1

codex δ

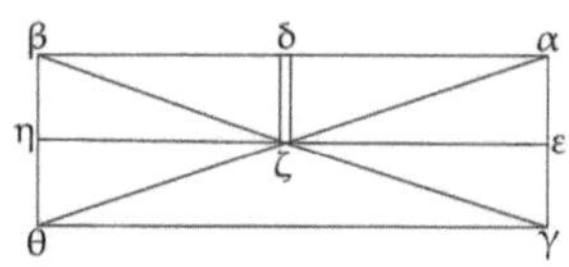

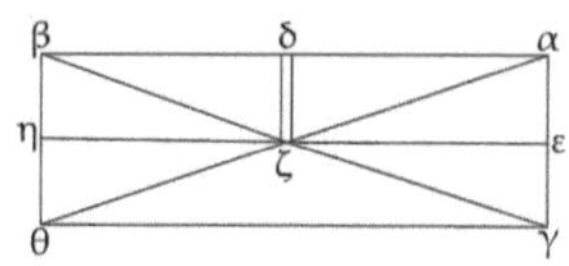

codex ε

codex ζ

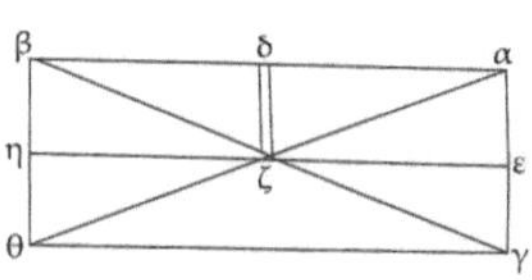

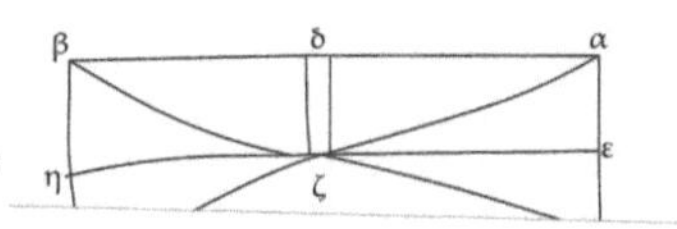

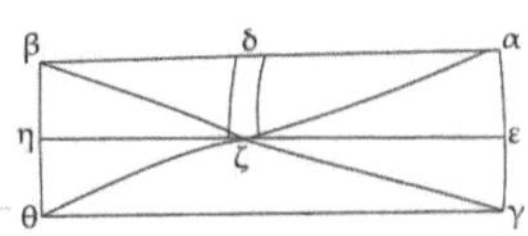

codex N^h

codex U^m

codex q

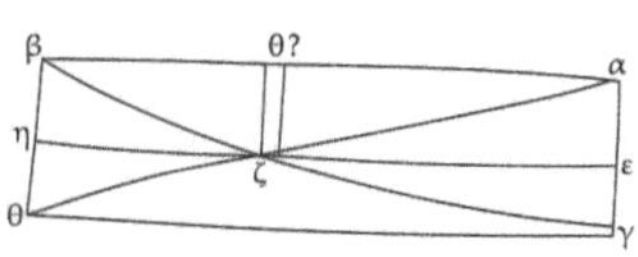

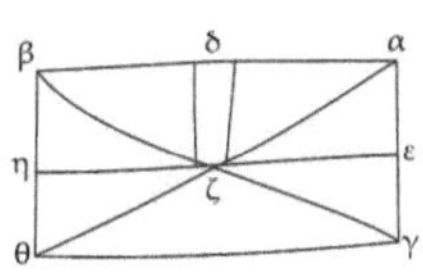

codex F^1

codex F^2

Problem 1.3

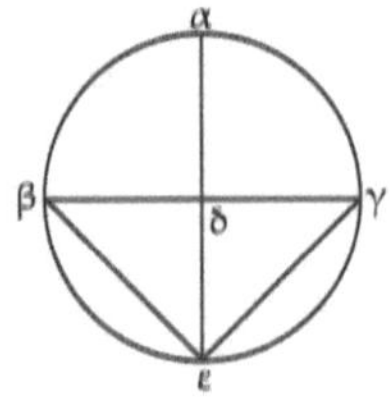

codex ω

Family *b*

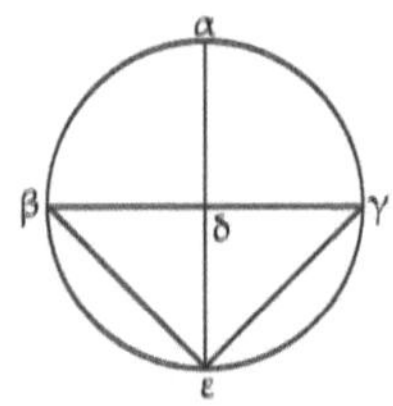

codex β

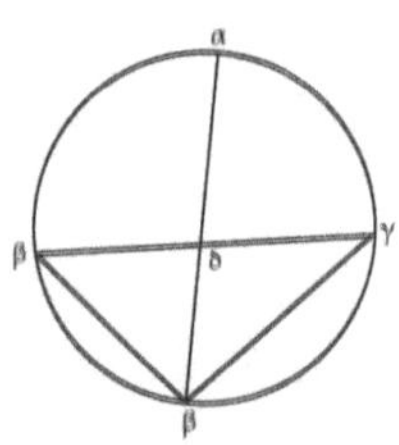

codex P

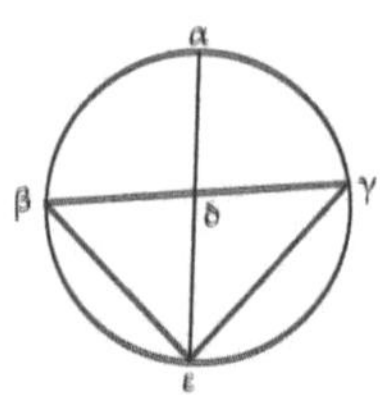

codex M^v

Family *c*

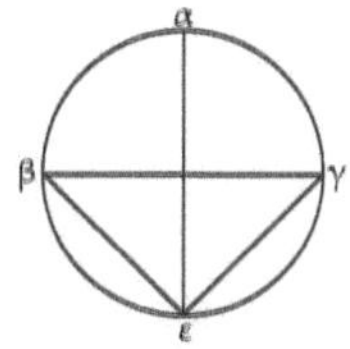

codex γ

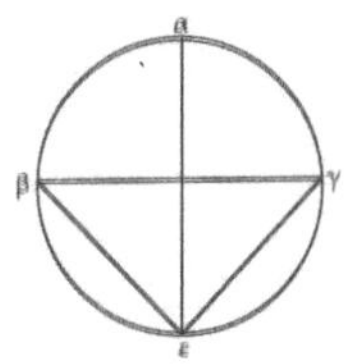

codex W^{a}

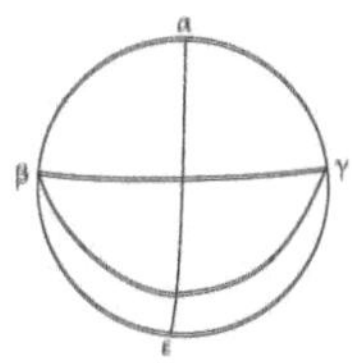

codex V^{1}

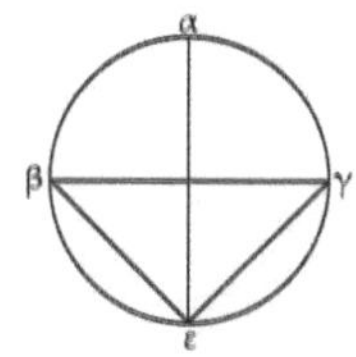

codex δ

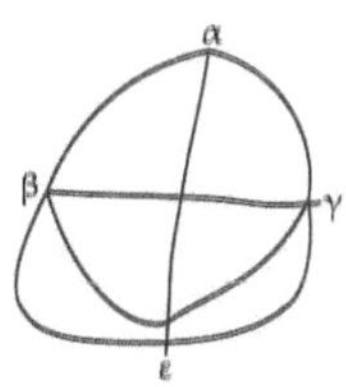

codex M^{1}

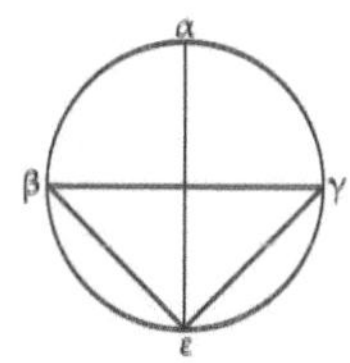

codex ε

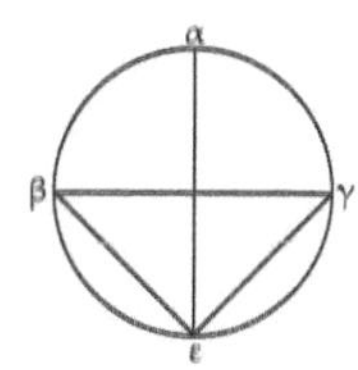

codex ζ

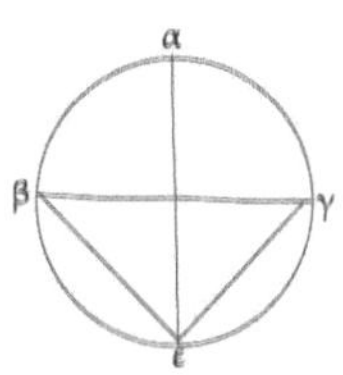

codex N^{h}

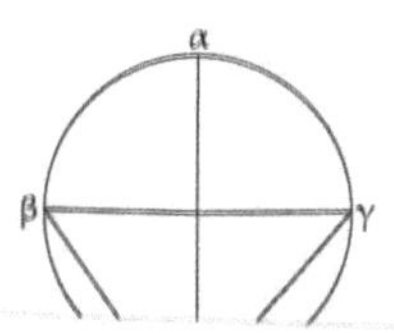

codex U^{m}

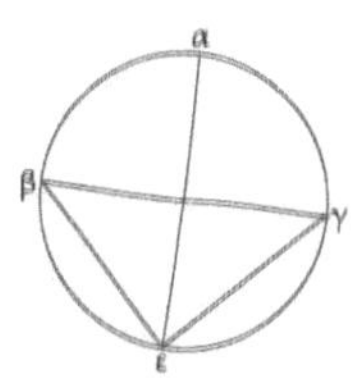

codex N^{b}

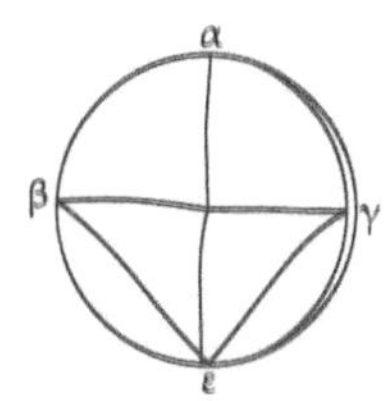

codex q

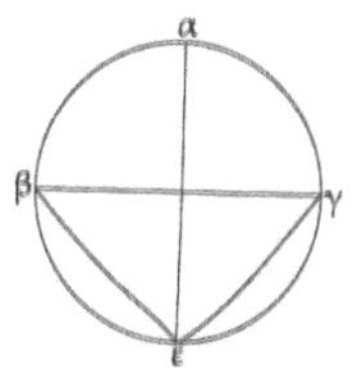

codex F^{1}

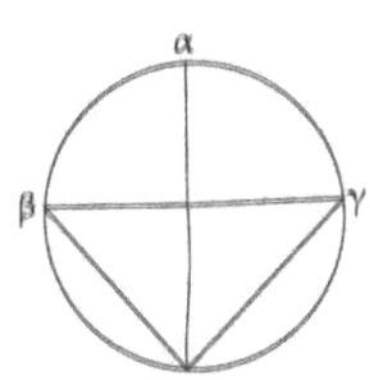

codex F^{2}

Problem 1.4

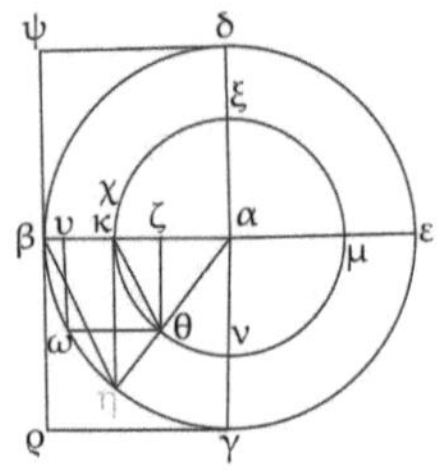

codex ω

Family *a*

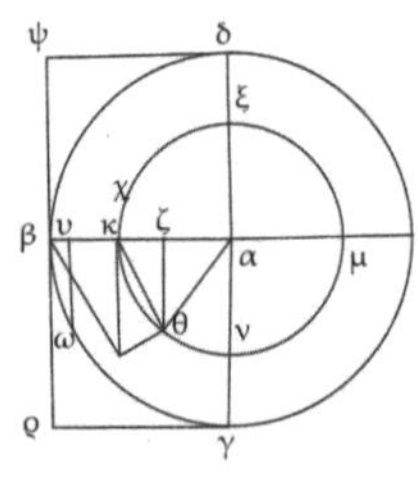

codex α

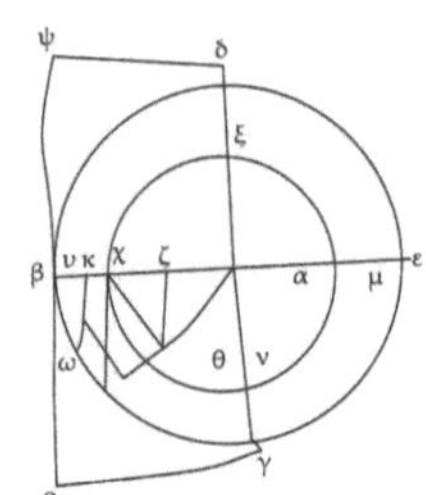

codex L

Family *b*

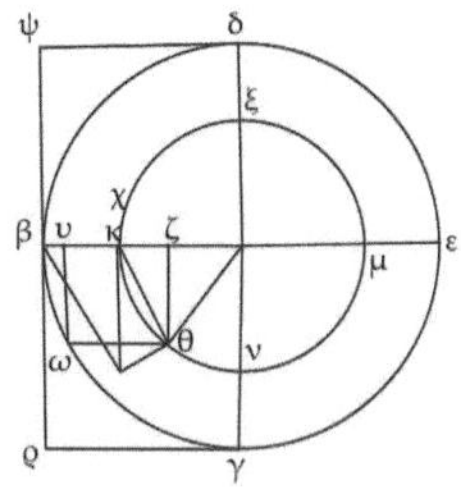

codex β

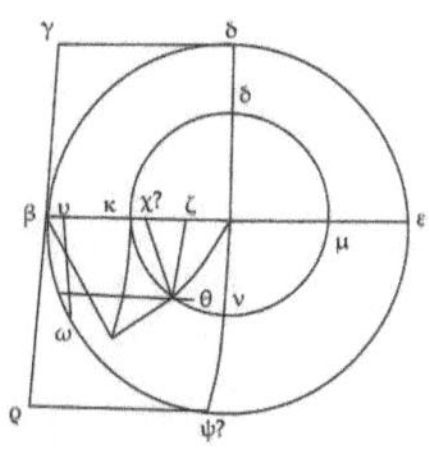

codex P

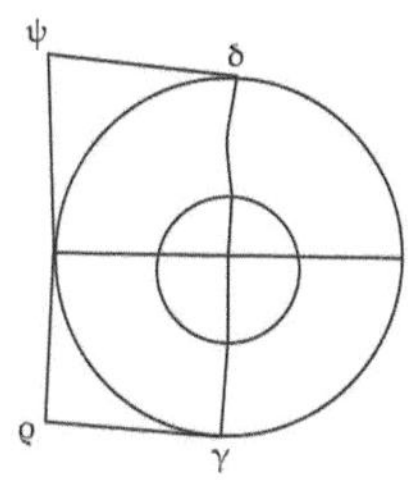

codex M^v

Family *c*

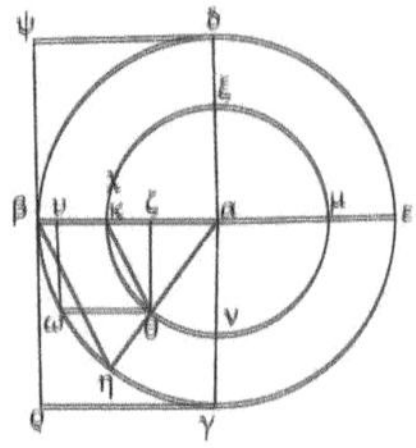

codex γ

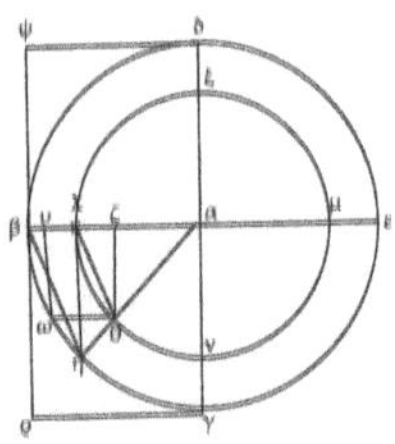

codex W[a]

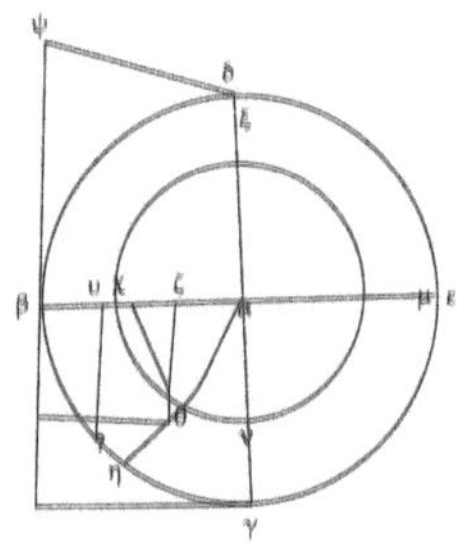

codex V[1]

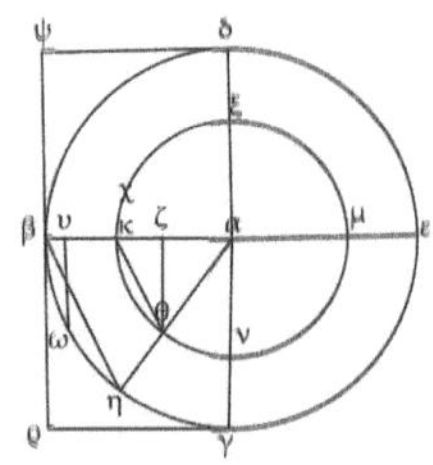

codex δ

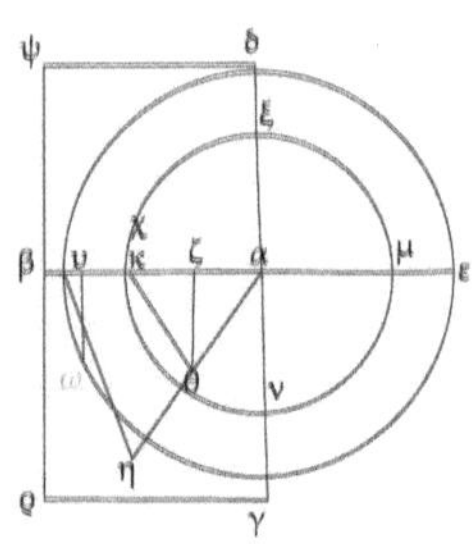

codex ε

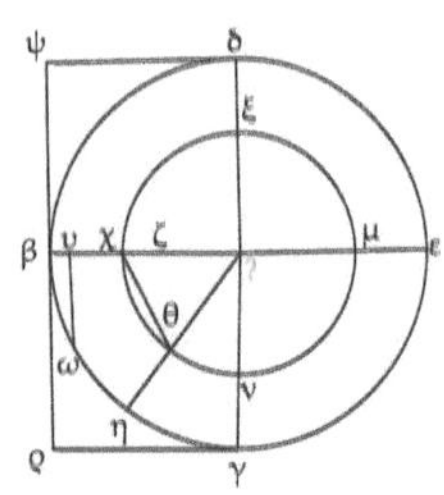

codex ζ

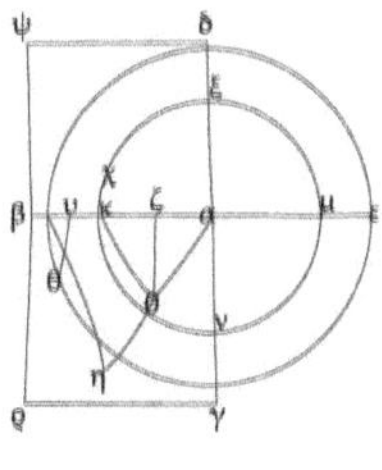

codex N^h

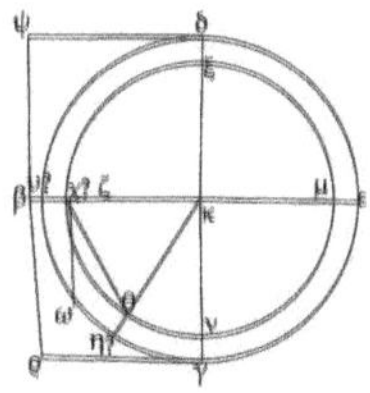

codex U^m

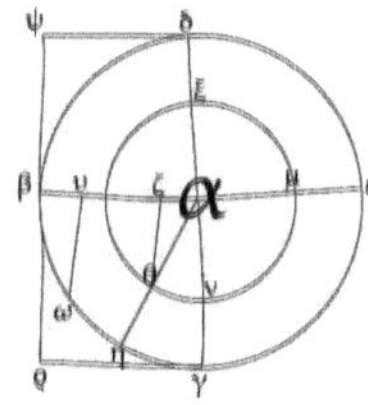

codex N^b

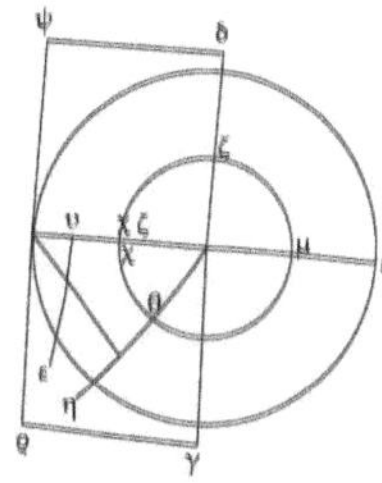

codex q

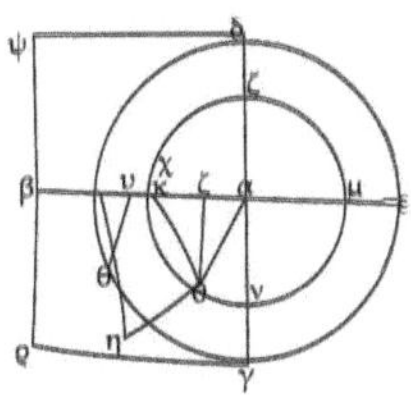

codex F^1

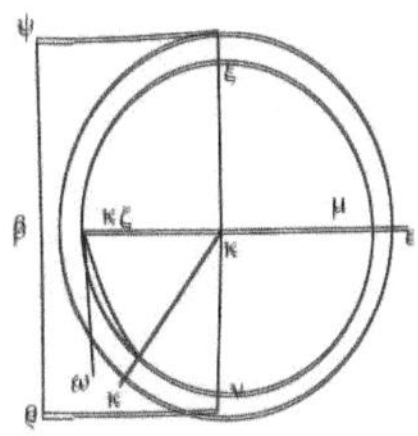

codex F^2

Problem 2

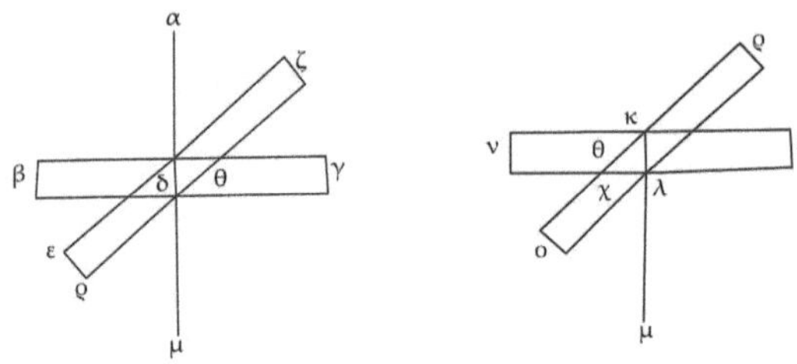

codex ω

Family *a*

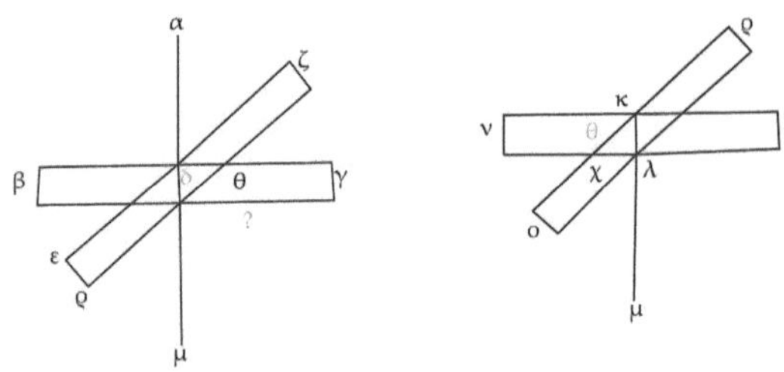

codex α

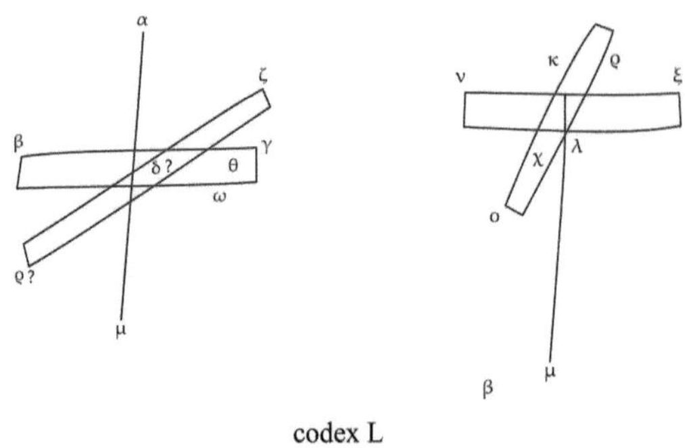

codex L

Family *b*

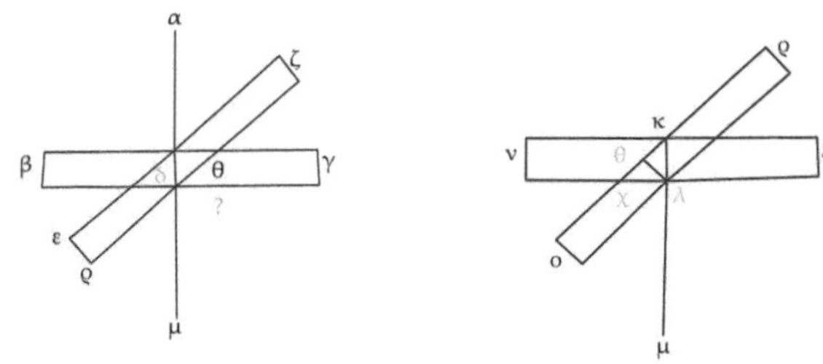

codex β

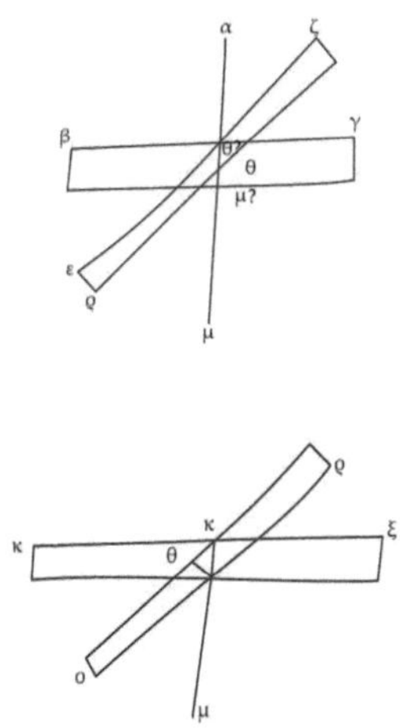

codex P

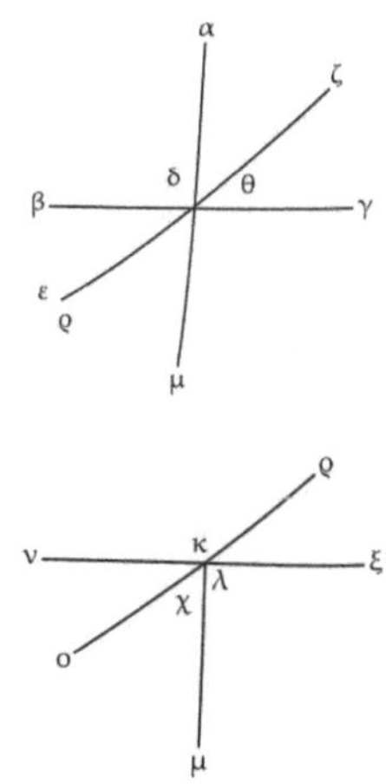

codex M^{v}

Family *c*

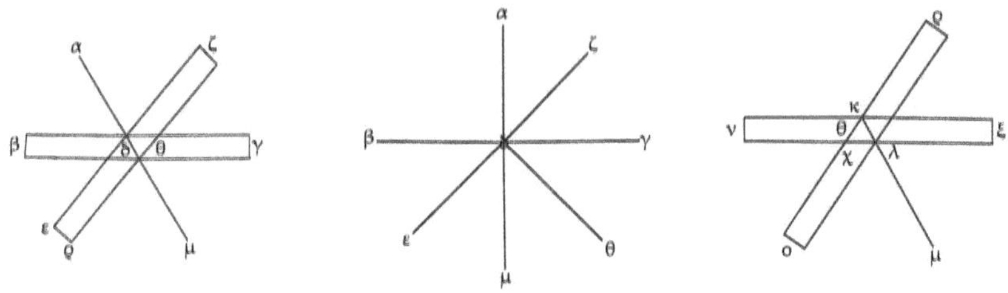

codex γ

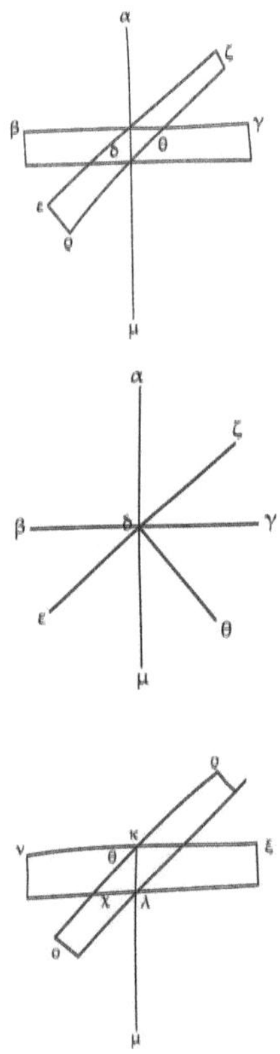

codex W[a]

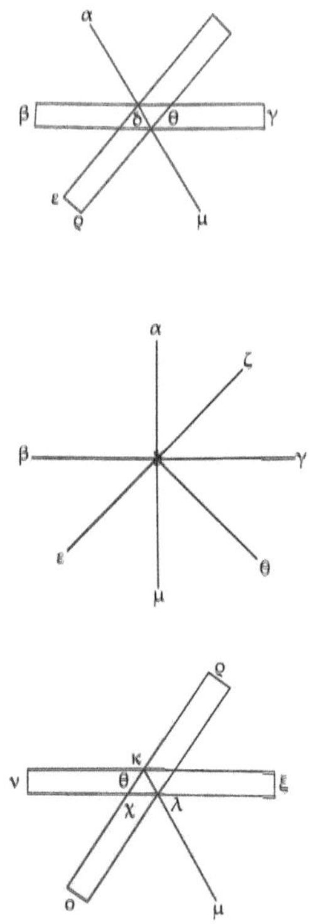

codex δ

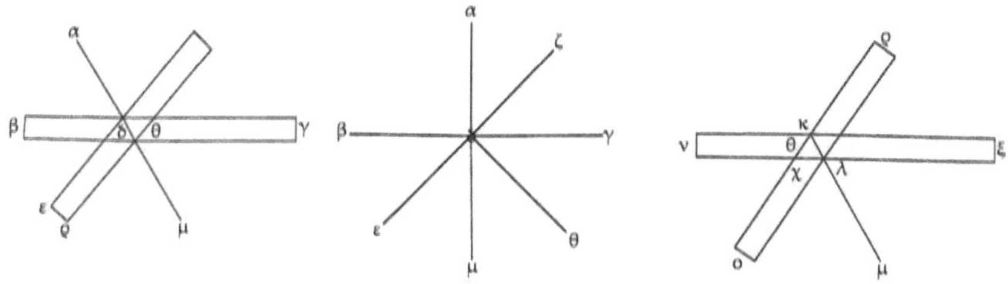

codex ε

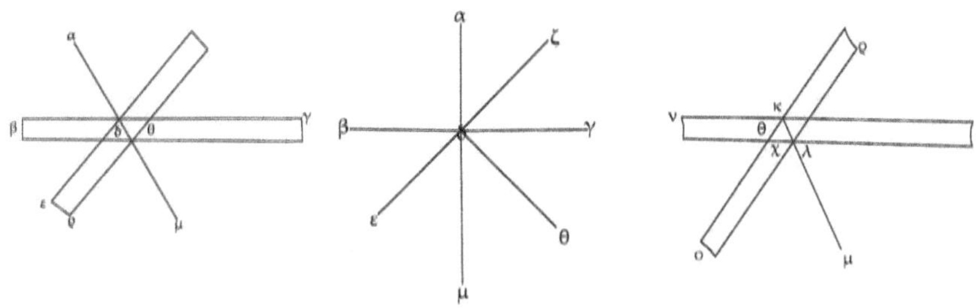

codex N^h

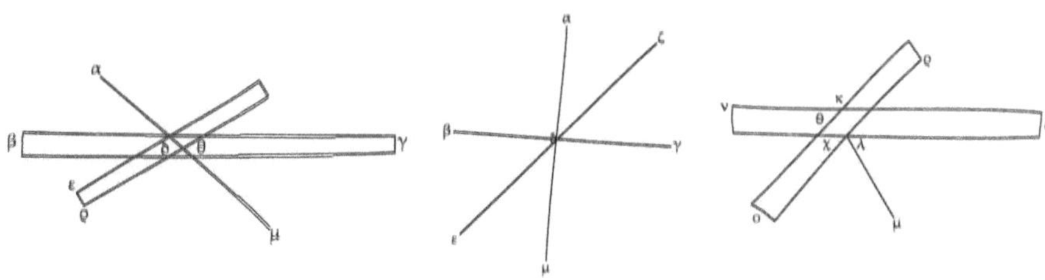

codex F^1

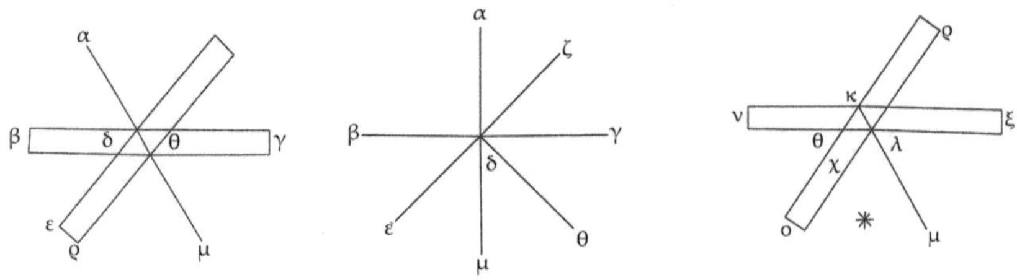

codex ζ

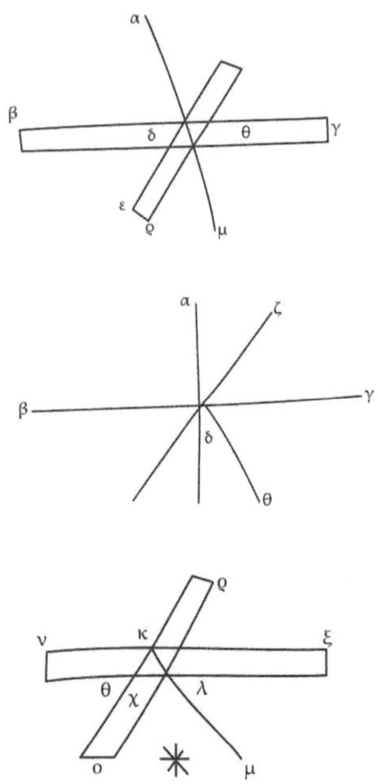

codex U^m

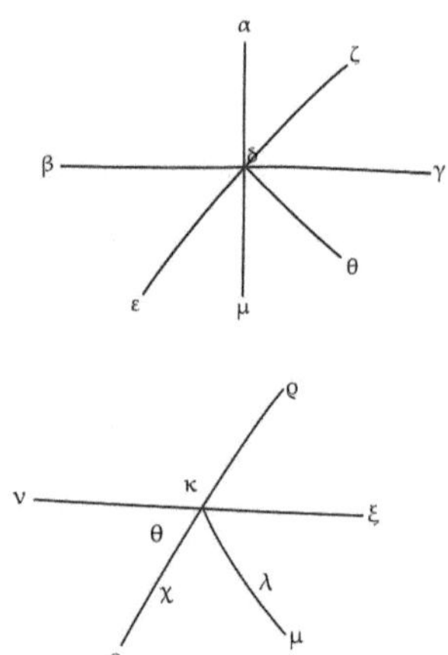

codex N^b

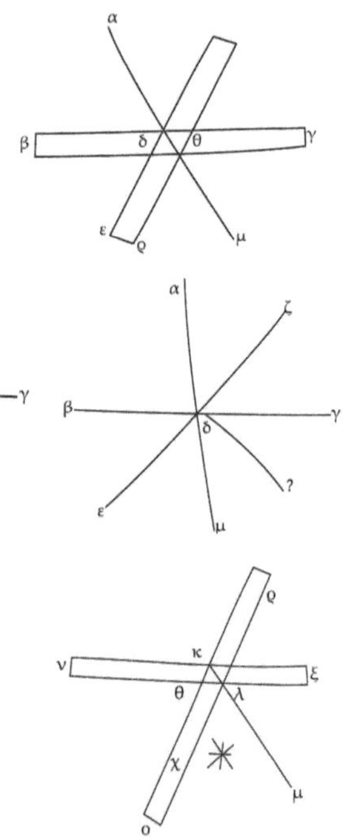

codex q

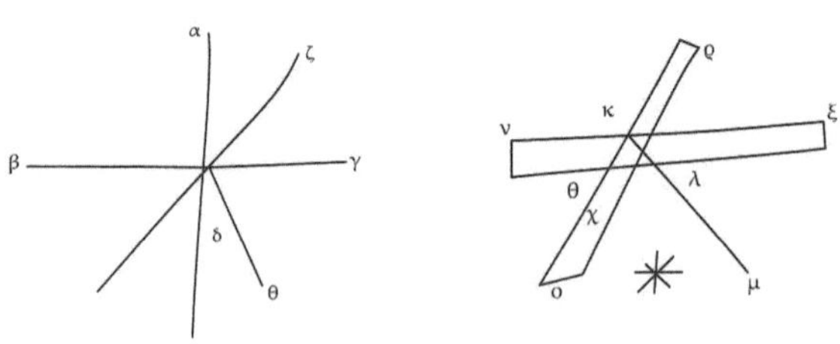

codex F²

Problem 3

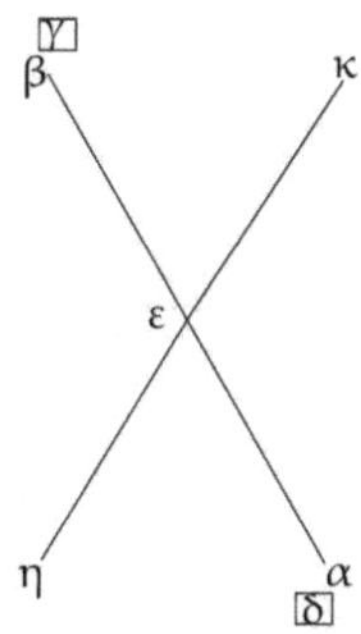

codex ω

Family *a*

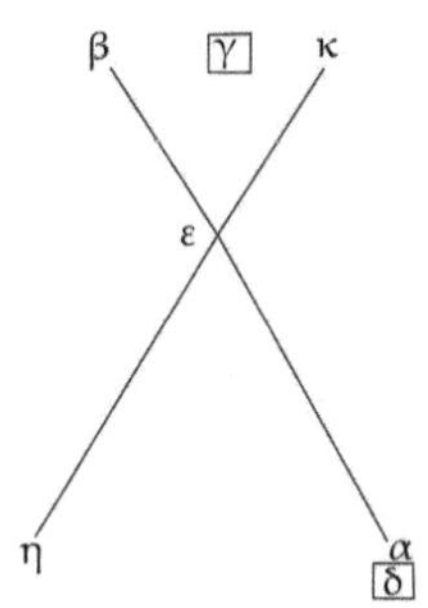

codex α

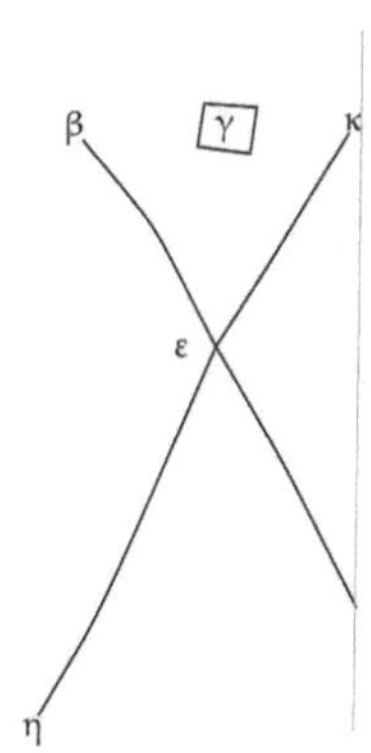

codex L

Family *b*

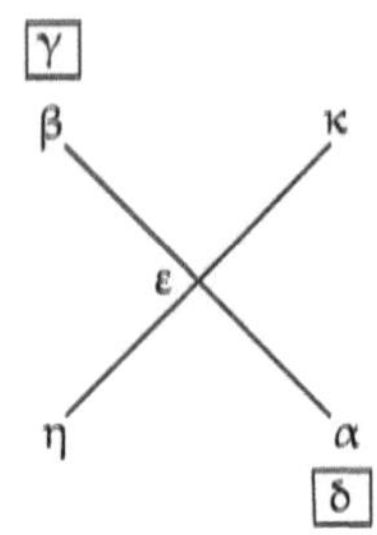

codex β

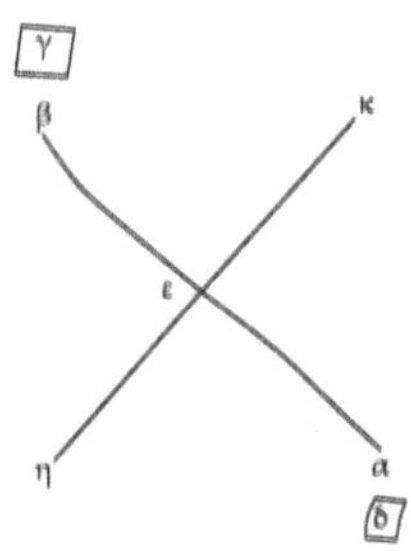

codex P

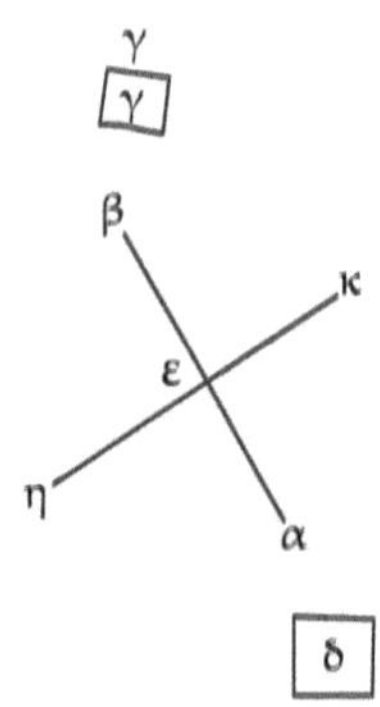

codex Mᵛ

Family *c*

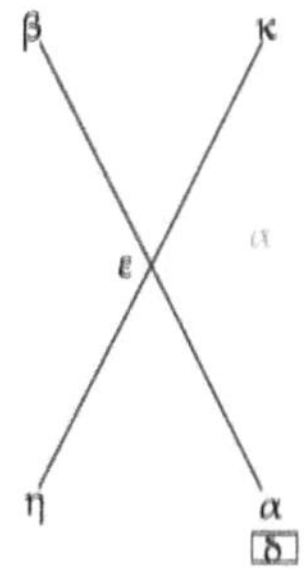

codex γ

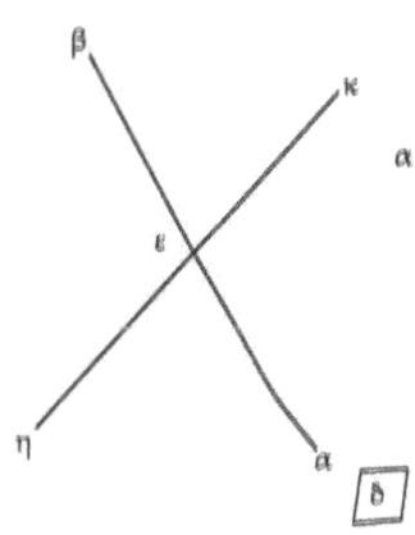

codex W[a]

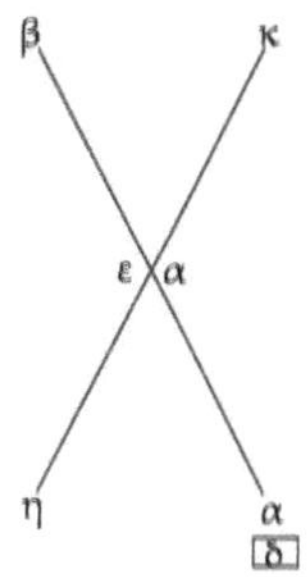

codex δ

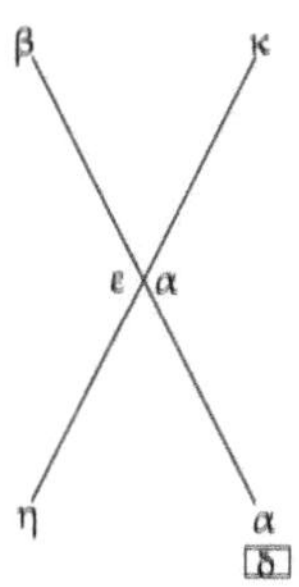

codex ε

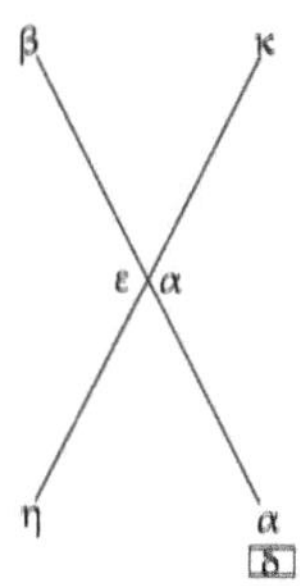

codex ζ

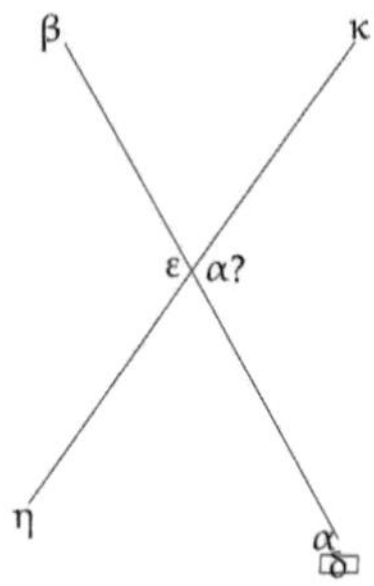

codex N^h

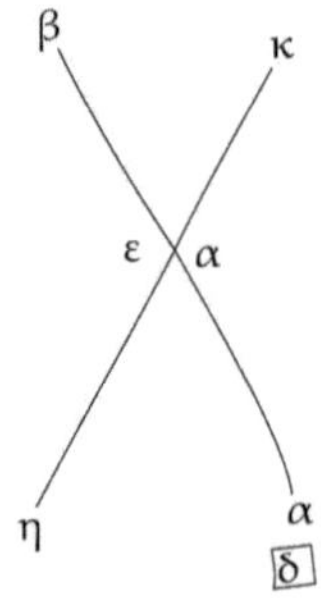

codex U^m

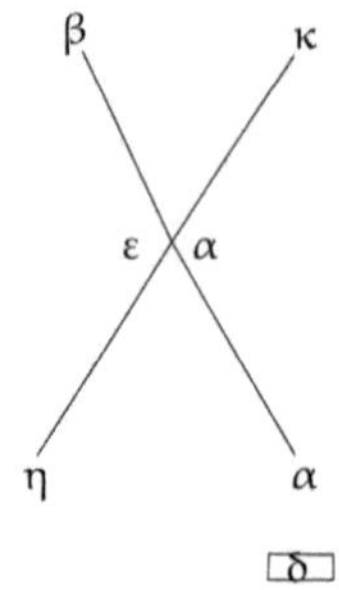

codex N^b

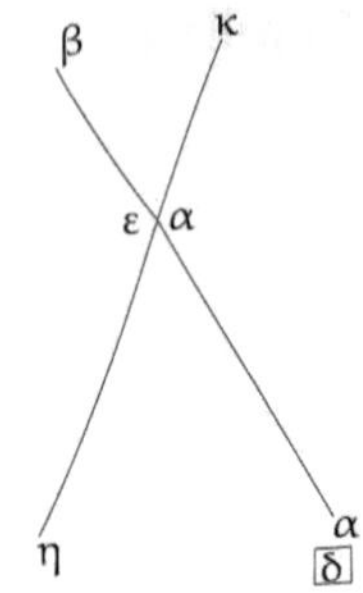

codex q

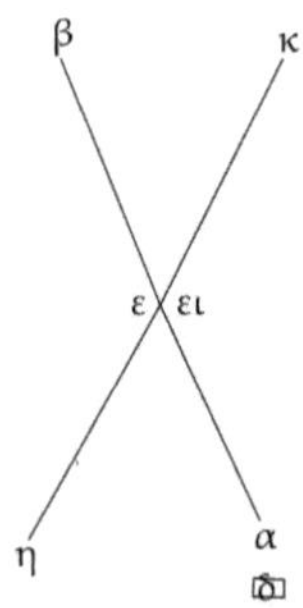

codex F^1

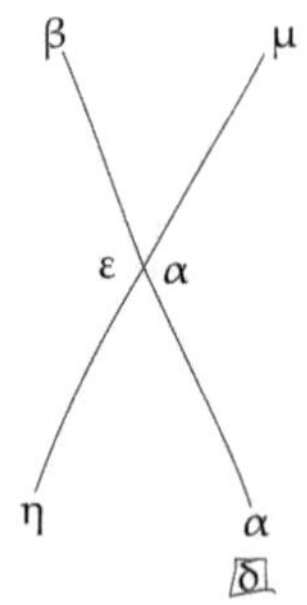

codex F^2

Problem 5

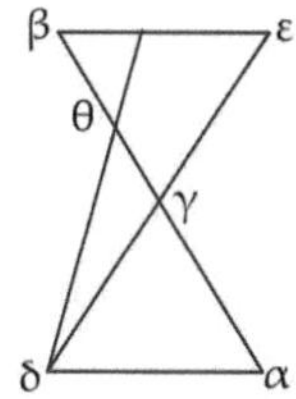

codex ω

Family *b*

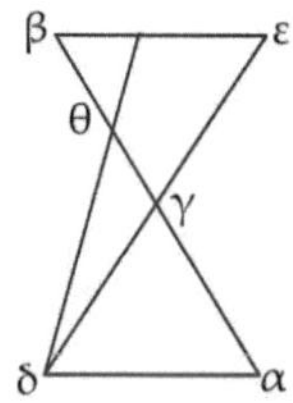

codex β

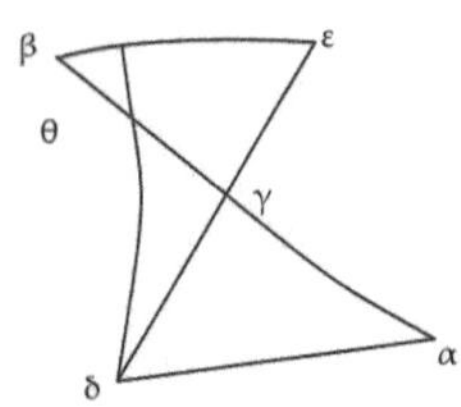

codex P

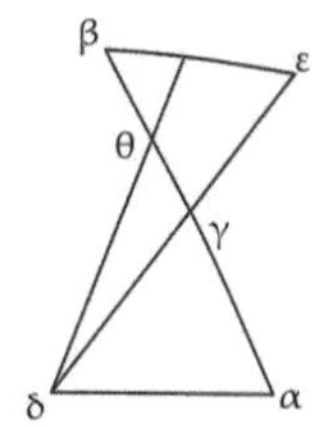

codex M[v]

Family *c*

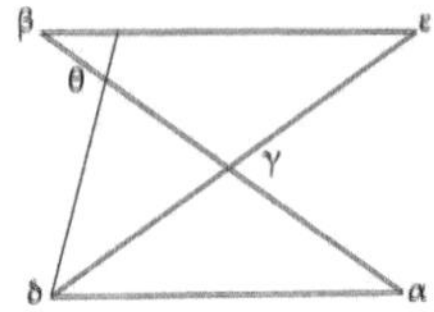

codex γ

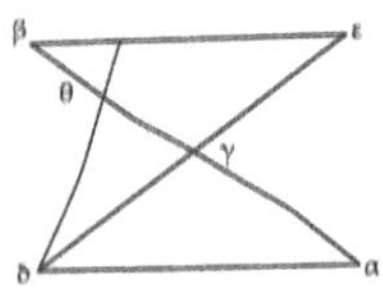

codex W^a

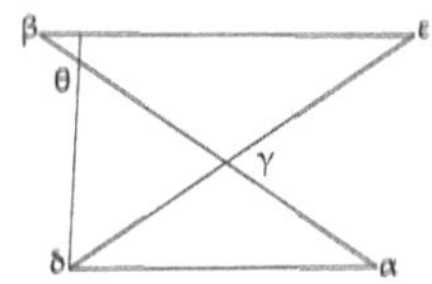

codex δ

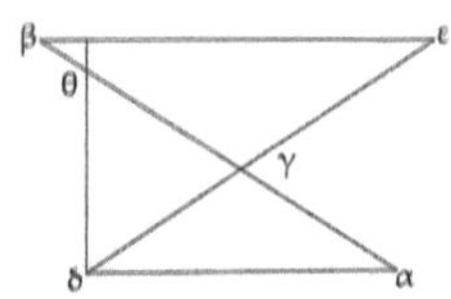

codex ε

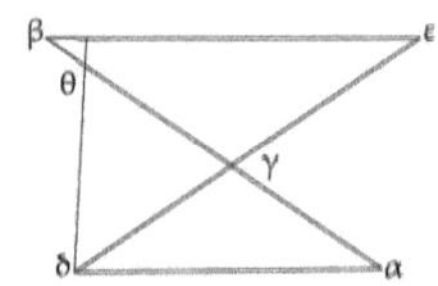

codex ζ

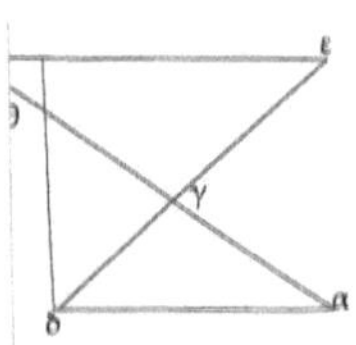

codex N^h

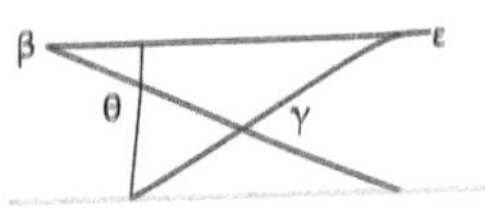

codex U^m

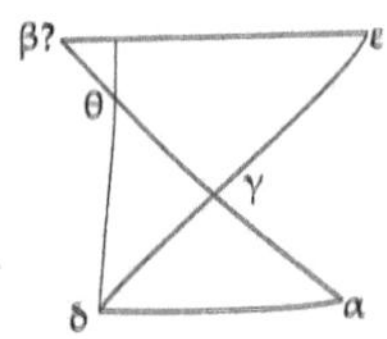

codex N^b

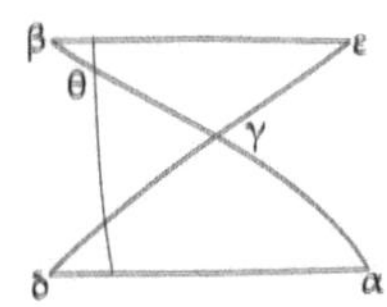

codex q

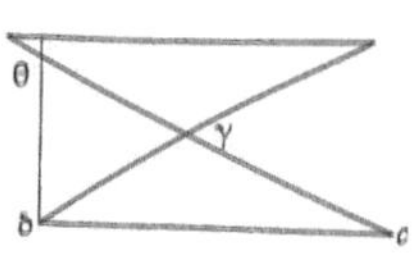

codex F^1

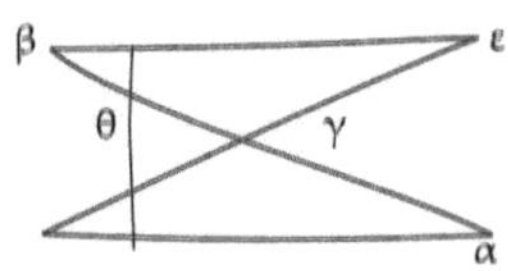

codex F^2

Problem 17

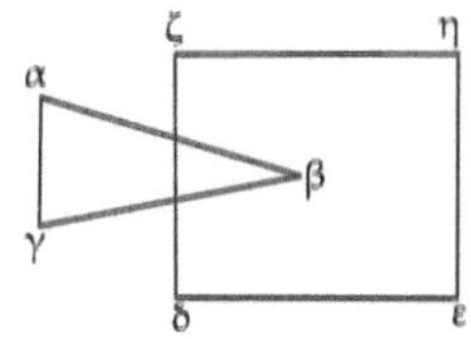

codex ω

Family *a*

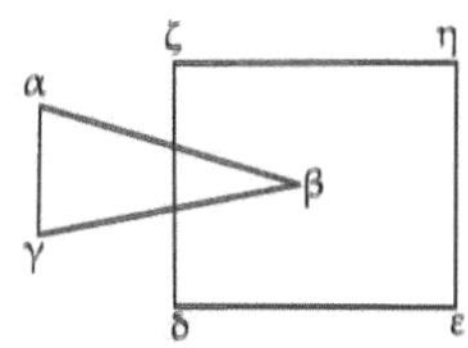

codex α

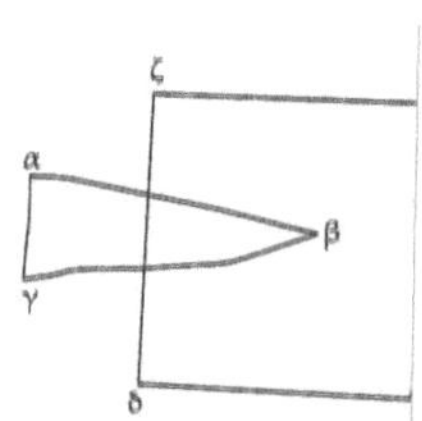

codex L

Family *b*

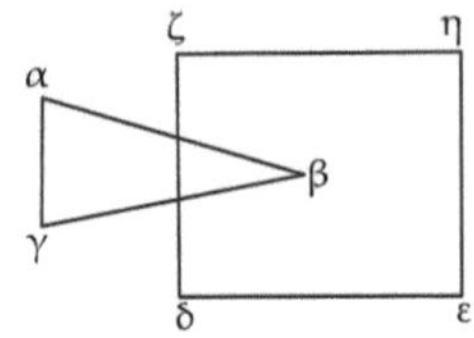

codex β

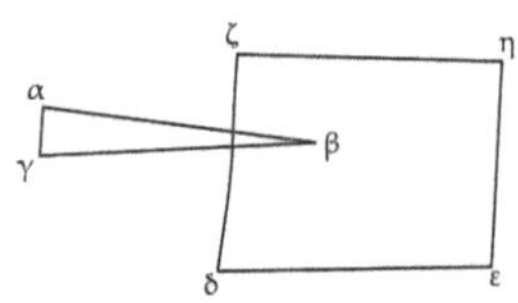

codex P

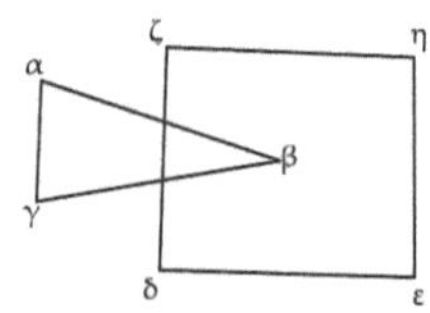

codex M[v]

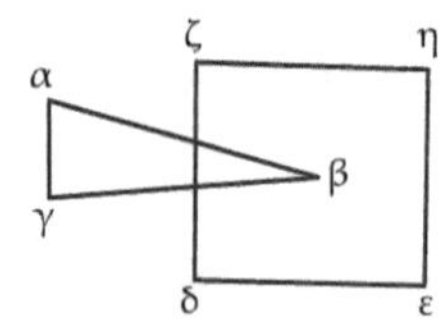

codex D[1]

Family *c*

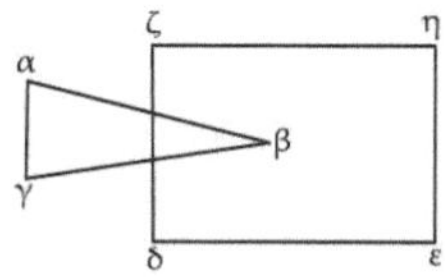

codex γ

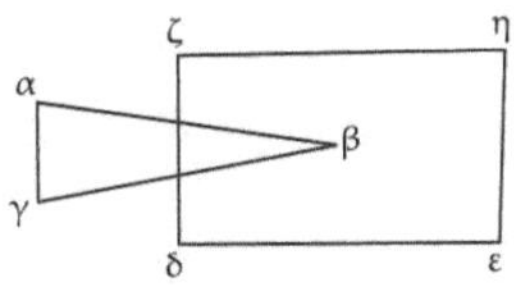

codex W^a

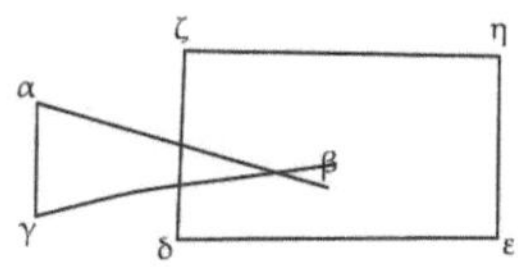

codex V^1

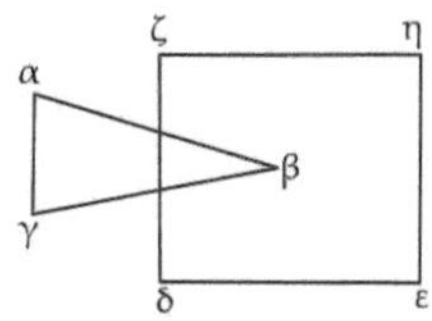

codex δ

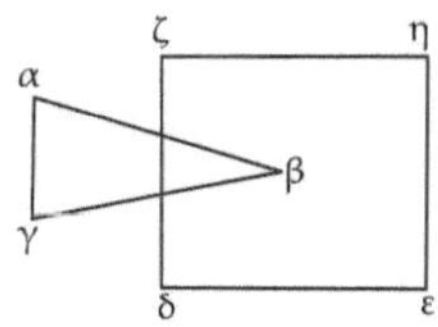

codex ε

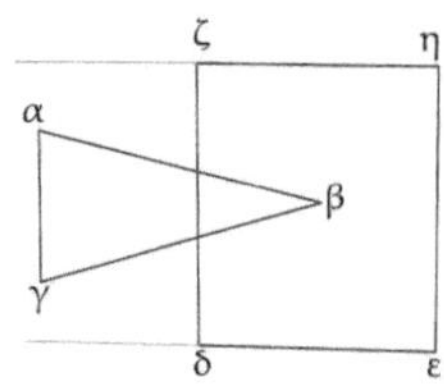

codex N^h

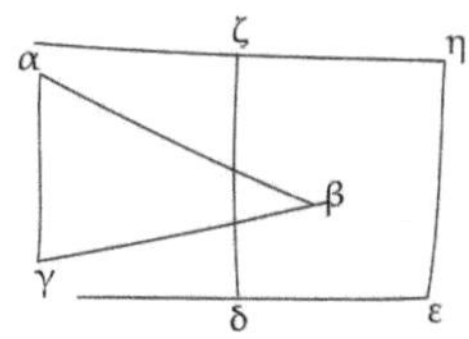

codex F^1

Problem 21

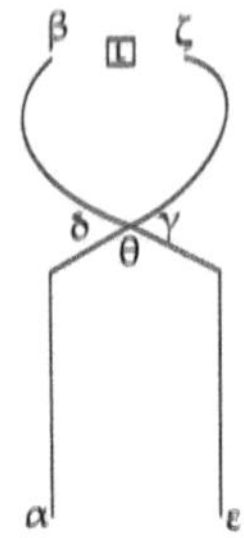

codex ω

Family *a*

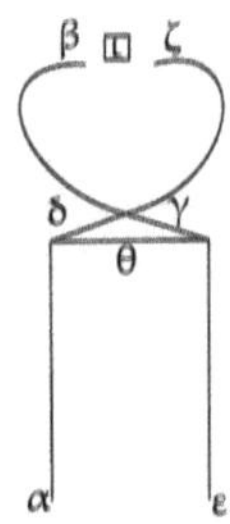

codex α

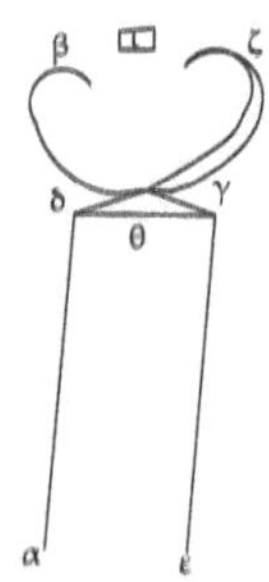

codex L

Family *b*

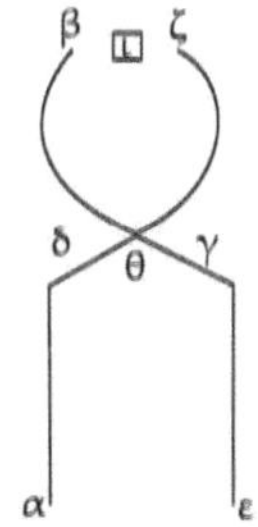

codex β

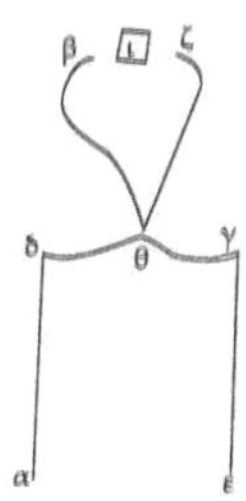

codex P

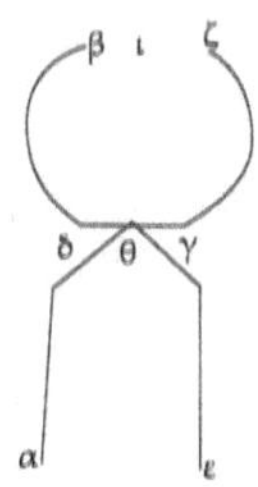

codex M^v

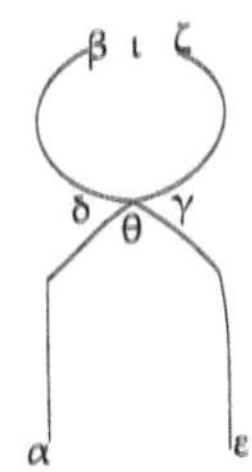

codex D^1

Family *c*

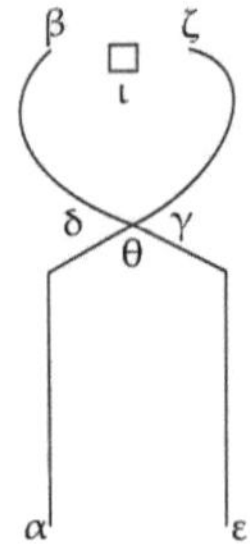

codex γ

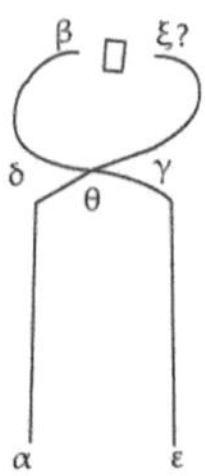

codex W^{a}

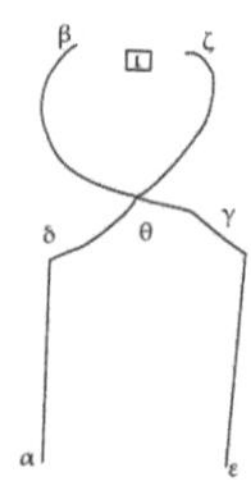

codex V^{1}

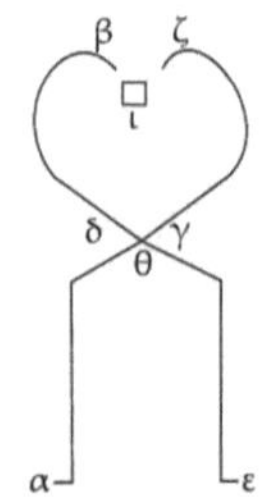

codex δ

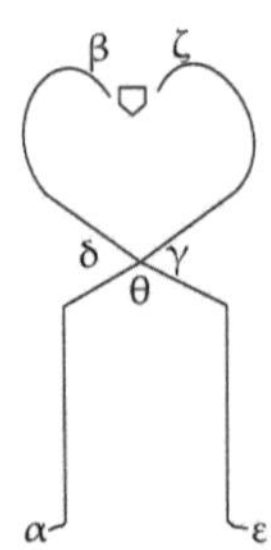

codex ε

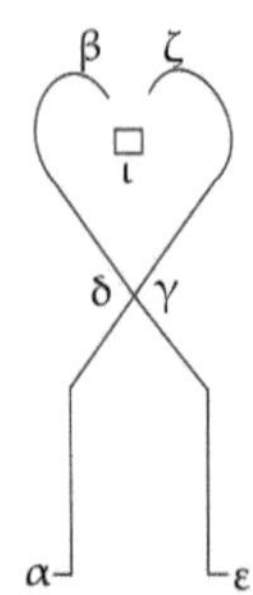

codex ζ

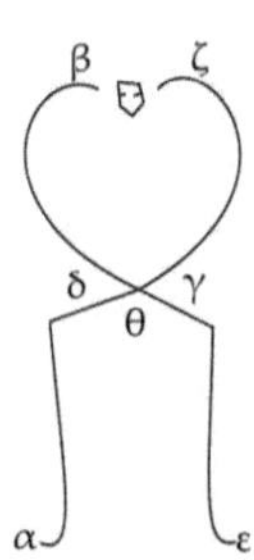

codex N^h

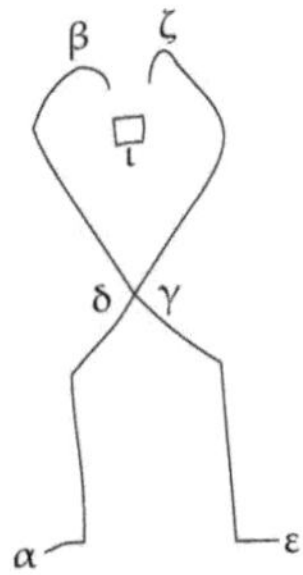

codex U^m

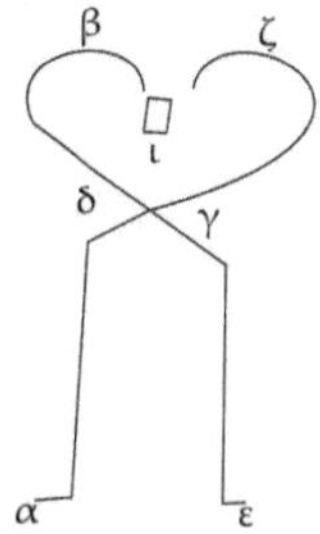

codex N^b

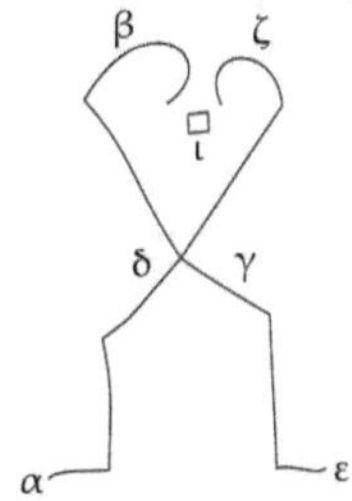

codex q

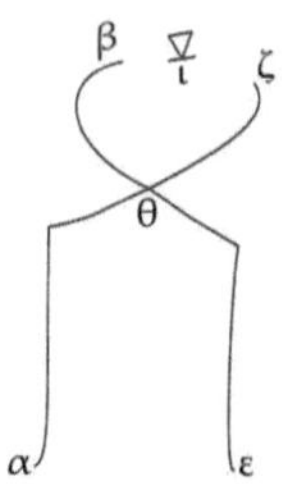

codex F^1

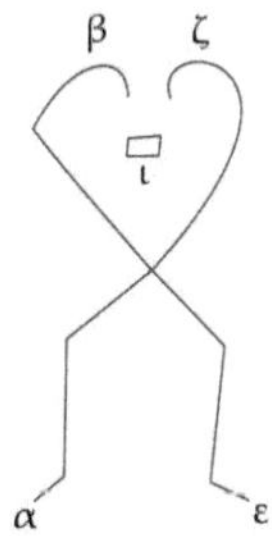

codex F^2

Problem 22

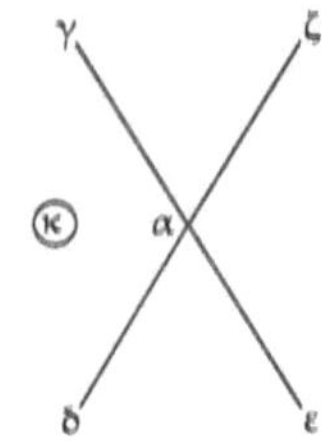

codex ω

Family *b*

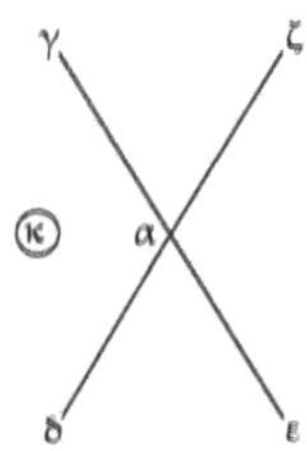

codex β

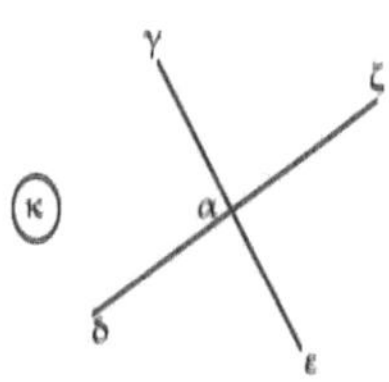

codex P

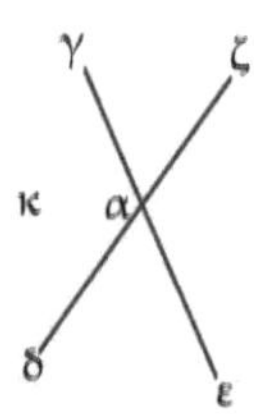

codex M^v

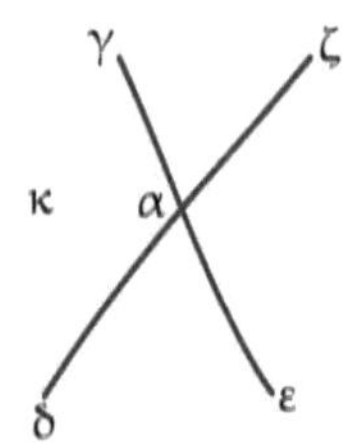

codex D^1

Family *c*

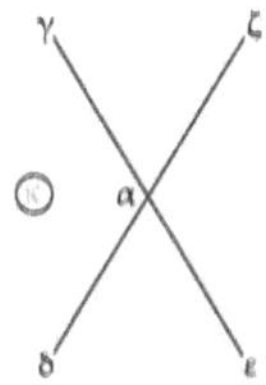

codex γ

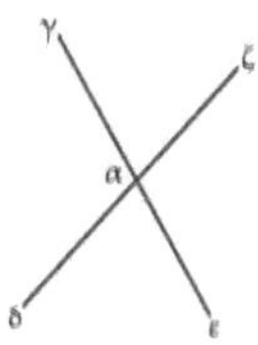

codex W^a

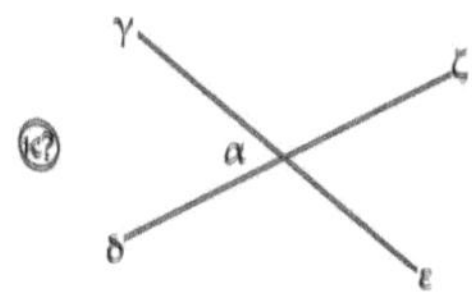

codex V^1

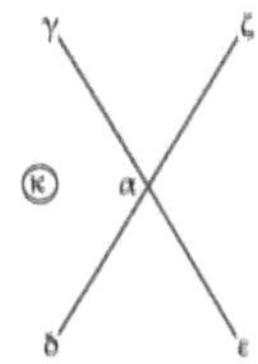

codex δ

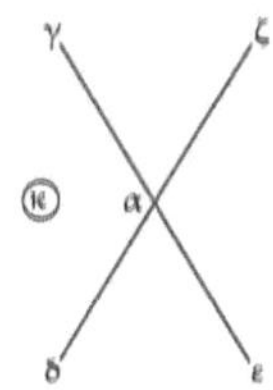

codex ε

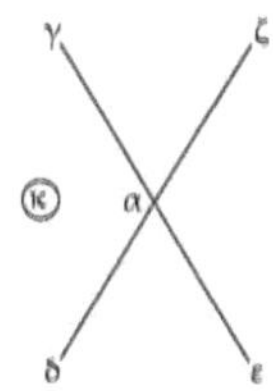

codex ζ

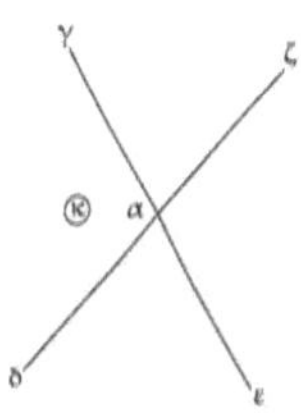

codex N^h

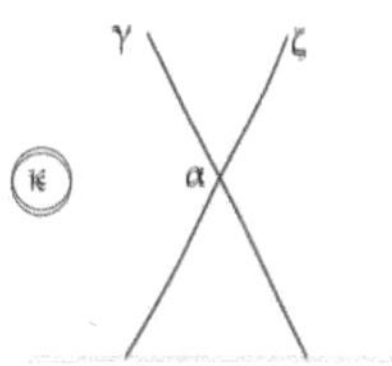

codex U^m

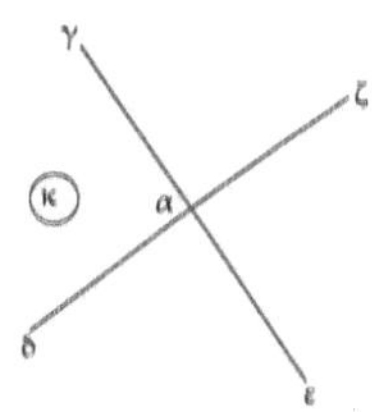

codex N^b

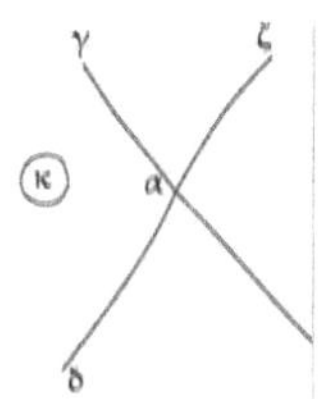

codex q

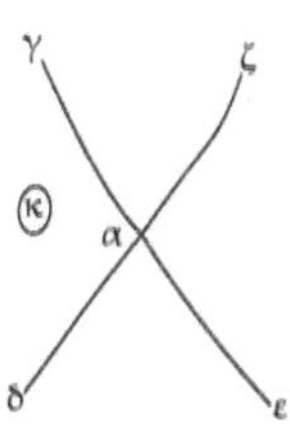

codex F^1

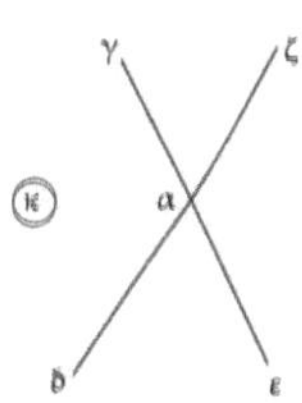

codex F^2

Problem 23

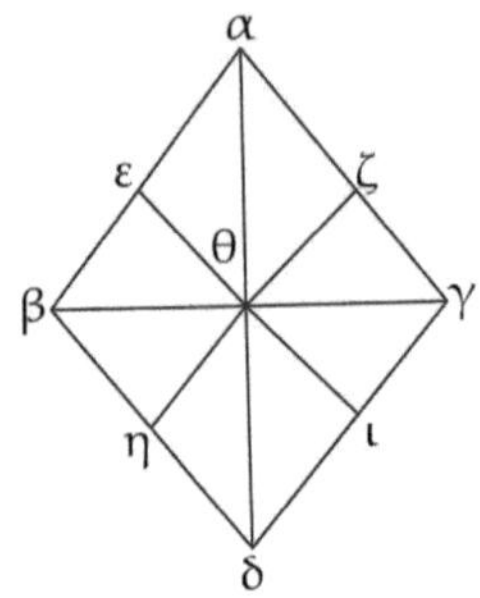

codex ω

Family *a*

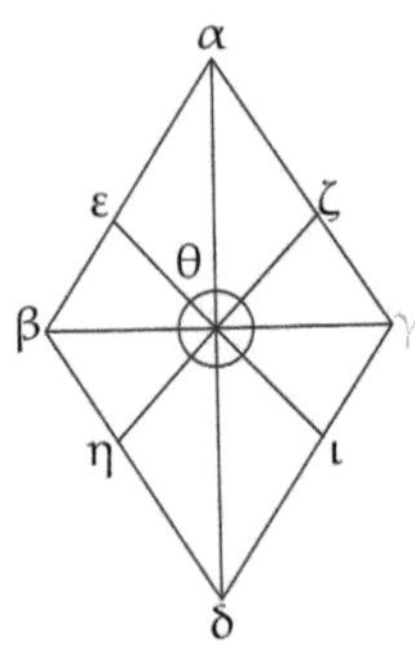

codex α

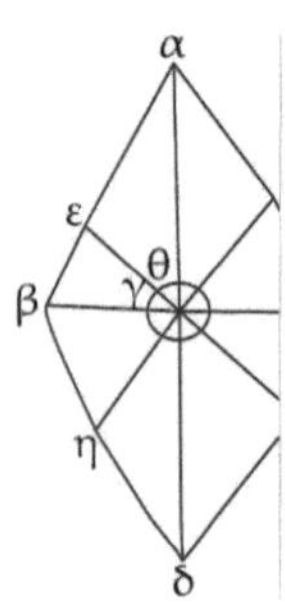

codex L

Family *b*

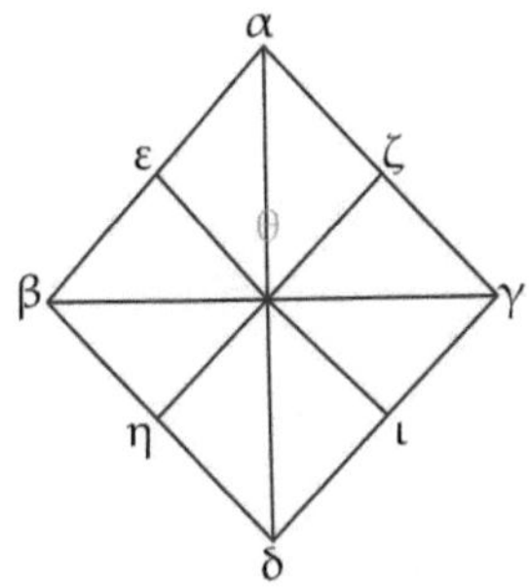

codex β

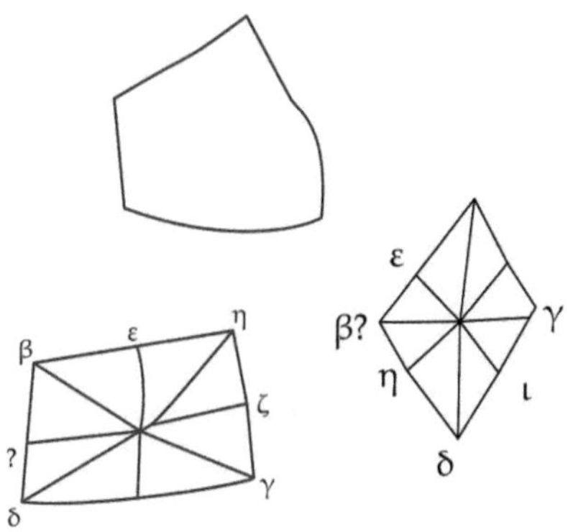

codex P

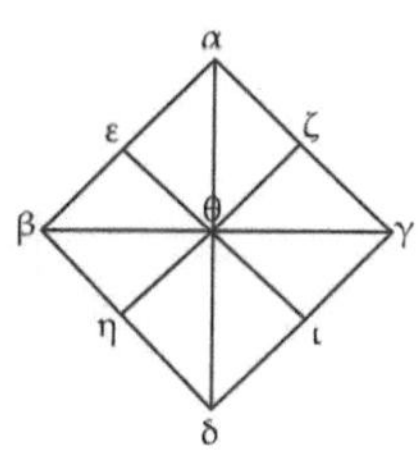

codex M^v

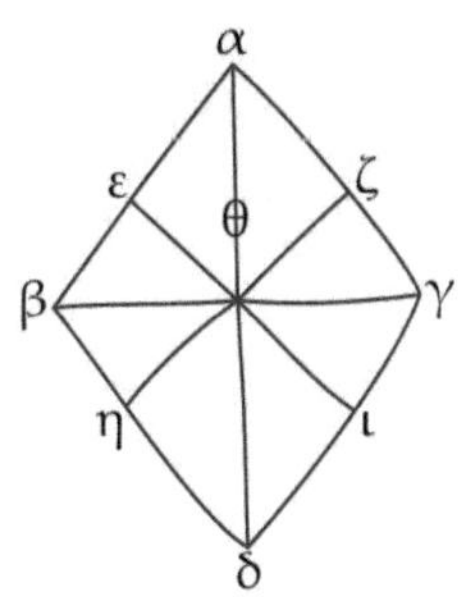

codex D^1

Family *c*

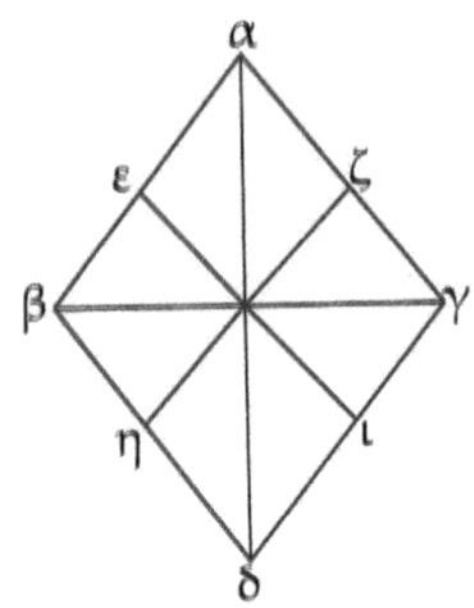

codex γ

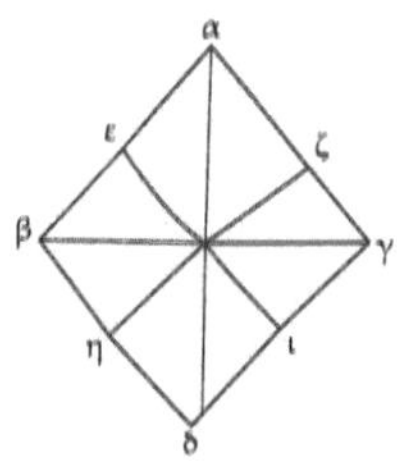

codex W^a

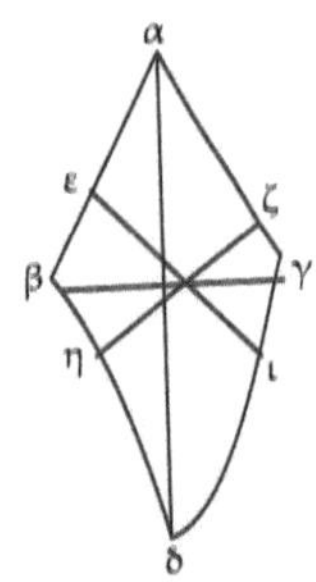

codex V^1

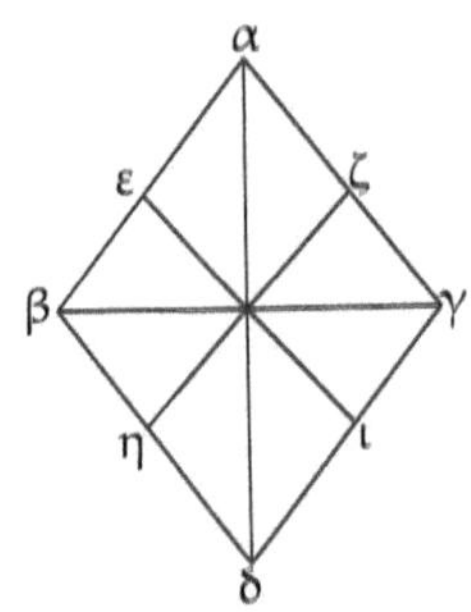

codex δ

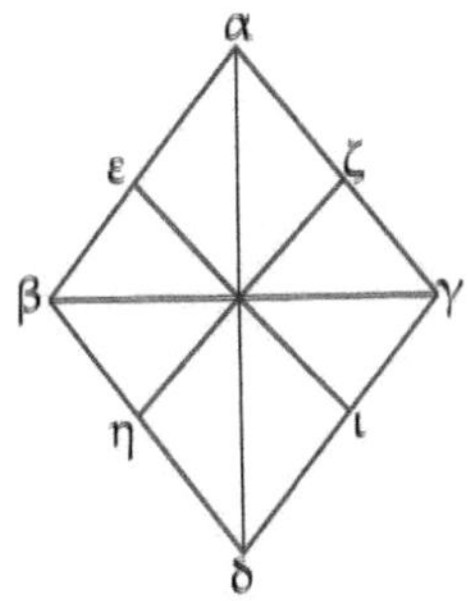

codex ε

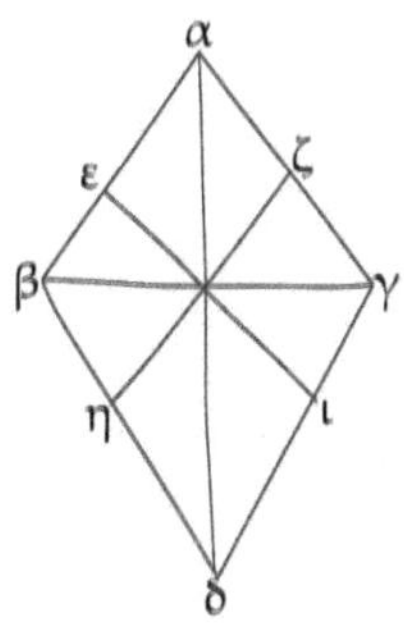

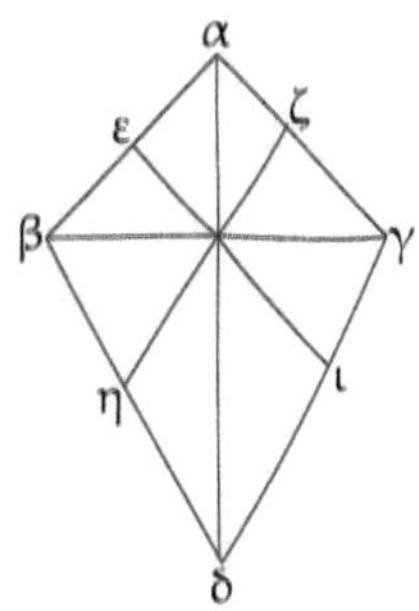

codex N[h]

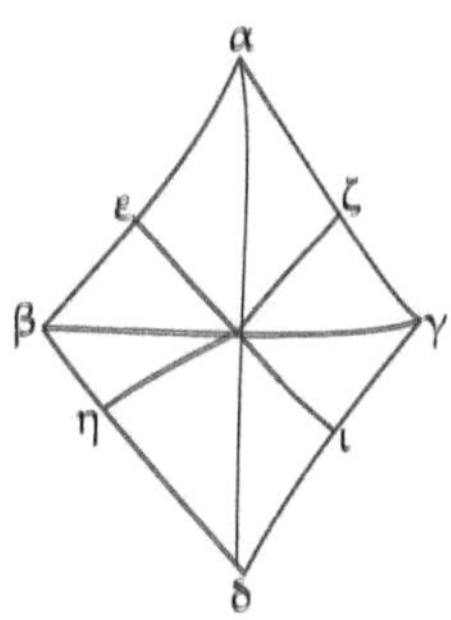

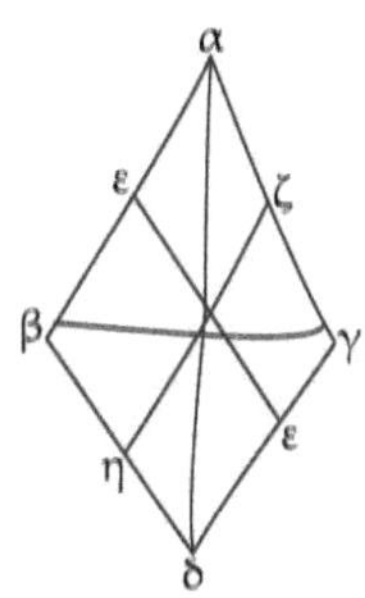

codex F[1]

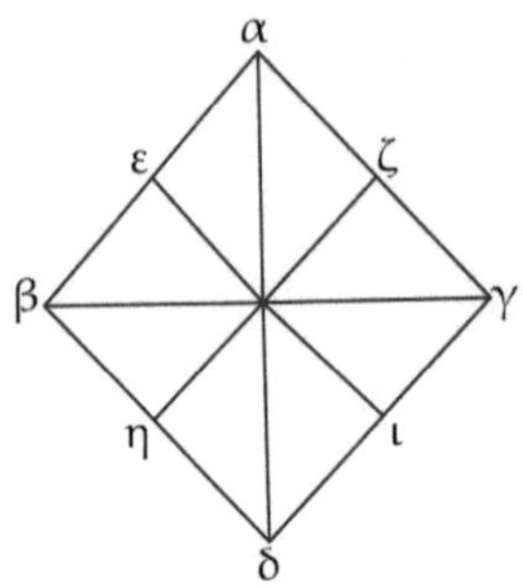

codex ζ

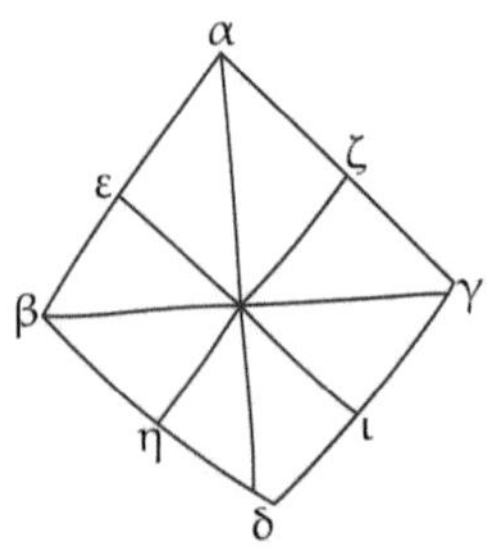

codex U^m

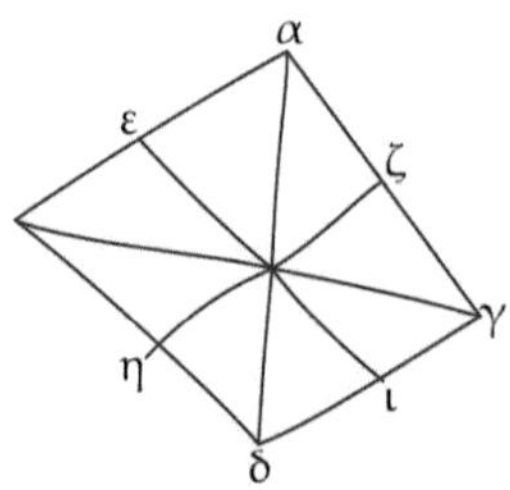

codex N^b

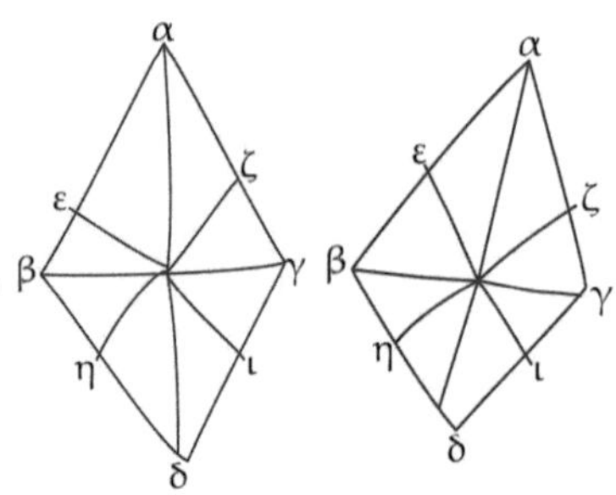

codex q

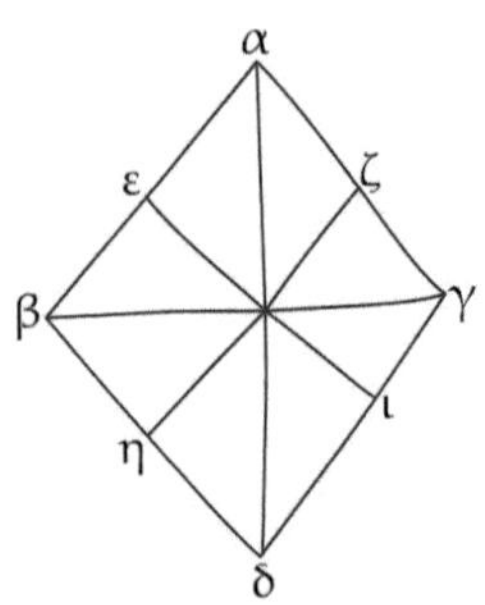

codex F^2

Problem 24.1

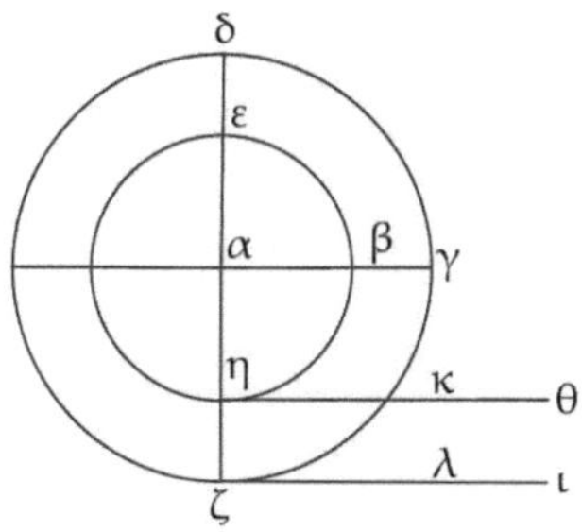

codex ω

Family *a*

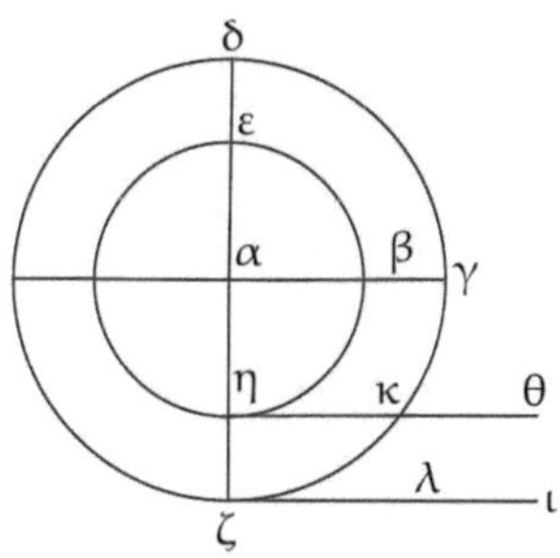

codex α

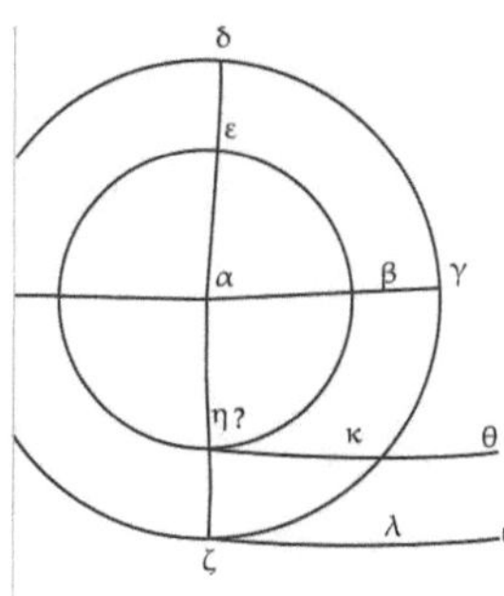

codex L

Family *b*

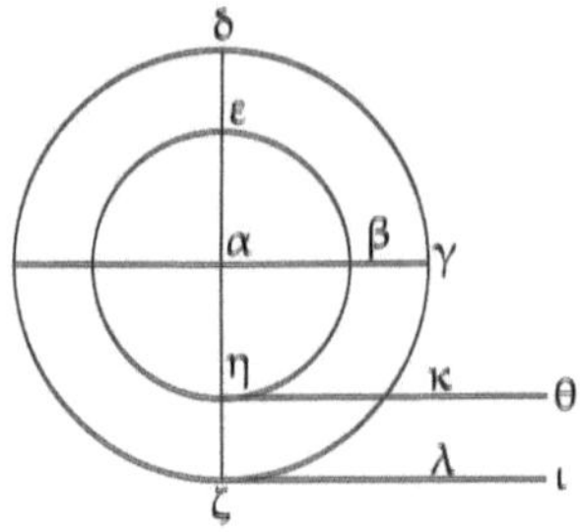

codex β

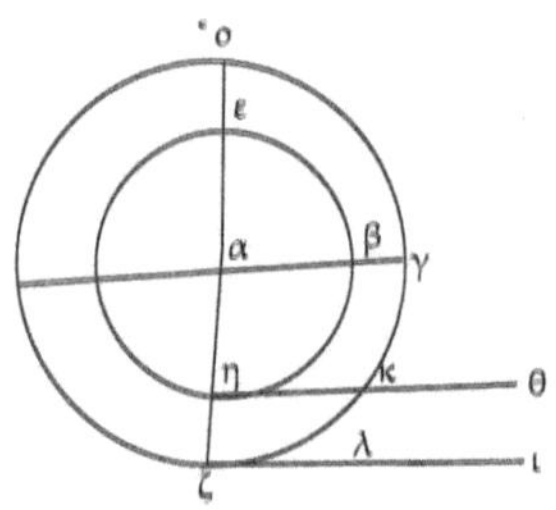

codex P

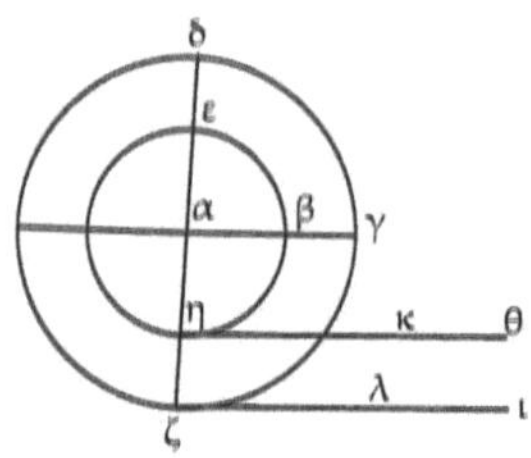

codex M^{v}

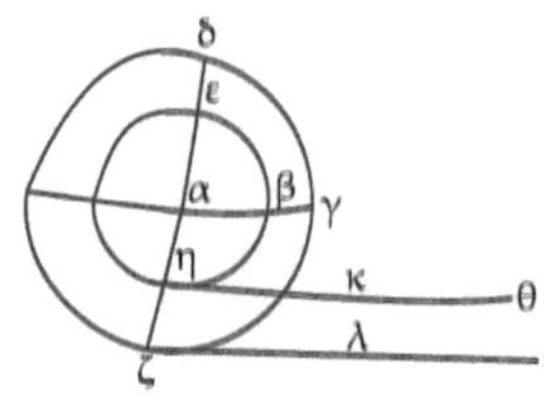

codex D^{1}

Family *c*

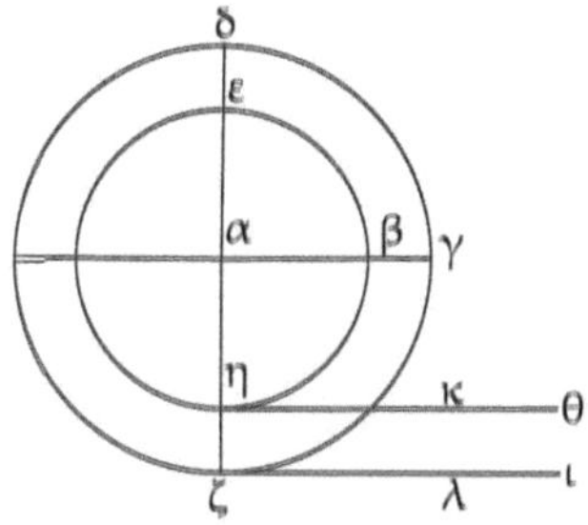

codex γ

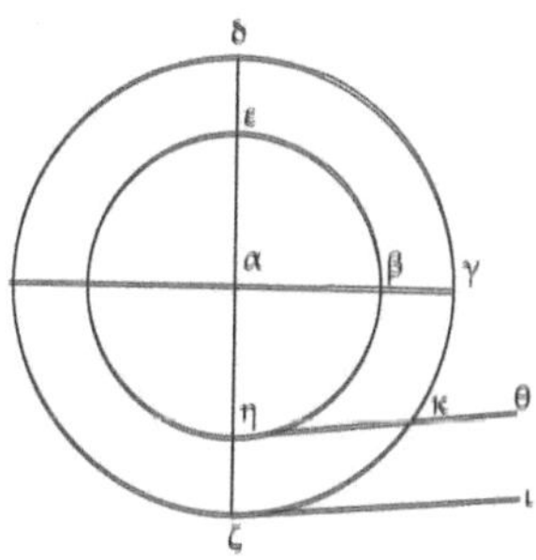

codex W^a

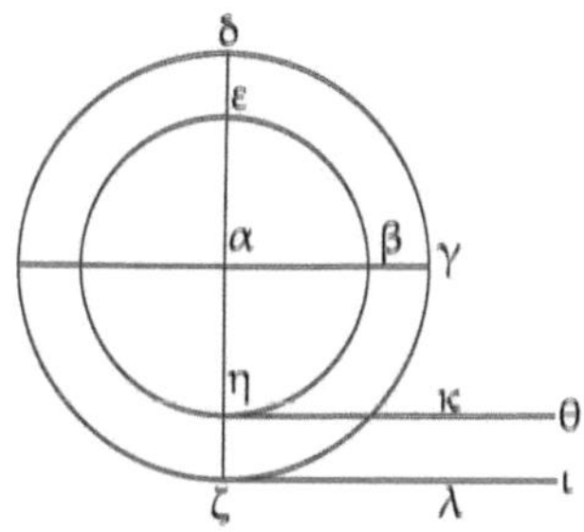

codex δ

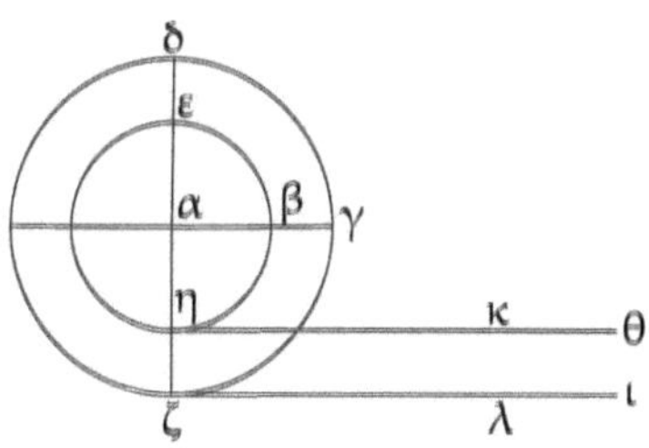

codex ε

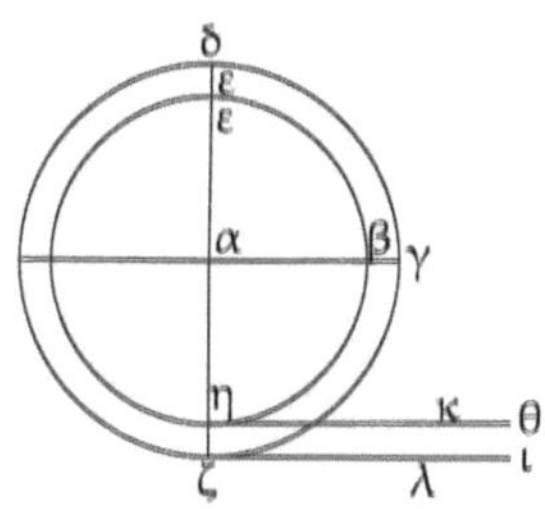

codex ζ

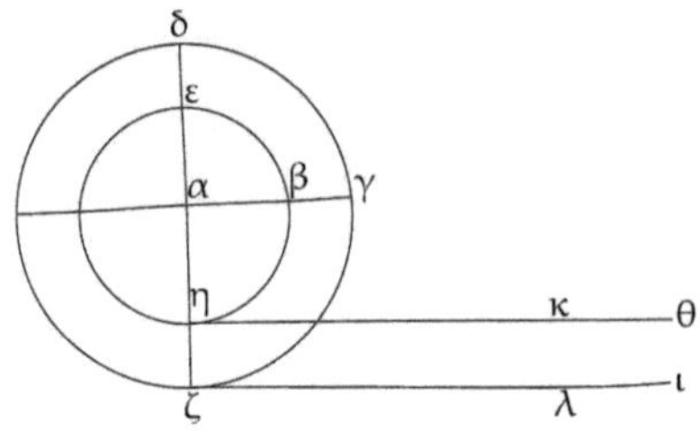

codex N^h

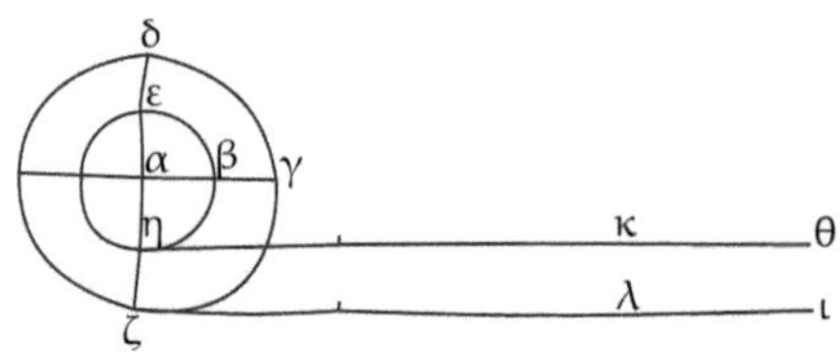

codex F^1

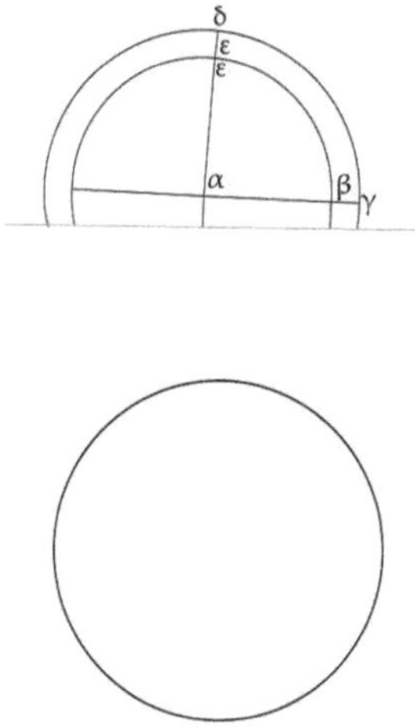

codex U^m

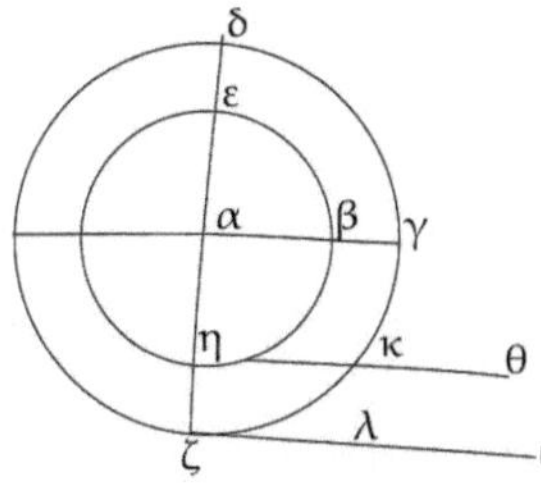

codex N^b

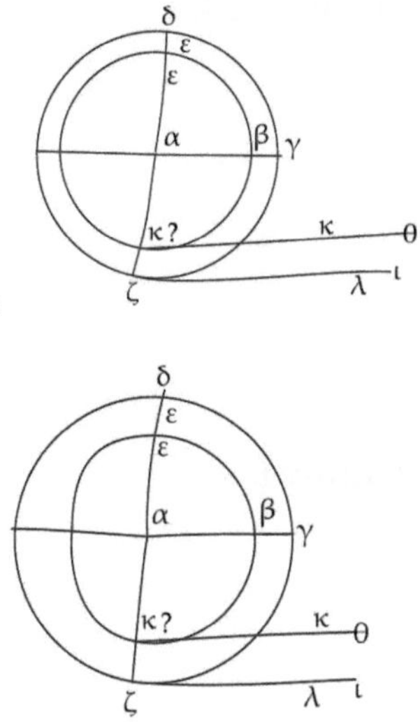

codex q

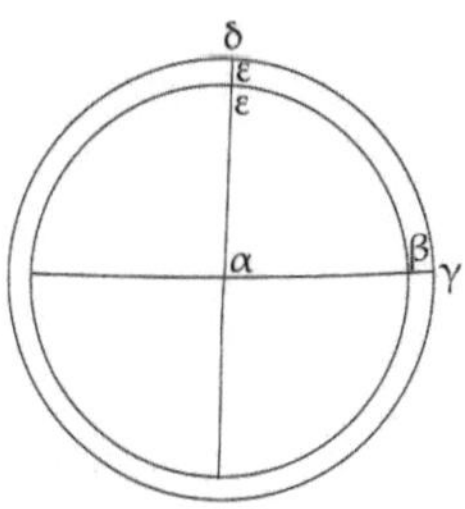

codex F^2

Problem 24.2

codex ω

Family *b*

codex β

codex P

codex M^v

codex D^1

Family *c*

codex γ

codex W^a

codex δ

codex ε

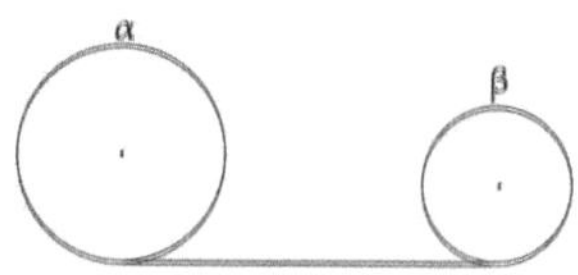

codex ζ

codex N^h

codex F^1

codex U^m

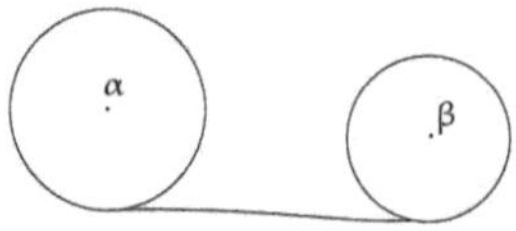

codex N^b

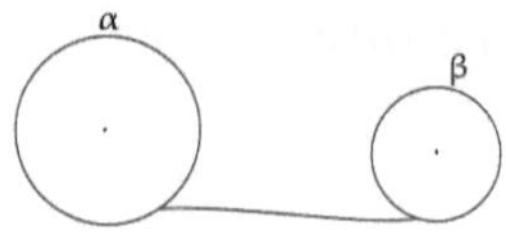

codex q

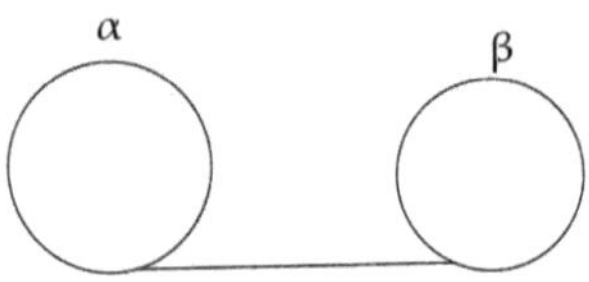

codex F^2

Problem 25

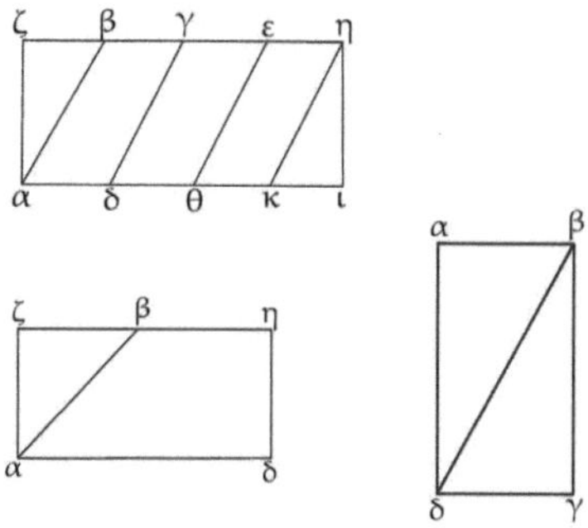

codex ω

Family *a*

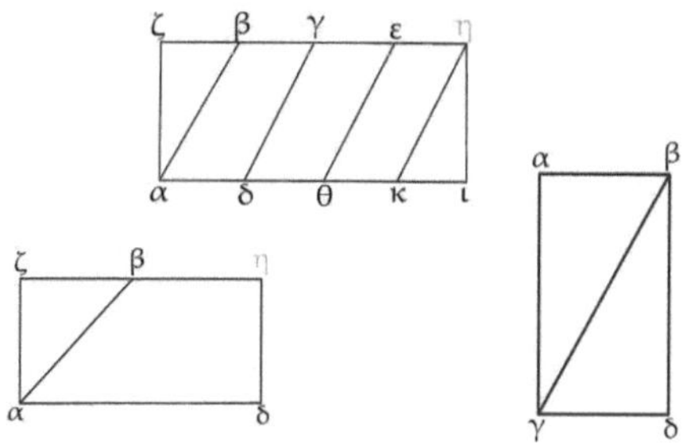

codex α

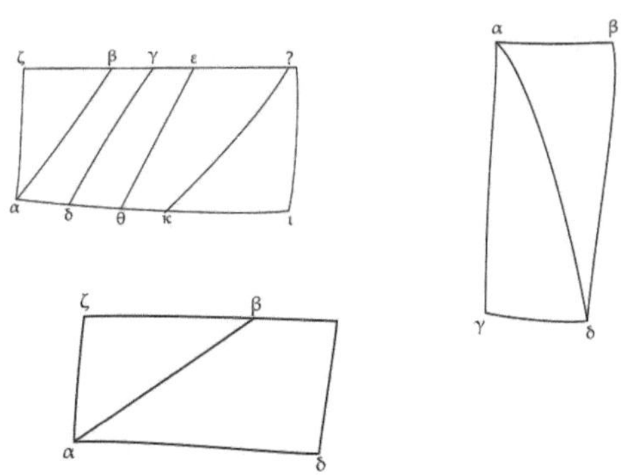

codex L

Family *b*

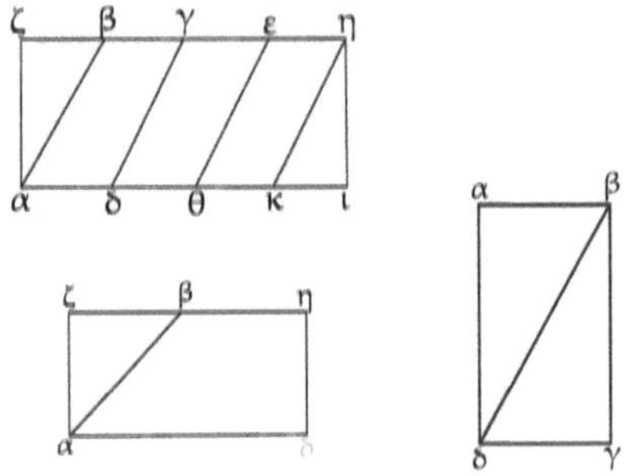

codex β

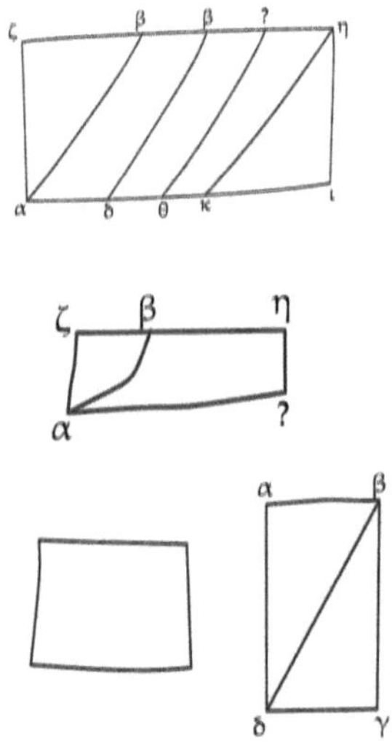

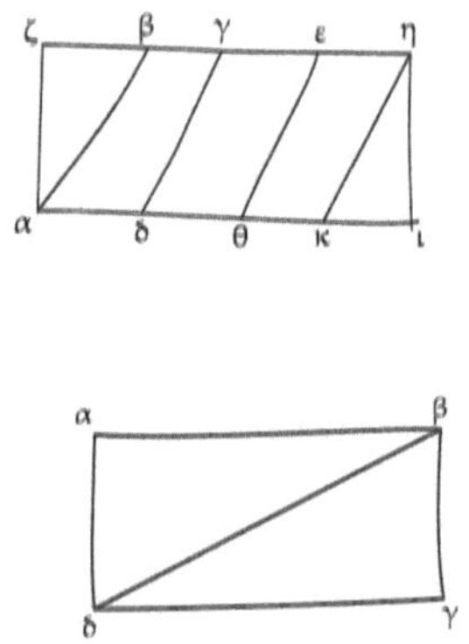

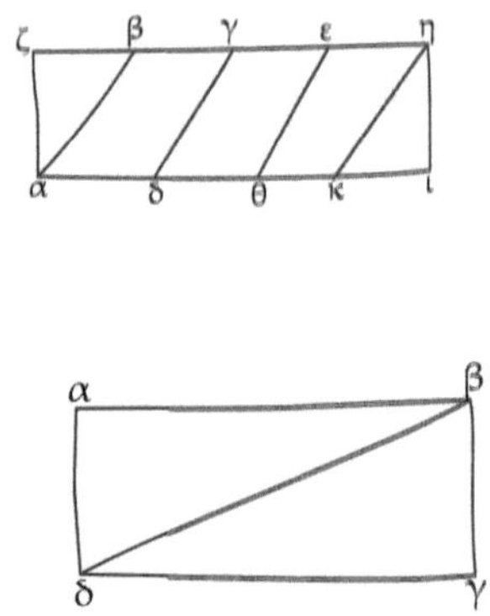

codex P

codex M[v]

codex D[l]

Family *c*

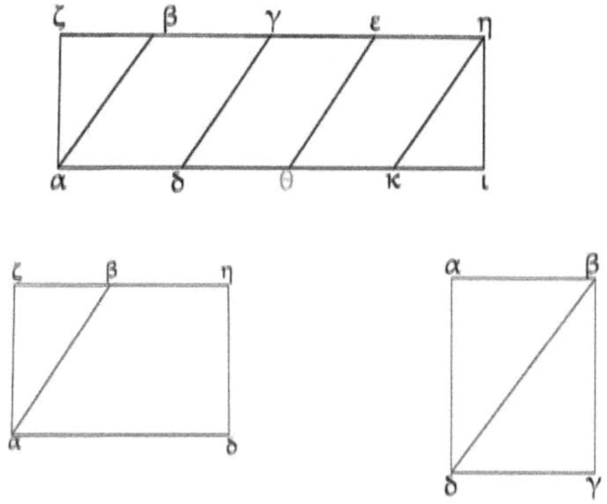

codex γ

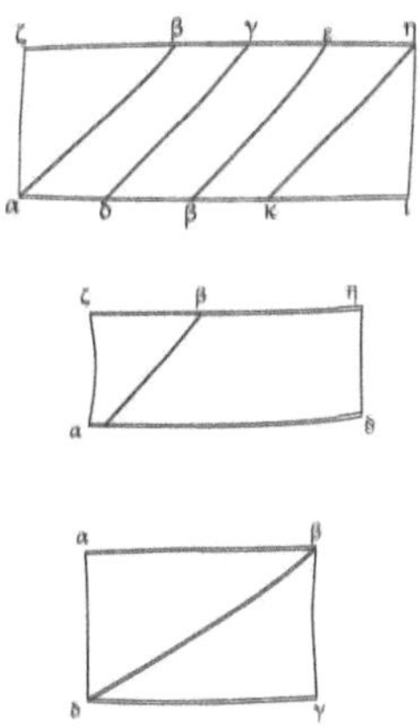

codex Wᵃ

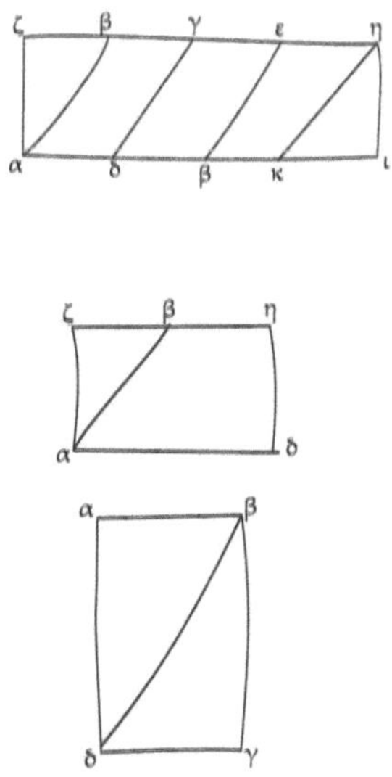

codex V¹

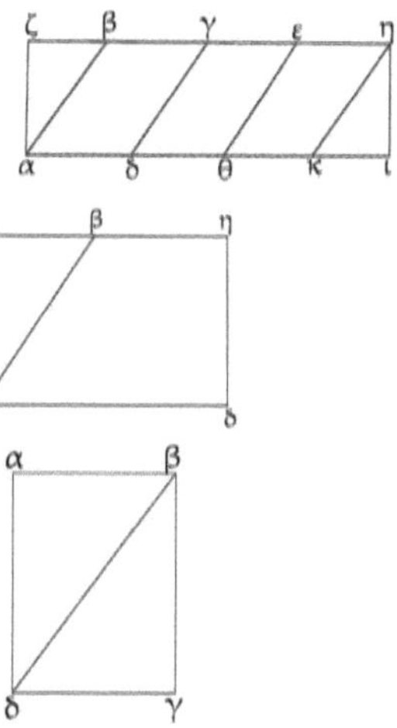

codex δ

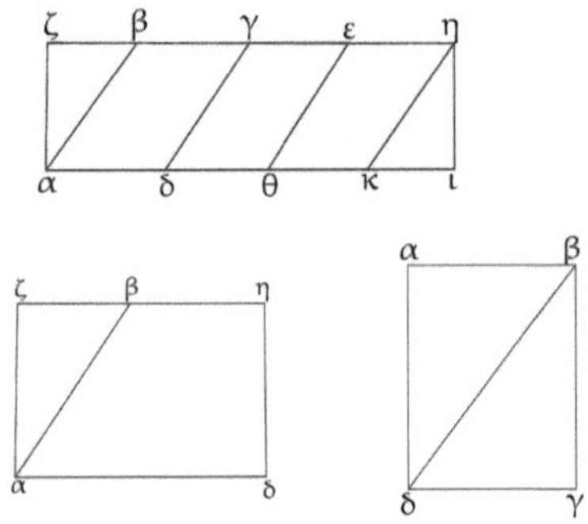

codex ε

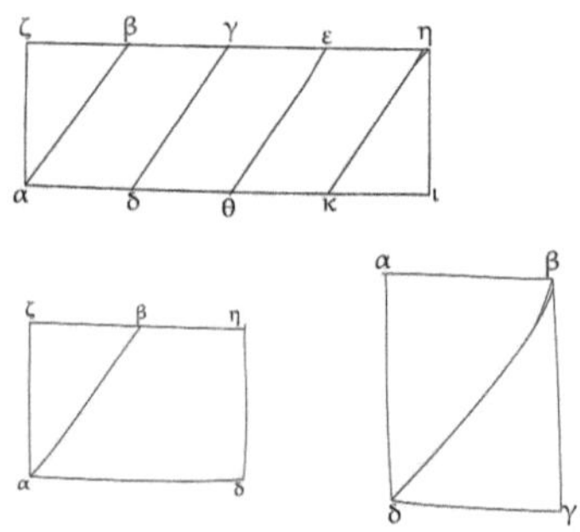

codex N^h

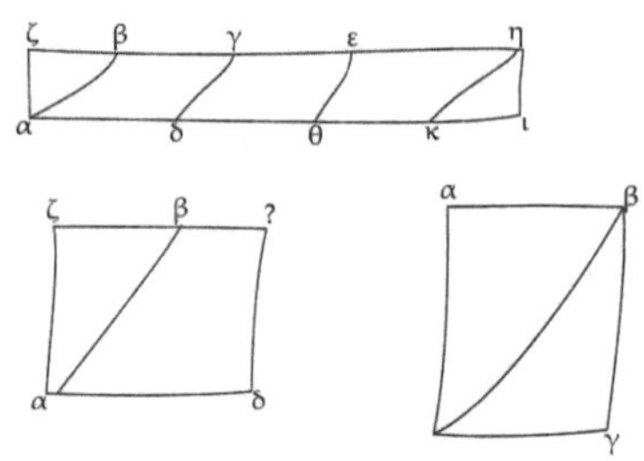

codex F^1

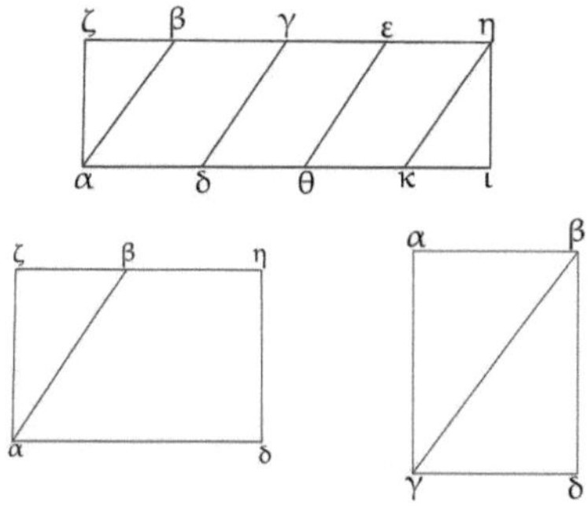

codex ζ

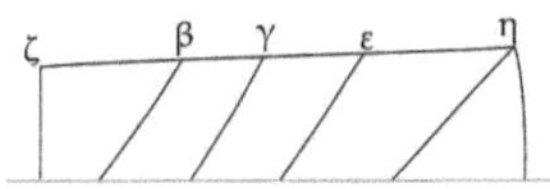

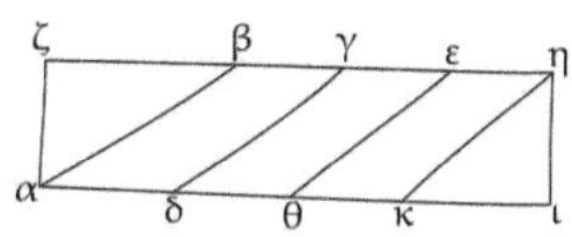

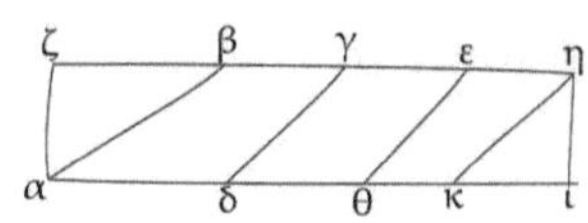

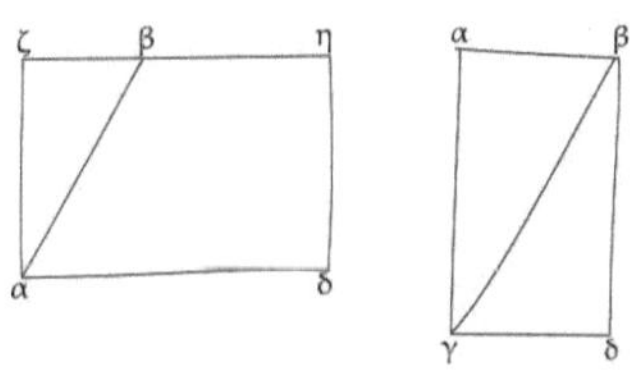

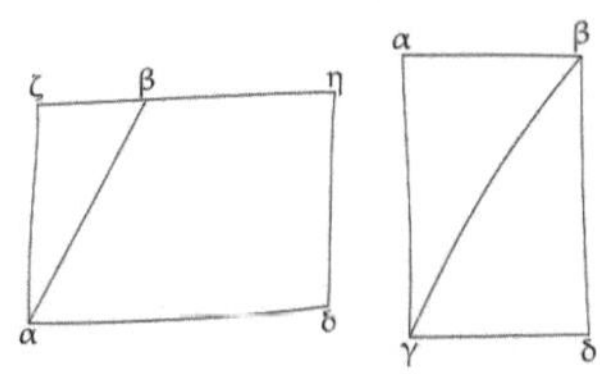

codex U^m

codex N^b

codex q

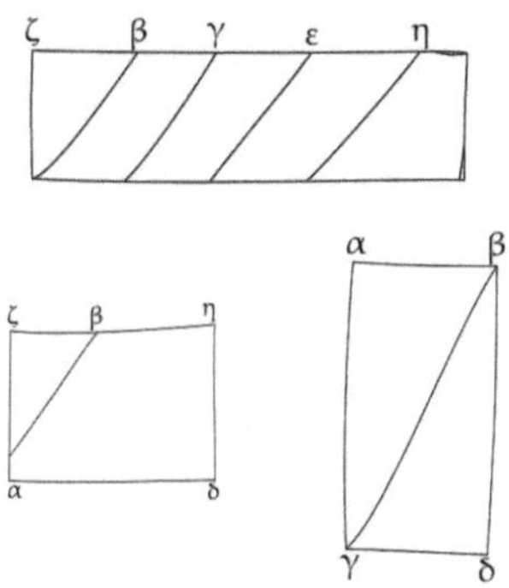

codex F^2

Problems 26–27

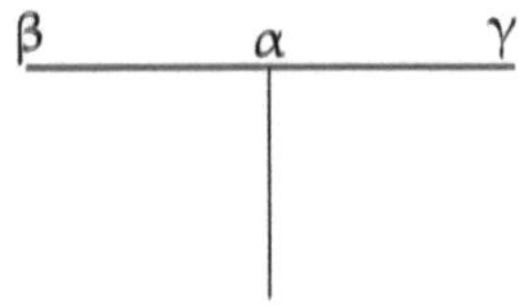

codex ω

Family *a*

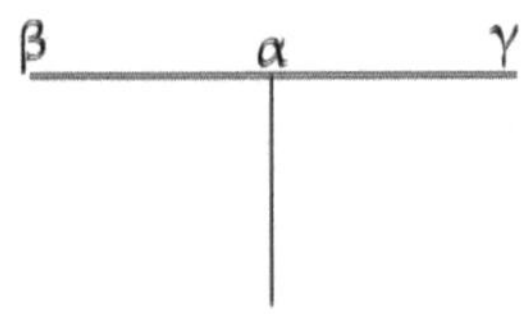

codex α

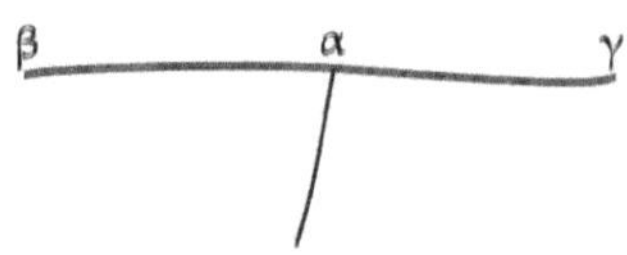

codex L

Family *b*

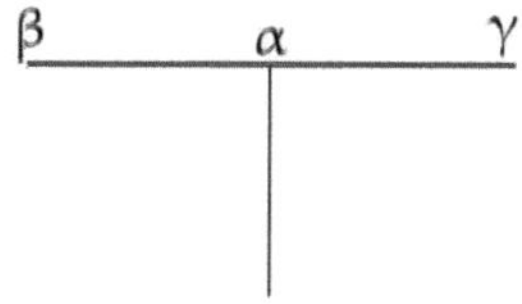

codex β

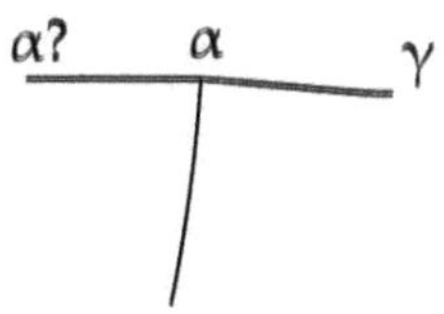

codex P

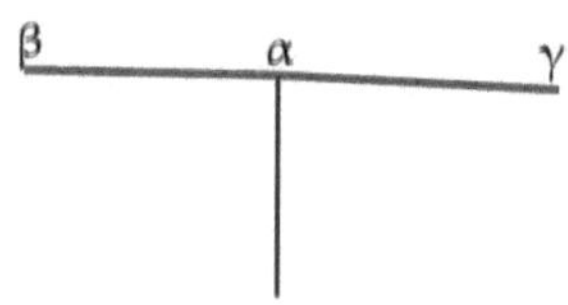

codex M^{v}

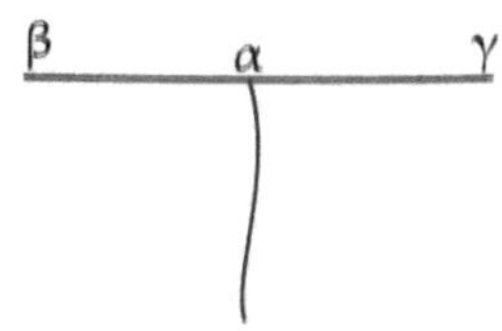

codex D^{1}

Family *c*

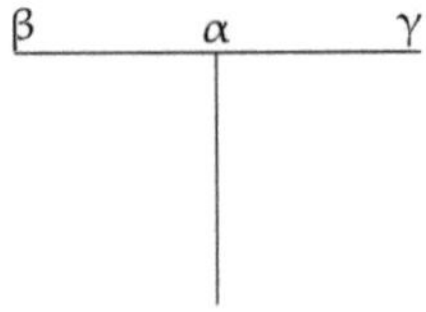

codex γ

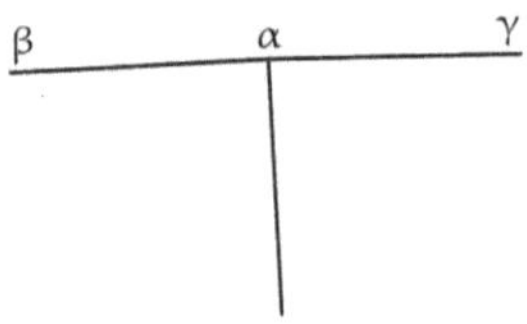

codex W^a

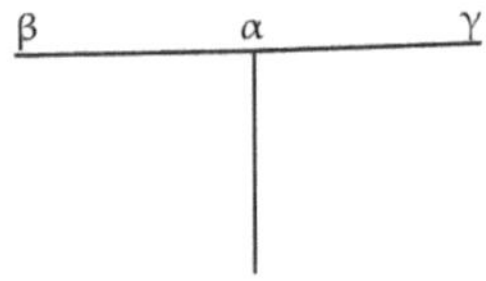

codex V^1

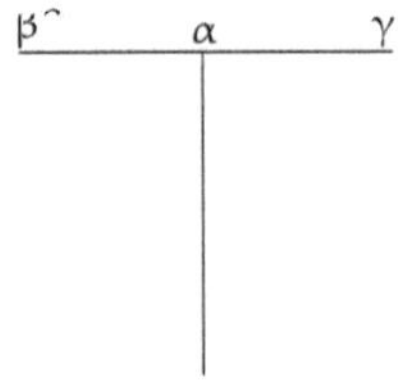

codex δ

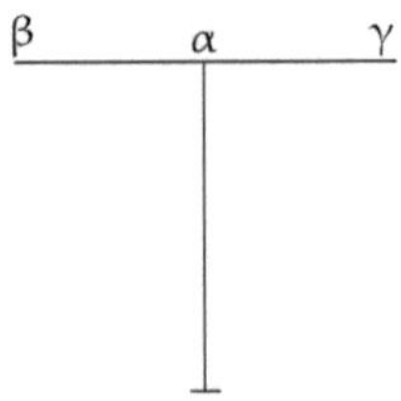

codex ε

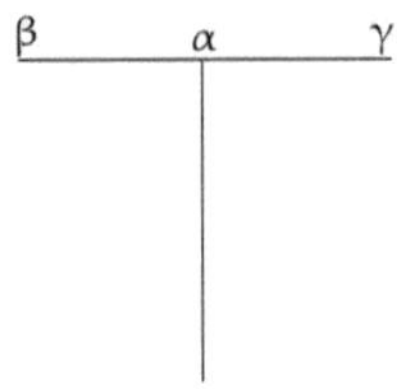

codex ζ

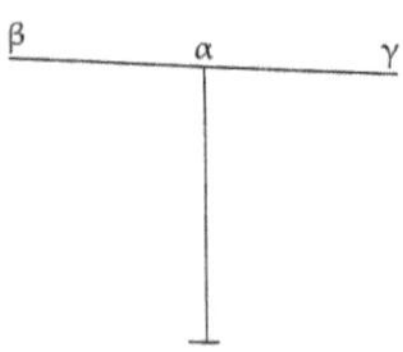

codex N^h

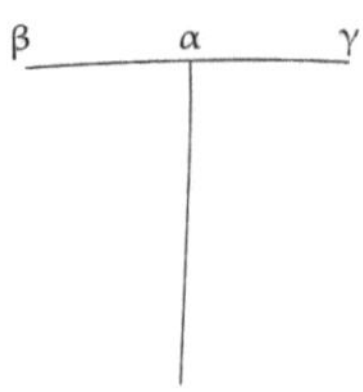

codex U^m

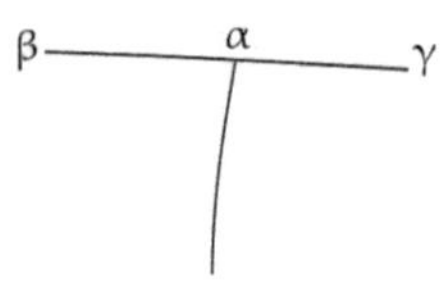

codex N^b

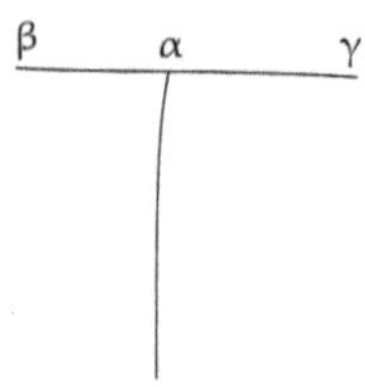

codex q

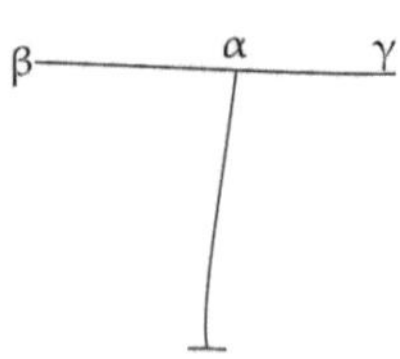

codex F^1

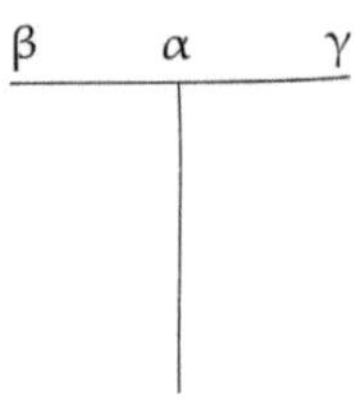

codex F^2

Problem 30

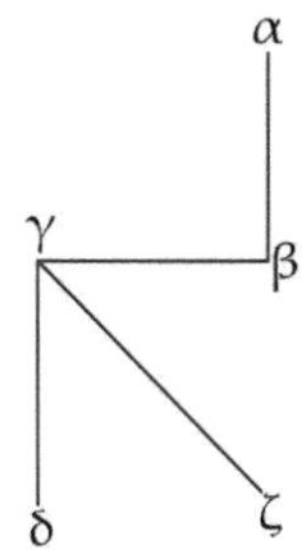

codex ω

Family *b*

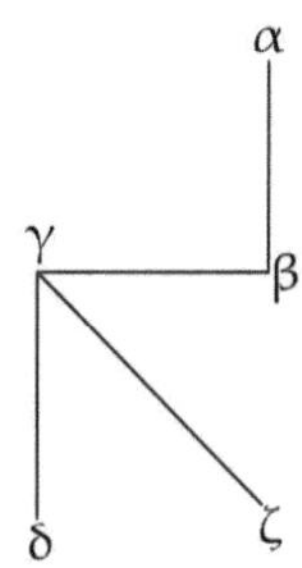

codex β

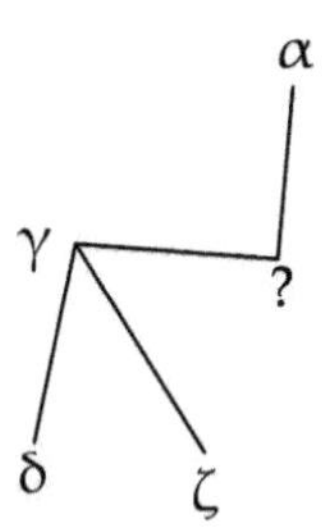

codex P

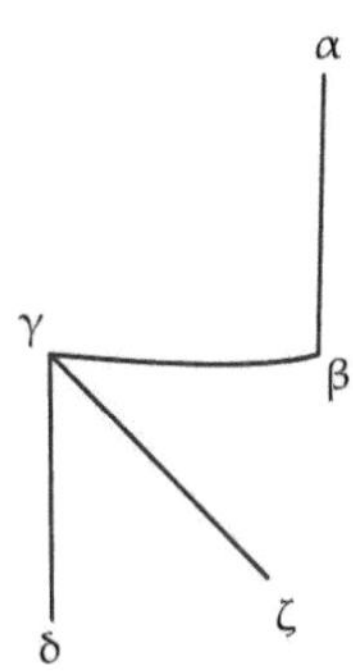

codex M^v

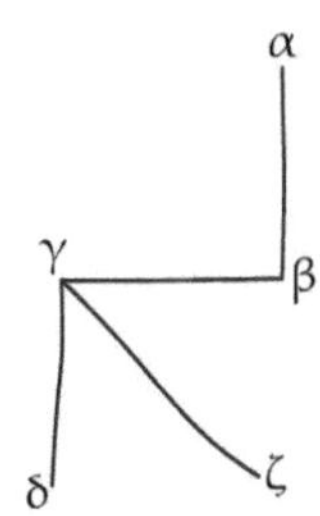

codex D^1

Family *c*

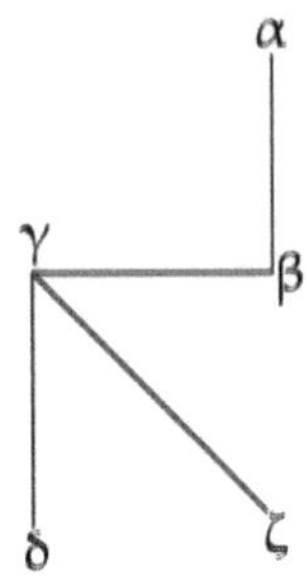

codex γ

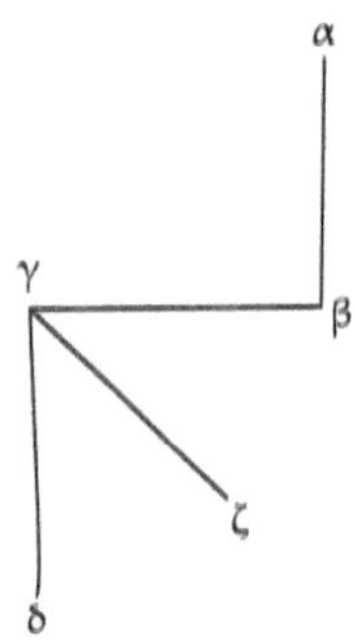

codex W^a

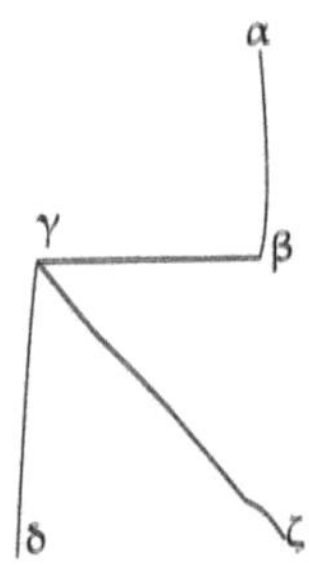

codex V^1

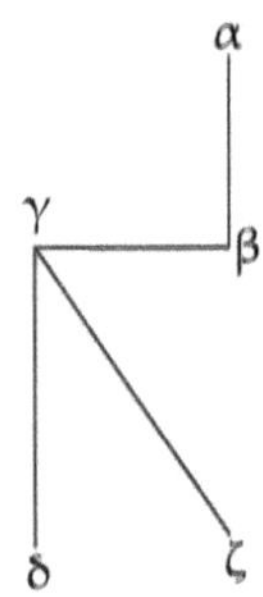

codex δ

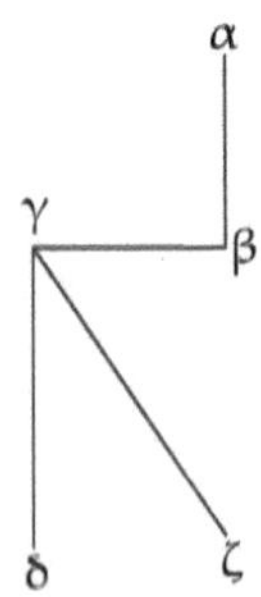

codex ε

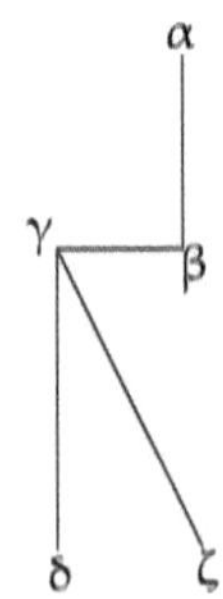

codex ζ

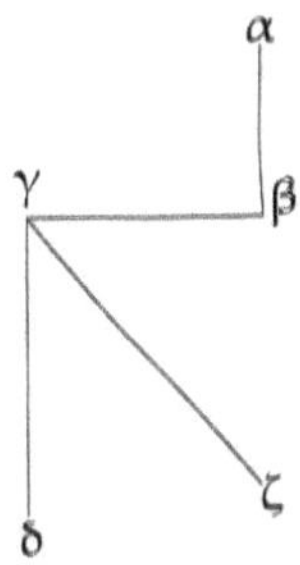

codex N^{h}

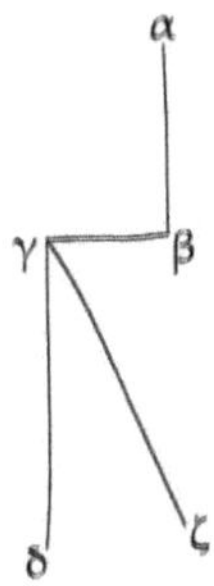

codex U^{m}

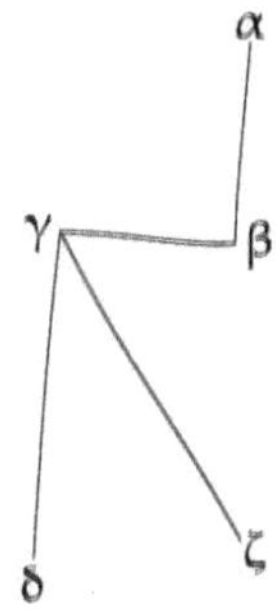

codex N^{b}

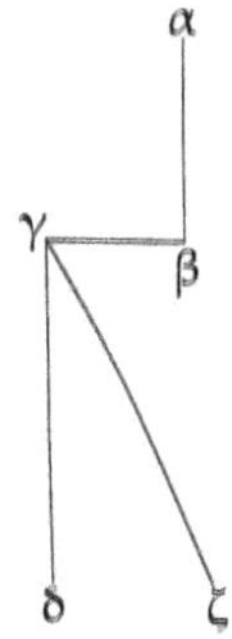

codex q

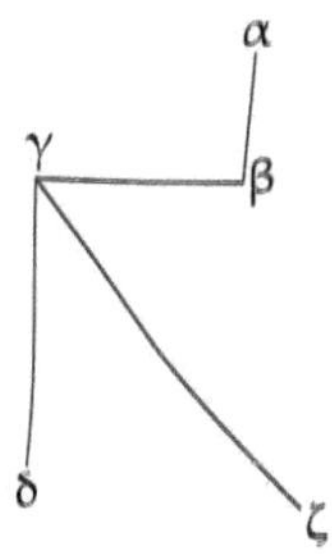

codex F^{1}

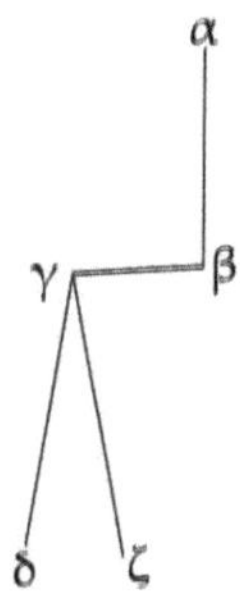

codex F^{2}

MIX
Papier aus verantwortungsvollen Quellen
Paper from responsible sources
FSC® C105338

If you have any concerns about our products,
you can contact us on
ProductSafety@springernature.com

In case Publisher is established outside the EU,
the EU authorized representative is:
Springer Nature Customer Service Center GmbH
Europaplatz 3, 69115 Heidelberg, Germany

Printed by Libri Plureos GmbH
in Hamburg, Germany